Recent Trends in Cold-Formed Steel Construction

Related titles

Transport Properties of Concrete
(ISBN 978-1-78242-306-5)

Advanced Composites in Bridge Construction and Repair
(ISBN 978-0-85709-694-4)

Rehabilitation of Metallic Infrastructure Using FRP Composites
(ISBN 978-0-85709-653-1)

Eco-Efficient Construction and Building Materials
(ISBN 978-0-85709-767-5)

Woodhead Publishing Series in Civil and Structural Engineering: Number 65

Recent Trends in Cold-Formed Steel Construction

Edited by

Cheng Yu

ELSEVIER

AMSTERDAM • BOSTON • CAMBRIDGE • HEIDELBERG
LONDON • NEW YORK • OXFORD • PARIS • SAN DIEGO
SAN FRANCISCO • SINGAPORE • SYDNEY • TOKYO
Woodhead Publishing is an imprint of Elsevier

WP
WOODHEAD PUBLISHING

Woodhead Publishing is an imprint of Elsevier
The Officers' Mess Business Centre, Royston Road, Duxford, CB22 4QH, UK
50 Hampshire Street, 5th Floor, Cambridge, MA 02139, USA
The Boulevard, Langford Lane, Kidlington, OX5 1GB, UK

British Library Cataloguing-in-Publication Data
A catalogue record for this book is available from the British Library

Library of Congress Cataloging-in-Publication Data
A catalog record for this book is available from the Library of Congress

ISBN: 978-0-08-100960-4 (print)
ISBN: 978-0-08-100157-8 (online)

Publisher: Matthew Deans
Acquisition Editor: Gwen Jones
Editorial Project Manager: Charlotte Cockle
Production Project Manager: Omer Mukthar
Designer: Maria Inês Cruz

Typeset by TNQ Books and Journals

Contents

List of contributors

J.C. Batista Abreu Bucknell University, Lewisburg, Pennsylvania, United States

D. Camotim Instituto Superior Técnico, Universidade de Lisboa, Lisbon, Portugal

H. Chen American Iron and Steel Institute, Washington, DC, United States

P.B. Dinis Instituto Superior Técnico, Universidade de Lisboa, Lisbon, Portugal

G.J. Hancock University of Sydney, Sydney, NSW, Australia

J. Leng Postdoctoral Fellow, Mechanical Engineering Department, McGill University, Montreal, Canada

Z. Li SUNY Polytechnic Institute, Utica, NY, United States

J.B.P. Lim The University of Auckland, Auckland, New Zealand

W. Lu Aalto University, Espoo, Finland

A.D. Martins Instituto Superior Técnico, Universidade de Lisboa, Lisbon, Portugal

D.J. Mynors University of Sussex, Brighton, United Kingdom

D.A. Nethercot Imperial College London, London, United Kingdom

C.H. Pham University of Sydney, Sydney, NSW, Australia

N.A. Rahman The Steel Network, Inc., Durham, North Carolina, United States

C.J. Wang University of Sussex, Brighton, United Kingdom

L.W. Williams Steel Framing Industry Association Falls Church, VA, United States

A.M. Wrzesien University of Strathclyde, Glasgow, United Kingdom

L. Xu University of Waterloo, Waterloo, ON, Canada

C. Yu University of North Texas, Denton, TX, United States

W. Zhang Tongji University, Shanghai, China

Woodhead Publishing Series in Civil and Structural Engineering

1 **Finite element techniques in structural mechanics**
C. T. F. Ross
2 **Finite element programs in structural engineering and continuum mechanics**
C. T. F. Ross
3 **Macro-engineering**
F. P. Davidson, E. G. Frankl and C. L. Meador
4 **Macro-engineering and the earth**
U. W. Kitzinger and E. G. Frankel
5 **Strengthening of reinforced concrete structures**
Edited by L. C. Hollaway and M. Leeming
6 **Analysis of engineering structures**
B. Bedenik and C. B. Besant
7 **Mechanics of solids**
C. T. F. Ross
8 **Plasticity for engineers**
C. R. Calladine
9 **Elastic beams and frames**
J. D. Renton
10 **Introduction to structures**
W. R. Spillers
11 **Applied elasticity**
J. D. Renton
12 **Durability of engineering structures**
J. Bijen
13 **Advanced polymer composites for structural applications in construction**
Edited by L. C. Hollaway
14 **Corrosion in reinforced concrete structures**
Edited by H. Böhni
15 **The deformation and processing of structural materials**
Edited by Z. X. Guo
16 **Inspection and monitoring techniques for bridges and civil structures**
Edited by G. Fu
17 **Advanced civil infrastructure materials**
Edited by H. Wu
18 **Analysis and design of plated structures Volume 1: Stability**
Edited by E. Shanmugam and C. M. Wang

19 **Analysis and design of plated structures Volume 2: Dynamics**
Edited by E. Shanmugam and C. M. Wang
20 **Multiscale materials modelling**
Edited by Z. X. Guo
21 **Durability of concrete and cement composites**
Edited by C. L. Page and M. M. Page
22 **Durability of composites for civil structural applications**
Edited by V. M. Karbhari
23 **Design and optimization of metal structures**
J. Farkas and K. Jarmai
24 **Developments in the formulation and reinforcement of concrete**
Edited by S. Mindess
25 **Strengthening and rehabilitation of civil infrastructures using fibre-reinforced polymer (FRP) composites**
Edited by L. C. Hollaway and J. C. Teng
26 **Condition assessment of aged structures**
Edited by J. K. Paik and R. M. Melchers
27 **Sustainability of construction materials**
J. Khatib
28 **Structural dynamics of earthquake engineering**
S. Rajasekaran
29 **Geopolymers: Structures, processing, properties and industrial applications**
Edited by J. L. Provis and J. S. J. van Deventer
30 **Structural health monitoring of civil infrastructure systems**
Edited by V. M. Karbhari and F. Ansari
31 **Architectural glass to resist seismic and extreme climatic events**
Edited by R. A. Behr
32 **Failure, distress and repair of concrete structures**
Edited by N. Delatte
33 **Blast protection of civil infrastructures and vehicles using composites**
Edited by N. Uddin
34 **Non-destructive evaluation of reinforced concrete structures Volume 1: Deterioration processes**
Edited by C. Maierhofer, H.-W. Reinhardt and G. Dobmann
35 **Non-destructive evaluation of reinforced concrete structures Volume 2: Non-destructive testing methods**
Edited by C. Maierhofer, H.-W. Reinhardt and G. Dobmann
36 **Service life estimation and extension of civil engineering structures**
Edited by V. M. Karbhari and L. S. Lee
37 **Building decorative materials**
Edited by Y. Li and S. Ren
38 **Building materials in civil engineering**
Edited by H. Zhang
39 **Polymer modified bitumen**
Edited by T. McNally
40 **Understanding the rheology of concrete**
Edited by N. Roussel
41 **Toxicity of building materials**
Edited by F. Pacheco-Torgal, S. Jalali and A. Fucic

42 **Eco-efficient concrete**
Edited by F. Pacheco-Torgal, S. Jalali, J. Labrincha and V. M. John

43 **Nanotechnology in eco-efficient construction**
Edited by F. Pacheco-Torgal, M. V. Diamanti, A. Nazari and C. Goran-Granqvist

44 **Handbook of seismic risk analysis and management of civil infrastructure systems**
Edited by F. Tesfamariam and K. Goda

45 **Developments in fiber-reinforced polymer (FRP) composites for civil engineering**
Edited by N. Uddin

46 **Advanced fibre-reinforced polymer (FRP) composites for structural applications**
Edited by J. Bai

47 **Handbook of recycled concrete and demolition waste**
Edited by F. Pacheco-Torgal, V. W. Y. Tam, J. A. Labrincha, Y. Ding and J. de Brito

48 **Understanding the tensile properties of concrete**
Edited by J. Weerheijm

49 **Eco-efficient construction and building materials: Life cycle assessment (LCA), eco-labelling and case studies**
Edited by F. Pacheco-Torgal, L. F. Cabeza, J. Labrincha and A. de Magalhães

50 **Advanced composites in bridge construction and repair**
Edited by Y. J. Kim

51 **Rehabilitation of metallic civil infrastructure using fiber-reinforced polymer (FRP) composites**
Edited by V. Karbhari

52 **Rehabilitation of pipelines using fiber-reinforced polymer (FRP) composites**
Edited by V. Karbhari

53 **Transport properties of concrete: Measurement and applications**
P. A. Claisse

54 **Handbook of alkali-activated cements, mortars and concretes**
F. Pacheco-Torgal, J. A. Labrincha, C. Leonelli, A. Palomo and P. Chindaprasirt

55 **Eco-efficient masonry bricks and blocks: Design, properties and durability**
F. Pacheco-Torgal, P. B. Lourenço, J. A. Labrincha, S. Kumar and P. Chindaprasirt

56 **Advances in asphalt materials: Road and pavement construction**
Edited by S.-C. Huang and H. Di Benedetto

57 **Acoustic emission (AE) and related non-destructive evaluation (NDE) techniques in the fracture mechanics of concrete: Fundamentals and applications**
Edited by M. Ohtsu

58 **Nonconventional and vernacular construction materials: Characterisation, properties and applications**
Edited by K. A. Harries and B. Sharma

59 **Science and technology of concrete admixtures**
Edited by P.-C. Aïtcin and R. J. Flatt

60 **Textile fibre composites in civil engineering**
Edited by T. Triantafillou

61 **Corrosion of steel in concrete structures**
Edited by A. Poursaee

62 **Innovative developments of advanced multifunctional nanocomposites in civil and structural engineering**
Edited by K. J. Loh and S. Nagarajaiah

63 **Biopolymers and biotech admixtures for eco-efficient construction materials**
Edited by F. Pacheco-Torgal, V. Ivanov, N. Karak and H. Jonkers

64 **Marine concrete structures: Design, durability and performance**
Edited by M. Alexander

65 **Recent trends in cold-formed steel construction**
Edited by C. Yu

66 **Start-up creation: The smart eco-efficient built environment**
F. Pacheco-Torgal, E. Rasmussen, C. G. Granqvist, V. Ivanov, A. Kaklauskas and S. Makonin

67 **Characteristics and uses of steel slag in building construction**
I. Barisic, I. Netinger, A. Fučić and S. Bansode

68 **The utilization of slag in civil infrastructure construction**
G. Wang

Preface

This book presents a collection of frontier research results and the latest code developments in cold-formed steel (CFS) applications in buildings. The content covers a large spectrum of knowledge: from basic design methodology to advanced computational modeling techniques, from structural behavior to the mechanical performance of CFS structures, and much more.

CFS has a long history of building applications, and in recent decades computer-assisted fabrication technologies have helped CFS to become an alternative material for load-bearing structures in low- and mid-rise constructions.

The wide variety of cross-section shapes, connection details, and assembly configurations make CFS a big challenge for structural design but also a great opportunity for construction industries, as the nature of CFS enables them to pursue endless optimal solutions.

Innovation keeps driving CFS to be more efficient, more sustainable, and more accessible. I hope this book will give readers a comprehensive introduction to the latest technical developments from the CFS research community.

I am thankful to the chapter authors who have made this book possible. We have a world-class team of experts in CFS research. Ten years hence the content of this book may no longer be at the cutting edge, but readers can always rely on our authors to present the latest trends in CFS construction.

Cheng Yu
May 15, 2016
Denton, Texas

Introduction to recent trends in cold-formed steel construction

1

L.W. Williams
Steel Framing Industry Association Falls Church, VA, United States

Cold-formed steel (CFS) as we know it today is one of the newest structural systems used in residential and nonresidential construction, but in a few short decades has grown into one of the most commonly used materials in developed economies around the world. Countries experiencing rapid economic and industrial development, including China (Fig. 1.1), India, and throughout the Middle East, South America, and Africa, are increasingly looking to CFS because it allows builders to erect new homes (Fig 1.2) and offices (Fig. 1.3) in a fraction of the time compared with traditional construction materials.

1.1 The history of cold-formed steel

The use of CFS as a construction material dates back into the 1800s, although in shapes and dimensions that are quite unlike the typical framing members we see today. Corrugated iron makers started using mild steel in the 1890s, and when cold-rolled and hot-dipped galvanized it became an ideal cladding and structural sheathing for a wide range of building types in a variety of conditions and climates around the world.

The transition to the current shapes and applications for steel in construction followed the revolutionary change that occurred when builders recognized that the heavy timber commonly used for building frames could be replaced by smaller closely

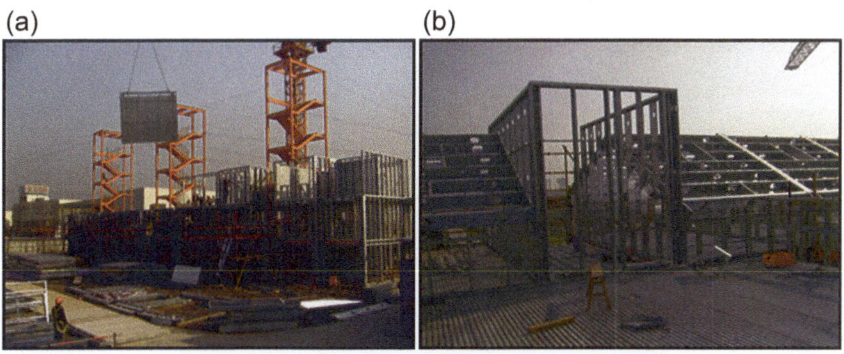

Figure 1.1(a,b) Cold-formed steel construction methods in China (shown here) follow the same general principles as in all other countries around the world.

Recent Trends in Cold-Formed Steel Construction. http://dx.doi.org/10.1016/B978-0-08-100160-8.00001-3

Figure 1.2 Residential.

Figure 1.3 Nonresidential.
Steel Framing Industry Association. cfsteel.org.

Figure 1.4 Virginia Baptist Hospital.

spaced wood members. This helped reduce the manpower required for assembling the building frame, but also made it possible to maximize the yield from a stand of trees which contributed to lower material costs for frames. Widespread adoption of this technique in the 1880s was further facilitated by the development of hydraulically powered sawmills and manufacturing techniques that could inexpensively produce wood framing and nails.

A recognizable form of CFS framing appeared in the early 20th century, most likely to mimic the dimensioned wood studs that had become the common construction method for shelter. While there are a few vague claims as to who built the first structure with CFS—an architect in Berlin… a house builder in upstate New York… an apartment complex in France—it appears that the first documented use of CFS as a building material is the Virginia Baptist Hospital (Fig. 1.4) built around 1925 in Lynchburg, Virginia. The walls were load-bearing masonry, with a floor system framed with double back-to-back CFS lipped channels. A site observation during a recent renovation confirmed that these joists from the "roaring twenties" are still supporting loads.

CFS finally made a grand debut in 1933 at the Chicago Century of Progress Exposition (World's Fair). Here the "home of the future" exhibit area featured three homes that either made extensive use of cold-formed shapers or were completely framed with steel: the General Houses home, the Armco-Ferro house (Fig. 1.5), and the Good Housekeeping/Stran-steel house (Fig. 1.6). CFS materials were also used as a nonstructural material in the Hall of Science. This new material was also touted in the official program of the World's Fair: "the genius of man has provided factory-made parts, wall materials pre-fabricated in shops, steel frames and clips and screws for quick assembly, and new compositions, all to permit the building of staunch structures…"

Figure 1.5 Armco-Ferro house.

Sadly, this enthusiasm for steel-framed homes never translated into new demand from the residential market. The primary reason was that costs were much higher than for wood-framed systems due to the difficulty in obtaining cold-formed studs and accessories and a total lack of any design, manufacturing, or installation standards.

The first true foundation stone for the CFS framing industry was laid in February 1939 when the American Iron and Steel Institute's (AISI) Committee on Building

Figure 1.6(a) Stran-steel house.

Figure 1.6(b) Cover of promotional World's Fair booklet.

Codes sponsored a research project at Cornell University that eventually resulted in the 1946 publication of the first edition of the AISI's "Specification for the Design of Light Gage Steel Structural Members" (Fig. 1.7).

The release of this important document coincided with the next major event in the development in the steel framing industry: the end of World War II.

1.1.1 The postwar boom

In the years immediately following World War II a number of countries around the world faced a housing crisis. Millions of homes and commercial buildings in Japan, Germany, France, and other countries had been destroyed in the global conflict. The United States and other nations faced a wave of returning soldiers who would soon be starting families and businesses.

Some countries, like Japan, did not have the wood to replace these houses. Postwar builders in Australia not only faced a shortage of building materials, but also the voracious Formosan subterranean termite that had immigrated into the northern region of the country. With steelmaking capacity no longer needed for war production, it occurred to more than one visionary that a noncombustible, termite-proof, engineered steel wall-framing system was the ideal solution.

In the United States the Lustron Corporation set out to mass produce prefabricated homes (Fig. 1.8), and between 1948 and 1950, completed 2500 structures in 36 states

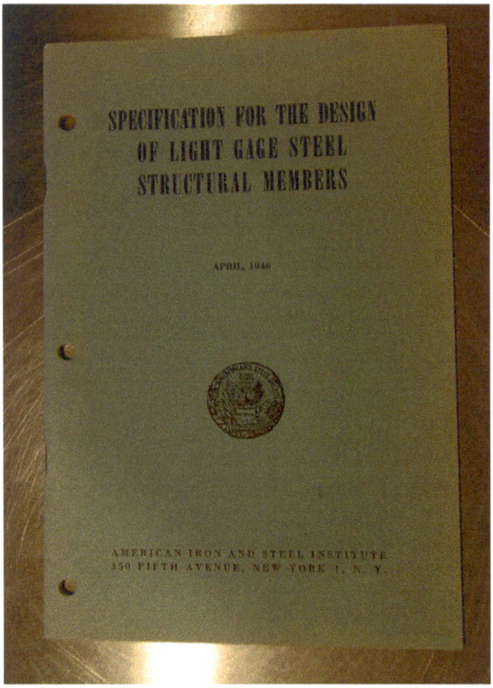

Figure 1.7 AISI CFS specification.

Figure 1.8(a) Lustron home on a trailer.

Figure 1.8(b) Original Lustron home.

and Venezuela. Unable to make a profit, however, the company eventually closed its doors. A number of other systems also came on the market during this period, including a housing development designed by architect Donald Wexler in the small community of Palm Springs, California.

In Australia the Econosteel system was first developed for widespread use, with bituminous paint providing a barrier for corrosion protection and a tab-and-slot connection method. More than 300 homes were built in the Canberra area during the 1950s, but the cost was six times higher than for a comparable wood-framing system. The experiment continued into the 1960s, when an American builder constructed about 50 houses on reclaimed land around the Gold Coast in Queensland, using galvanized frames fabricated similarly to timber and finished with stucco. In the late 1960s the first steel-framed houses were also built in New Zealand.

1.1.2 CFS and the commercial market

While interest in steel-framed homes waxed and waned, CFS really made lasting headway as a mainstream building material in nonresidential construction in the United States. During the 1950 and 1960s the construction of taller buildings where safety and constructability were primary concerns created new demand for lightweight, noncombustible CFS. New technologies were being introduced that made steel-framing construction easier and faster, including the self-drilling screw to replace "nailable" studs (Fig. 1.9) and wire ties for metal lath. The parallel development of tools to drive the screws during the 1950s made CFS even more attractive, and steel-framing manufacturers and suppliers began to pop up across North America.

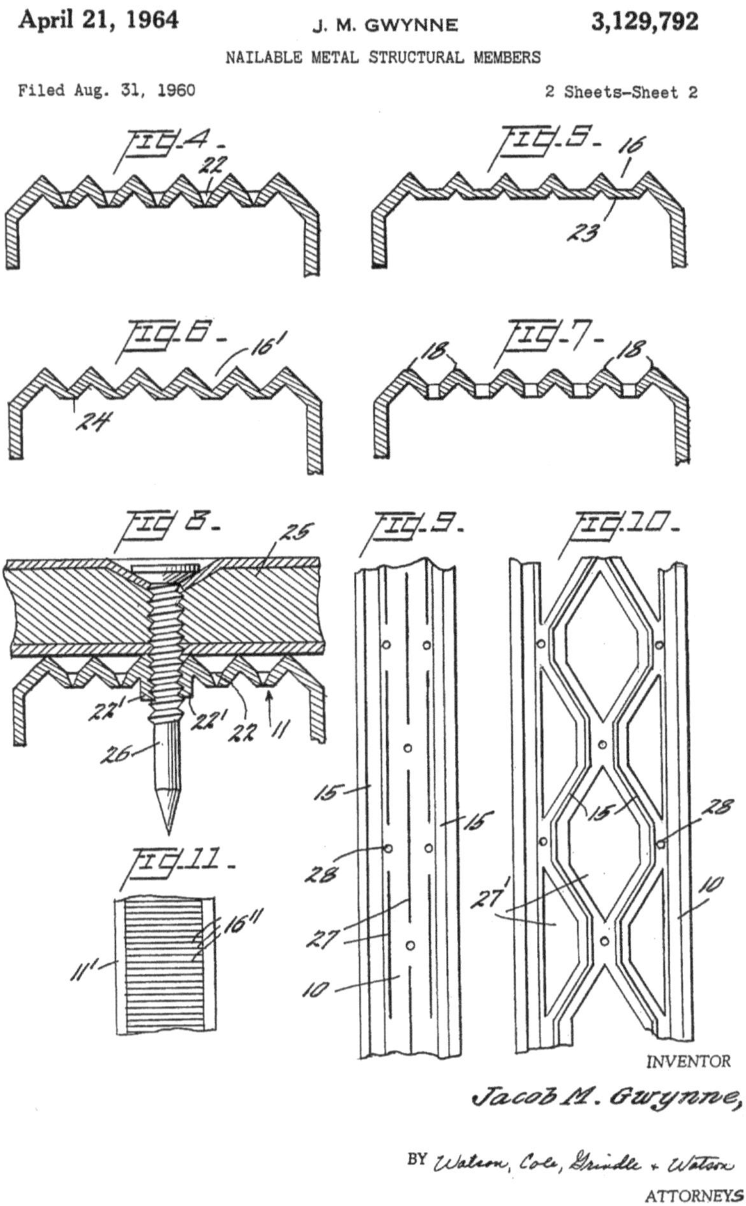

Figure 1.9 Nailable stud.

During the 1960s CFS was used in new systems like curtain walls, exterior framing with brick veneer, and interior shaft walls.

The initial gains in CFS market share occurred in nonstructural applications like partition walls within buildings. In 2004 the Steel Framing Alliance reported that 81% of interior walls built in the United States used CFS framing. With greater familiarity in the market, the establishment of building codes and standards, and availability of fire-rated assemblies, the use of structural studs also picked up steam and by 2011 had surpassed the tons of steel used to manufacture nonstructural studs.

Today, the Steel Framing Industry Association reports that 30–35% of all nonresidential buildings in the United States use CFS structural and nonstructural framing.

1.1.3 The residential market (1990–early 2000s)

Even with the success in nonresidential construction, the big prize has long been the potentially enormous demand for steel that home building represents. In 2007 the Steel Framing Alliance estimated that the US homebuilding market alone could consume up to 20 million tons of steel—or 20% of all the steel produced in the United States.

The first focused industry efforts were organized in the 1990s with the formation of groups like the Light Gauge Steel Engineers Association (USA), Steel Framing Alliance (USA), National Association of Steel Housing (Australia and New Zealand), AISI/Committee on Framing Standards (USA), Steel House Club (Korea), and the South African Steel Framing Association. These organizations had the common objective of setting technical standards, encouraging new technologies, and developing the training, marketing, and supply chains needed to enable and encourage growth in the residential CFS framing industry.

1.1.4 The Living Steel global project

In 2005 the International Iron and Steel Institute (IISI, since renamed the World Steel Association) launched a global marketing effort, Living Steel, to promote interest in the use of CFS, primarily for the housing market. The centerpiece of the effort was an annual architectural contest to identify specifiers and provide opportunities for IISI member companies to engage decision-makers. Demonstration projects were attempted in India, China, and Russia.

Steel producers introduced new capabilities and product lines to capitalize on the residential market opportunity, such as Arcelor's Styltech system (Fig. 1.10) in France and Nucor's Nuconsteel group, which helped develop innovative new automated manufacturing processes to simplify and speed design, manufacturing, and assembly. Important work was also being accomplished in other parts of the world, including Sweden, Turkey (Fig. 1.11), Japan, the UK, and Germany.

The initial success of this industry-wide effort was seen in Hawaii (Fig. 1.12), where it was aided by the Formosan subterranean termite (Fig. 1.13), one of the hungriest of wood-consuming insects and responsible for billions of dollars in structural damage each year. In the 1990s termite damage was costing builders up to $50,000–60,000 per project for repairs, in addition to endless lawsuits from residents. Ultimately, both homeowners and builders supported new building code regulations that required any wood framing to be treated for termites, and immediately alternative building materials that once were slightly more expensive, like CFS framing, were on a level playing field. By the early 2000s, 72% of all homes in Hawaii were framed with steel.

Figure 1.10 Styltech.
ArcelorMittal.

Figure 1.11 House01—Turkey.

Figure 1.12 House on the hill in Hawaii.

Success in other parts of the world was mixed. In Australia, residential market share for steel framing reached 12% in 2005. On the US mainland, residential market share reached 16% in multifamily walls and 3% in the single-family segment. However, the Living Steel project showed no appreciable return on investment.

The economic crisis of the last decade and economic and demographic trends have reshaped many of the markets for building products around the world. There is an acute need for housing and business structures in developing economies like China, Brazil, and India, giving framing systems that can be installed more quickly a competitive advantage over traditional concrete construction. In countries with aging populations, building types more suitable for urban environments are increasingly in demand. Adaption to climate change will increasingly require materials that are strong, ductile,

Figure 1.13 Formosan subterranean termite.

Figure 1.14 Blast furnace.
Steel Framing Industry Association. cfsteel.org.

and durable. These all infer increasing demand for strong, adaptable, and sustainable building materials like CFS.

1.2 What is cold-formed steel?

Metal studs. Light-gauge metal. Light-gauge steel. Lightweight framing. All these terms have been part of the vernacular over the years, and they may be correct to some degree. However, the common name for our material has evolved to what is now the most proper term: cold-formed steel.

Many steel products are produced or processed with the introduction of varying degrees of heat, ranging from molten steel that is cast into slabs or hot band coils to extruded products like structural shapes, bars, and wire, or hot-milled products like plate. As the name suggests, CFS products require no additional heat when being formed into the desired shape.

Like other steel products, CFS starts with the production of raw steel, which is made by combining iron ore or steel scrap with small amounts of carbon in a basic oxygen furnace (Fig. 1.14) or electric arc furnace. Molten steel is poured into slabs that are reduced to thinner strips of steel, called "hot band." In the steel finishing process, the hot band is reduced once again into "cold-rolled steel." A protective coating of zinc is then added through the galvanization process. The final product is called a "coil" (Fig. 1.15). To convert coils into CFS framing members, they are first slit into widths that match the intended dimensions of the final product.

The slit coils (Fig. 1.16) of CFS are formed into C-sections and other shapes, usually by roll forming (Fig. 1.17) the steel through a series of dies. Penetrations for plumbing and electrical runs are punched at predetermined locations, helping reduce installation times. A variety of steel thickness is available to meet a wide range of structural and nonstructural applications.

Figure 1.15 Coils.
Steel Framing Industry Association. cfsteel.org.

Figure 1.16 Slit coils.
Steel Framing Industry Association. cfsteel.org.

To meet custom material requirements, studs, track, and joist material can be cut within 1/8 in. of specification, and length is restricted only by the mode of physical transportation—typically 40 feet for containers and flatbed trucks. This ability to order material cut to length has several advantages for the builder, including less field cutting, and less labor and waste on the job site, which in turn reduce costs and noise during the construction process.

Modern automated roll-forming equipment (Fig. 1.18) goes one step further by directly integrating CAD designs into the manufacturing process to produce framing

Figure 1.17 Roll forming.
Steel Framing Industry Association. cfsteel.org.

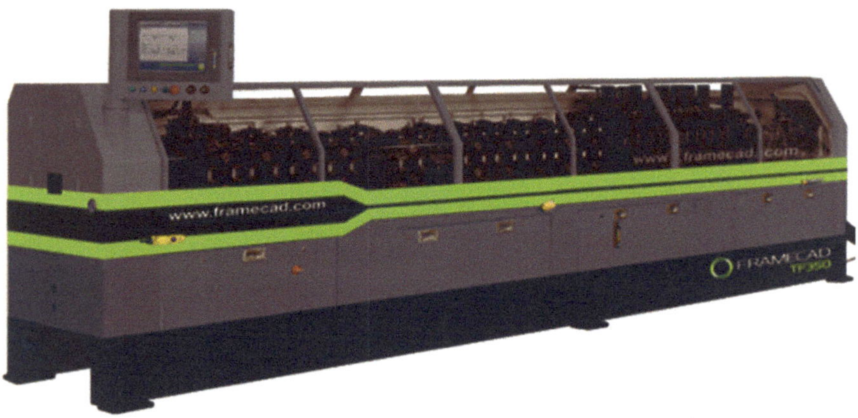

Figure 1.18 Automated roll former.
FrameCAD, Limited.

members that are precut, punched, and sized for specific locations in the wall, floor, or roof. Members can then be efficiently assembled into panels in a controlled environment. This new technology can help reduce the training time needed for installation crews, and almost completely eliminate waste in the factory and on the job site. Some manufacturers even provide roll-forming equipment that can be stationed at a job site for the duration of a project.

The finished product is sold to contractors and builders, and many distributors have become full-service materials providers selling framing members, screws, tools, and other steel-framing accessories. These suppliers have made steel-framing products

Figure 1.19 Bundled panels.

more accessible. In addition, many traditional building materials suppliers also stock CFS framing materials.

Finally, CFS framing materials are then either used to build wall and floor panels and roof trusses, or delivered to the job site in bundles (Fig. 1.19).

CFS can be found in all durable goods, like appliances and automobiles, but is most commonly used to describe products in the construction sector.

1.2.1 Physical dimensions

In the United States, steel thickness is expressed as "mils" or thousandths of an inch. When defining CFS, the accepted range of thickness can be from 0.0147 in (0.373 mm) to about $^1\!/_4$ in (6.35 mm). Steel plates and bars as thick as 1.0 in (25.4 mm) can also be cold-formed successfully into structural shapes, but this special thickness may entail additional considerations in design and installation.

1.2.2 Cold-formed steel framing shapes

Typically, CFS framing is "C" shaped with the three primary surfaces being the web, the flange, and the return lip. The web depth typically has the greatest variability, and is sized to meet the structural requirements of a specified design condition and to accommodate wall or floor sizes. In wall framing, structural members (called "studs"—Fig. 1.20) carry axial loads and thus are made with thicker steel than nonstructural studs, which are used for interior walls and expected to provide only a surface for attaching wall finishes.

The horizontal boundary of a wall is provided by a section called a "track" (Fig. 1.21), and the vertical framing members ("studs") are seated into the track at a

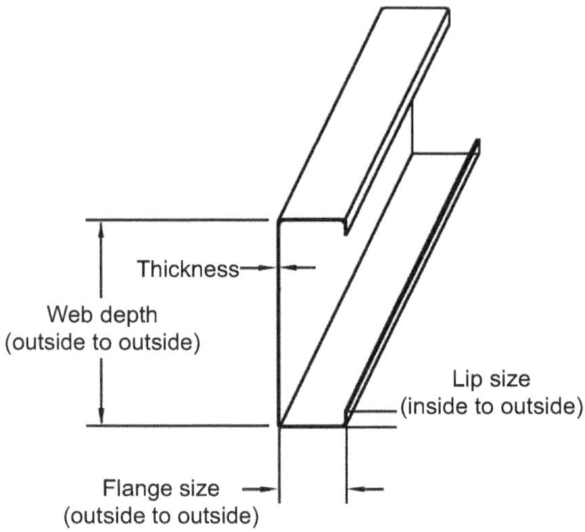

Figure 1.20 Stud.

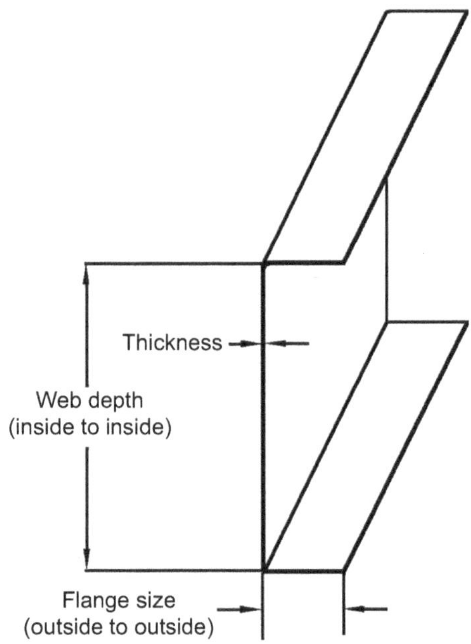

Figure 1.21 Track.

perpendicular angle. In floor and roof framing the load-bearing members are called "joists" (Fig. 1.22), and the track is typically referred to as the "rim joist."

CFS members can be used to serve a variety of purposes, and the most common members and uses include the following.

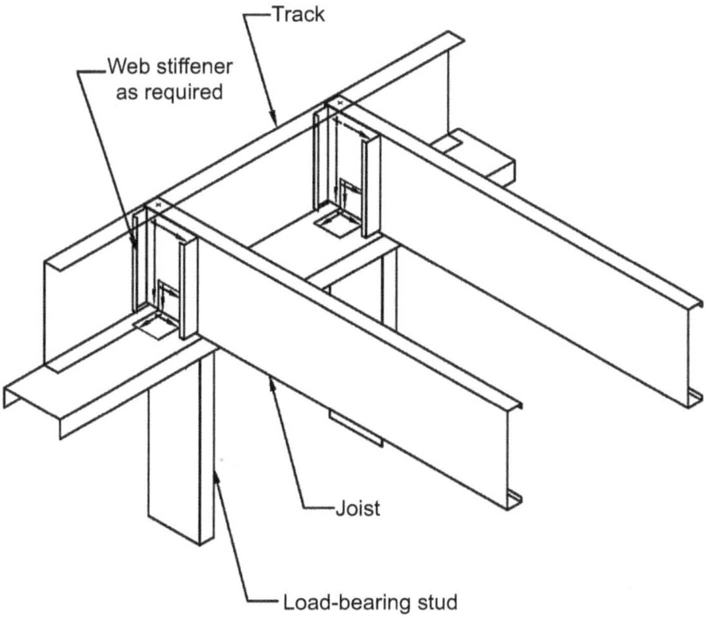

Figure 1.22 Joists.

- A **U-channel** (Fig. 1.23) serves as bridging material, backing for intersecting walls, and an attachment surface for cabinets.
- A **furring channel** (Fig. 1.24), or "hat" channel, may be used to brace walls or as a component in sound-attenuating assemblies.
- Headers can be constructed with joist material. An increasingly popular option is to fasten an L-shaped strip of CFS to the top plate above a wall opening. This is called an **L-header** (Fig. 1.25). In some engineered designs this can also effectively serve as a load distribution plate.

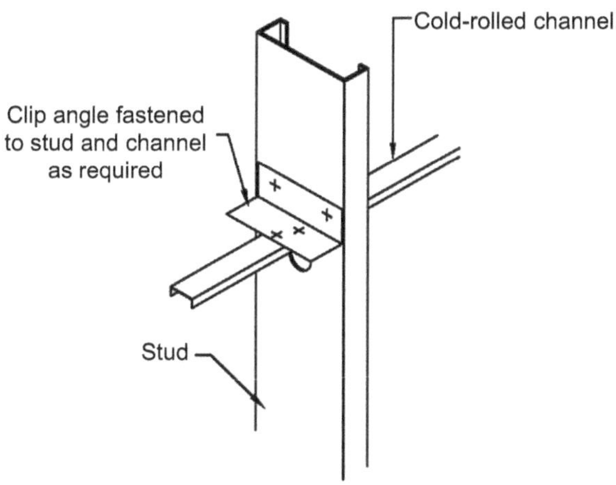

Figure 1.23 Cold-rolled channel.

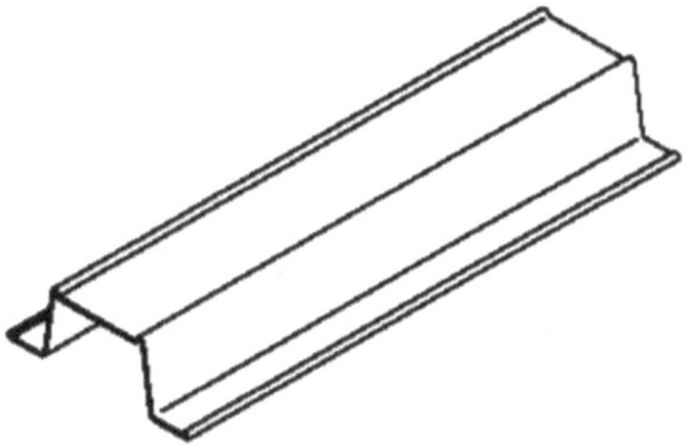

Figure 1.24 Hat (furring) channel.

CFS strapping is used to provide bracing, blocking, and wall bridging, and there are dozens of specialty connectors and framing products available to meet almost any structural or assembly requirements.

CFS is an engineered material, meaning that such characteristics as strength and ductility can be manipulated through slight alterations in the chemical composition and the manufacturing process. In many parts of the world CFS framing has a tensile strength of around 33 ksi (Kips per square inch) (228 MPa). In Australia,

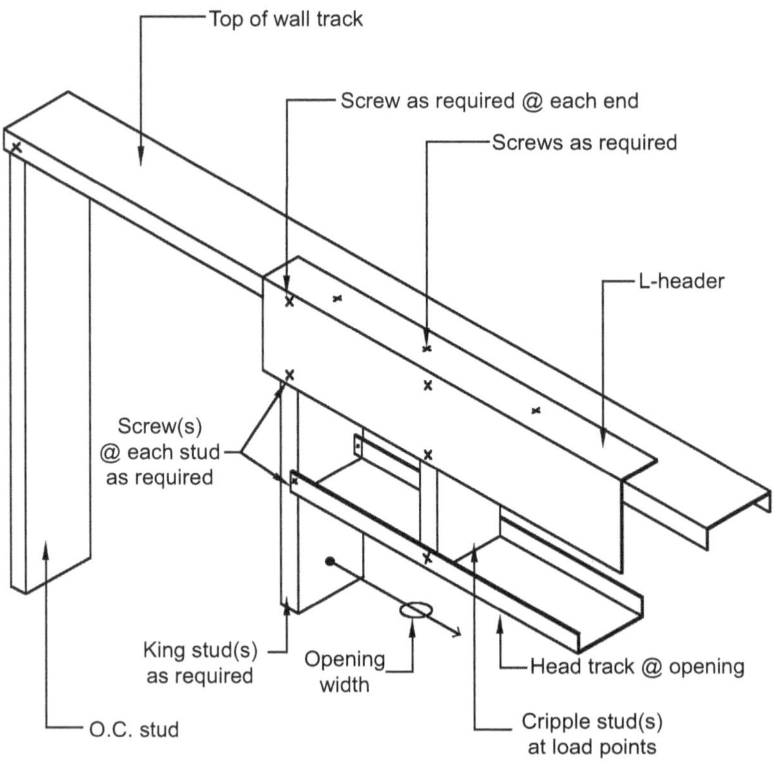

Figure 1.25 L-header.

Figure 1.26 EQ stud.

New Zealand, and South Africa, however, the use of high-strength steel of 80 ksi (550 MPa) is standard. Stronger steel members allow flexibility in stud spacing and roof truss placement, as well as studs that are somewhat lighter than those common in other countries.

1.2.3 EQ studs

Over the past 15 years the ability to use the properties of steel to produce more competitive products has led to the development of new framing products. Perhaps the most widely adopted product is the EQ (equivalent gauge) stud (Fig. 1.26) for nonload-bearing walls.

As with high-strength steel, EQ studs are formed from steel with higher yield strength than traditional framing studs typical in such markets as North America, Japan, and Europe. EQ studs also take advantage of stud manufacturing methods that introduce ribs and other devices to stiffen and strengthen the flanges and webs. The higher-strength steel also offers the advantage of higher pullout and shear values.

These new products are termed "equivalent" because the combination of the stiffening elements and higher yield strength enables the production of studs with thinner steel than traditional studs, yet able to carry the same loads. There are a number of proprietary EQ studs on the market today, accounting for 90% of all CFS studs used in the United States.

1.2.4 Designator systems

Since 2001 the United States has used a designator system (Fig. 1.27) that has become the industry standard for identifying the most commonly used CFS framing members. The designator consists of five sequential codes. The first is a three- or four-digit numeral indicating the member web depth (D) in 1/100 in. The second is a single letter indicating the type of member, as follows:

S = stud or joist framing member with lips
T = track section

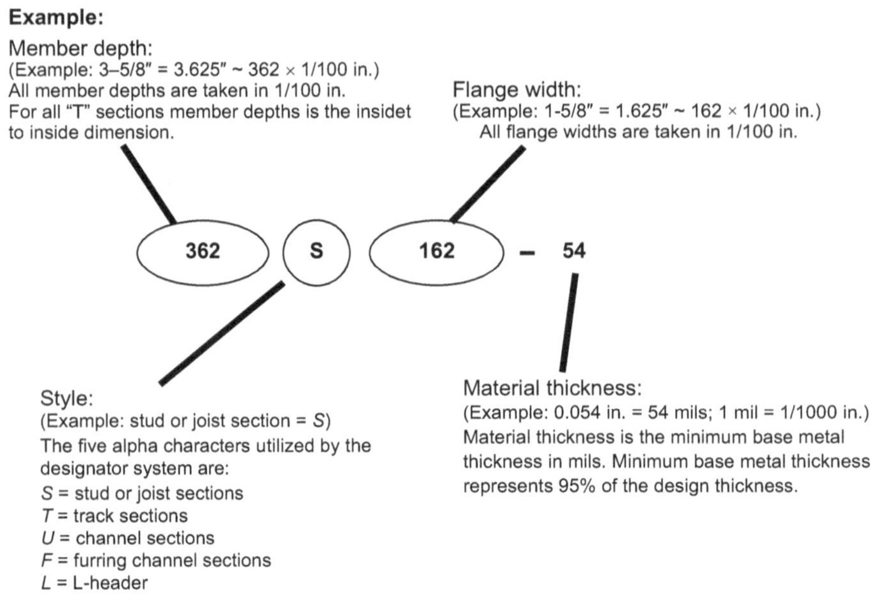

Figure 1.27 Designation system.

U = channel or stud framing section which does not have lips
F = furring channel
L = angle or L-header.

The third code is a three-digit numeral indicating flange width (*B*) in 1/100 in., followed by a dash. The fourth is a two- or three-digit numeral indicating the base steel thickness in 1/1000 in. (mils).

As an example, the designator for an 8″ 16-gauge C-shape with 1−5/8″ flanges is **800S162-54**:

800 = 8″ member depth expressed in 1/100th in. (outside to outside dimension)
S = stud or joist with flange stiffeners
162 = 1−5/8″ flange in 1/100th in.
54 = minimum base steel thickness in mils (0.054 in = 54 mils).

1.2.5 Corrosion resistance

When moisture comes into contact with bare steel, it creates a chemical reaction called corrosion that over time will degrade the structural properties of the metal. CFS members are typically protected from moisture by a thin coating of zinc that is applied in a process called galvanization.

The zinc protects the steel in two ways. First, it provides an excellent barrier to moisture because galvanized coatings (Fig. 1.28) have excellent adhesion

Figure 1.28 Spangle typical of a galvanized coating.

characteristics and are extremely resistant to abrasion. Unlike paint, a galvanized coating will not crack, peel, fade, or mechanically degrade over time.

Also, when the base steel is exposed as a result of a cut, scratch, or other type of surface damage, the steel is cathodically protected by the sacrificial action of the zinc. This occurs because zinc is more electronegative (more reactive) than steel in the galvanic series (Fig. 1.29), which means that steel cannot corrode as long it is

Galvanic series of metals and alloys (in seawater)

Corroded end – anodic
(Electronegative)
Magnesium
Zinc
Aluminum
Cadmium
Iron or steel
Tin
Copper
Lead
Silver
Stainless steel (passive)
Gold
Protected end – cathodic or most noble
(Electropositive)

Note: any one of these metals and alloys will theoretically corrode while protecting any other that is lower in the series as long as both form part of an electric circuit.

Figure 1.29 Galvanic series.

adjacent to zinc. This is also the reason why additional corrosion protection is not required on edges of metallic-coated steel framing members, even if they are shop or field cut, punched, or drilled.

In enclosed locations (walls, attics, floors) zinc coatings can protect the steel for as long as 1150 years. For exterior exposed or semiexposed locations in an aggressive environment subject to higher humidity and exterior pollutants, the higher corrosion rates can still be extrapolated to over 150 years of coating life, well beyond the serviceable life of modern buildings.

1.2.6 Coating designations

Each coated steel sheet product has its own coating weight designation system, which is defined in the appropriate American Society for Testing Materials (ASTM) standard. For example, the most widely used ASTM metallic-coated sheet standard is A653/A653M, which covers hot-dip galvanized products.

One of the coating weight designation systems in this standard uses descriptors such as G40, G60, and G90. The "G" means the coating is galvanized (zinc), and the numbers refer to the weight of zinc on the surface of the steel sheet in inch-pound (UK/imperial) units. Taking G90 as an example, the coating weight on one square foot of sheet (total on both sides of the sheet) shall have a triple spot test average minimum of 0.90 oz.

The other measurement system in widespread use today is the SI (metric) system. The conversion from the inch-pound weight in ounces per square foot (oz/ft^2) to the SI mass in grams per square meter (g/m^2) is:

1.00 oz/ft^2 = 305 g/m^2
To convert from oz/ft^2 to g/m^2, multiply by 305.
Example: G90 (0.90 oz/ft^2) = Z275 (275 g/m^2).

Structural and nonstructural CFS framing members used in construction are required to have a minimum metallic coating that complies with ASTM A1003/A1003M, as follows.

Structural members—G60 minimum.
Nonstructural members—G40 or equivalent minimum.
Harsh environments (see definitions in glossary)—G90.

1.2.7 Equivalent coatings

Most industry members are familiar with popular, standard coatings like galvanized and galvanealed, but new "EQ" or equivalent coatings already in use by the automotive and appliance industries are finding a market.

Commonly, EQ coatings are the combination of a base metallic coating and an outer coating that chemically bonds to the base. The base metallic coatings are generally standard coatings that do not meet the minimum weight requirements of ASTM A1003. To improve the performance of the base coating, a supplemental coating is added to increase the coating corrosion resistance.

These new coatings are produced by a number of manufacturers and vary in both appearance and performance. ASTM 1003 provides guidance on the minimum performance requirements.

1.2.8 Thermal conductivity

Energy use, as a key component of sustainability, has risen to the top of the regulatory agenda around the world, and buildings have become a priority target for proponents of energy efficiency. New codes and standards are continuing to evolve, and building officials and code development organizations are much more informed about energy use and related issues in buildings.

In wall construction, energy efficiency is typically managed through mitigating thermal transfer of heat or cold from one surface of the wall through to the other surface (interior and exterior, or both sides of an interior wall). All objects in a wall, include wood or steel studs, joists, trusses, concrete beams, masonry block, brick, and even nails and screws, can provide a bridge or pathway for heat and cold to transfer from one side to the other. In some cases the extra heat loss amounts to very little in terms of actual energy use. In other cases, especially in colder climates, it can be significant.

Steel is a highly efficient conductor of heat and cold, and one of the most common methods for reducing thermal transfer in a CFS framed wall is the use of continuous insulation (Fig. 1.30). When applied to the exterior side of the CFS studs, the sheathing creates a thermal break and allows the system to achieve the desired energy performance requirements.

An alternative solution that has been studied and seen some limited use in Scandinavia and cold climates is the slotted stud (Fig. 1.31), where staggered longitudinal

Figure 1.30 Installing continuous insulation.

Figure 1.31 Slotted stud.

penetrations are introduced into the web of the stud during manufacturing. These slots create a longer path for the thermal energy to travel, which reduces the amount of heat or cold that reaches the flange on the opposite side of the stud. This product has had limited use due to the additional costs of manufacturing and the reduced load capacity resulting from the web penetrations.

1.3 Advantages of cold-formed steel

Light frame systems, like those that use CFS, offer a number of inherent advantages over heavy frame construction, including design flexibility and shorter construction cycles that enable owners to enjoy faster occupancy or rental revenue. CFS also has unique advantages over other light frame materials, particularly wood, in a number of other ways.

Lateral load resistance. Structures are designed to absorb energy produced by ground movement and wind by "flexing" or "deflecting" in varying degrees, depending upon the construction materials, design of the structure, quality of construction, level of engineering, and the applicable building code requirements. CFS is an optimal material for this purpose because it is ductile, making it more forgiving than other more brittle materials in earthquakes and high winds, and has inherent strength in uplift and gravity loading.

Consistent performance. Steel behaves in a highly predictable manner when subjected to the structural loads and movements imposed by high wind and seismic events. This is because steel is an inherently stable, manufactured material with consistent chemical and mechanical properties: once a steel stud has been formed, it will remain straight with virtually no change to the strength, stiffness, thickness, width, or other dimensions. Likewise, fasteners used to join steel framing members retain their strength and reliability over time.

Strength-to-weight ratio. This relatively easy way to compare the merits of several different materials is determined by dividing the maximum imposed load by the weight of the material. Of all commonly used construction materials, steel has the highest strength-to-weight ratio. When CFS sheet is formed into a C-shape, like a stud, the bends act as stiffeners and increase the strength of the steel sheet dramatically, providing a strength-to-weight ratio that is up to seven times greater than that of dimensional lumber.

Noncombustible. Steel-framed structures are inherently noncombustible, and do not burn or contribute to the spread or intensity of a fire. Noncombustible CFS construction also makes sense from a cost-saving standpoint, as insurers traditionally offer lower builders' risk and general liability premiums compared to wood.

Connection strength. Because the material and geometric properties of a steel-framing member are stable, the overall strength of the structure will depend upon the quality of connections between the studs. Steel framing typically uses screws that provide a mechanical locking connection where the load is carried in shear. This is in direct contrast to wood, where connection strength is often limited not by the strength of the fastener, but by the resistance of the wood in bearing or withdrawal.

The sustainability of steel

> **Recycling.** The Steel Recycling Institute reports that steel is recycled more than paper, plastic, glass, copper, lead, and aluminum combined. All steel products, including steel framing, contain recycled steel. Steel framing contains on average a minimum of 25% recycled steel and is 100% recyclable at end of life. Using recycled steel takes the pressure off renewable resources: about six scrapped cars are needed to build a typical 2000 ft^2 home with steel framing. Finally, in contrast to many other building materials, steel is routinely collected in aggregate quantities from construction and demolition sites and recycled into new steel products.
>
> **Energy and emissions.** The World Steel Association states that the steel industry worldwide has reduced energy consumption since the 1970s in the manufacture of steel by 50%. This directly relates to a reduction in greenhouse gas emissions. The US Environmental Protection Agency documents that the North American steel industry has reduced greenhouse gas emissions by 47%.

Termite-proof. Each year termite infestations cost billions of dollars in damage, and represent a threat to buildings in warmer climates. CFS is one of the few materials that can resist termites in nearly any climate or building type. Even though it seems like termite damage would be a long-term issue, when a hurricane or other high wind strikes, it pays to have a building that performs as designed versus one that may be weakened by termite damage. The Formosan termite poses a unique threat because, unlike the more traditional subterranean termite that attacks from the ground up, the Formosan termite can establish colonies even on the roof due to its ability to attack aerially. In these cases using CFS framing for the entire structure will increase the building's resilience.

Mold-proof. The use of CFS framing mitigates two primary causes of mold growth. Since steel will not absorb water, as wood does, it will not retain dampness in the space it encloses.

In effect, steel framing aids in drying out the space faster as it will not serve as a moisture reservoir. Steel is inorganic. It will not function as a food source for mold. Further, steel is dimensionally stable in a moist environment. It will not warp. Walls and floors remain plumb and level in a wet environment.

1.4 Applications for cold-formed steel framing in buildings

CFS can be used as the sole material in a framing system, or used in hybrid systems that combine several materials to take advantage of the unique structural properties of each.

1.4.1 Wall systems: interior/nonbearing/nonstructural

In high-rise commercial and multifamily residential construction, or structures that are eight stories and taller, CFS framing is typically used for interior partitions (Fig 1.32) and to support exterior walls and cladding. It provides the noncombustibility often required by local building codes, and also the lightweight that is desirable when lifting building materials to the higher levels of a building.

1.4.2 Wall systems: exterior/load-bearing/structural

Structural CFS-framed wall assemblies carry gravity loads (live, snow, and dead loads) and lateral loads. They are commonly used for commercial structures of all heights, and increasingly as the structural system in multistory retail, mixed-use, and

Figure 1.32 Interior partition.

apartments. Typically four to seven stories in height, these are generally classified as midrise buildings (Fig. 1.33).

The basic elements of a structural steel wall include structural studs, track, bracing and bridging, headers or load distribution members, and fasteners and connectors. When shear walls are part of the element design, they are typically framed within the wall assembly.

In a typical wall-framing plan, the load path is established by aligning the structural studs over the joists, which are then positioned over the studs below, and so on. This is called "in-line" framing and applies to all members, from the roof rafters or trusses through the wall studs and joists to the foundation.

1.4.3 Floors and roofs

Over the past decade the use of CFS framing in roof systems has grown, along with the use of structural studs for load-bearing walls. There are a number of reasons for this, including the similarity of framing principles, durability, and the need for noncombustible construction.

Conventional, or rafter, framing is an assembly of joists, rafters, bracing, and blocking constructed in place. The main advantage of rafter framing is that it typically requires fewer steel members—which, in turn, decreases the number of connections and amount of labor needed. The disadvantages are that it takes longer to frame rafters than to set preengineered trusses. In addition, as rafters are framed in place, framers must operate from ladders and scaffolds for longer periods of time.

Truss framing is becoming increasingly popular because it offers all the advantages of a lightweight, noncombustible structure combined with prefabrication, which can shorten overall construction times. Using a wide range of depths and material

Figure 1.33 Midrise building.

Figure 1.34 Wide-span trusses.

Figure 1.35 Hollow-core plank.

thicknesses, trusses can be designed for up to 70 ft clear span (Fig. 1.34), and for a variety of spacing.

The trusses are fabricated from CFS components such as the typical C-shaped member, "Z" and "hat" channels, and special tube-shaped members. Truss members cannot be notched, cut, or altered in any manner without design consideration, and should be designed in accordance with design standards, local building code requirements, or an approved design using engineering principles.

1.4.4 Hybrid systems

Buildings today must meet increasingly complex design and performance criteria, including combinations of stiffness, strength, ductility, construction efficiency, and cost. Hybrid structures allow for a more efficient design of concrete and steel components because the material performance profiles can be tailored to match the capacity of the materials.

Several common hybrid structural systems that use CFS are:

- CFS with precast concrete plank (Fig. 1.35)
- CFS with long-span deck (Fig. 1.36)
- CFS with wood framing (Fig. 1.37)
- CFS with bar (or composite) joists (Fig. 1.38)
- CFS with structural steel (Fig. 1.39).

1.5 Construction methods

There are three ways to build structural walls, floors and roofs.

Stick building. A process whereby walls are framed on the job site, one stud at a time. Walls can be built on a flat surface at the site (Fig. 1.40), like a concrete slab, and then stood in place.
Preengineered system. The structural elements of the building are engineered for specific design conditions, to optimize the use of material fully or achieve another use or design

Figure 1.36 Long span.

Figure 1.37 CFS and wood framing.

Figure 1.38 CFS with bar joists.

Figure 1.39 CFS with structural steel.

Figure 1.40 Worker fastening studs together.

objective. Preengineered systems typically enable the use of specialty members or increase the spacing between studs, sometimes to as much as 4 ft (600 cm) and 6 ft (800 cm) on center.

Panelized components. Walls, floors, and roofs are preassembled off site and then installed as components. This method is effective at reducing construction time while improving the efficiency and quality of the steel-framing assembly. While trusses can be built on the job site, it is more common for them to be designed and assembled at a plant, then installed on site.

1.5.1 Panelization

One of the most important trends in modern light-frame construction is the rapid growth in the use of panelization, or assembling the components of the house (walls, floors, roofs) in a controlled manufacturing environment (Fig. 1.41). In a typical panelization project, 75–80% of the work is complete when the panel assembly is made available to the cold-formed framing trade (Fig. 1.42). Panel installation (Fig. 1.43) only takes 20–25% of the time it takes to fabricate the panels. This cuts traditional on-site building schedules for this scope of work by approximately 75%. While these improvements may not reduce the subcontractor's costs or prices, schedule improvements for the general contractor and/or owner have many economic values, including reduced delays for inclement weather conditions, more efficient job site supervision, lower safety risk, and faster turnover of completed projects.

Figure 1.41 Panel tables.

Figure 1.42 Panels on the job site.

CFS framing is particularly suited for panelization because it is manufactured to precise and consistent tolerances, and its lightweight allows easier handling of assembled components than is the case for other light framing materials. The relatively lightweight of CFS also makes it feasible to ship completed panels over long distances.

Figure 1.43 Panel being lifted into place.

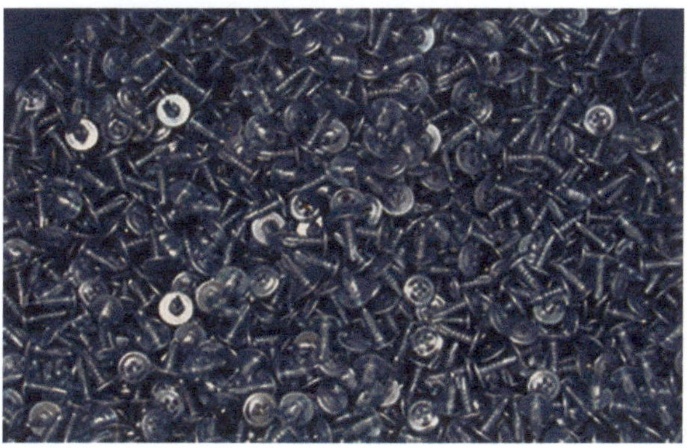

Figure 1.44 Screws.

1.5.2 Connection methods

Welds, bolts, screws, and powder-actuated fasteners are commonly used to connect elements in CFS structures. Typically screws (Fig. 1.44) and welds (Fig. 1.45) are used to attach bridging, studs to the bottom and top tracks, built-up jamb studs, built-up window headers, sill track to jamb studs, built-up headers to jamb studs, and connectors. Bolts (Fig. 1.46) are usually used to attach the bottom track to concrete or masonry, and powder-actuated fasteners are used to attach CFS members to structural steel, concrete, or masonry.

Figure 1.45 Welds.

Figure 1.46 Bolts/bottom track.

Sources of further information and advice

American Iron and Steel Institute, www.steel.org
Association of Wall and Ceiling Industries, www.awci.org
Cold-Formed Steel Engineers Institute, www.CFSEI.org
National Association of Steel Housing, www.NASH.asn.au
Southern Africa Light Steel Frame Building Association, www.SASFA.co.az
Steel Construction Institute, www.steel-sci.com
Steel Framing Industry Association, www.cfsteel.org
Wei-Wen Yu Center for Cold-Formed Steel Structures, www.ccfssonline.org
World Steel Association, www.worldsteel.org

Further reading

American Iron and Steel Institute, 2007. Commentary on North American Specification for the Design of Cold-formed Steel Structural Members. Washington, DC, Published.

ASCE Manuals and Reports on Engineering Practice No. 84, Mechanical Connections in Wood Structures, ASCE.

Cable, K., 2005. Western Metal Lath, Oral History.

GalvInfo Center, GalvInfoNote 1.1: Understanding Coating Weight Designations for Zinc-based Coatings on Sheet Steel.

"Official Guide Book of the Fair", A Century of Progress Administration Building Chicago, 1933.

IZA Research Program ZC4, Galvanized Steel Framing for Residential Buildings, January 2009. Ten Year Report, prepared by, NAHB Research Center, Inc., 400 Prince George Blvd., Upper Marlboro, MD 20774 USA.

Kosny, J., Christian, J., Desjarlais, A., 2001. Improving Energy Performance of Steel Stud Walls, Steel Framing Can Perform as Well as Wood. Oak Ridge National Laboratory. http://web.ornl.gov/sci/roofs+walls/research/detailed_papers/steel_frame/index.html.

Lustron Homes: The History of a Postwar Prefabricated Housing Experiment by Tom Fetters.

http://www.nash.asn.au/nash/nash-and-steel-framing/nash-milestones.html.

National Association of Steel Housing, www.NASH.asn.au.

Ryan Smith, P.E., December 2011. LEEP AP, Know Your Equivalent Coatings, Construction Dimensions.

Steel Framing Industry Association, 2015. Cold-formed Steel and Resilience.

Ten Year Corrosion Data of Zinc Coated Materials, June 2008. Corus Research, Development & Technology, Swindon Technology Centre, Moorgate, Rotherham, South Yorkshire S60 3AR UK.

The Steel Framing Alliance, Residential Steel Framing National Training Curriculum, http://store.steelframingalliancestore.com/restfrnatrcu1.html.

Thomson, S., 2005. Wrinkly Tin: The Story of Corrugated Iron in New Zealand, ISBN 978-1877338700.

Yu, W.-W., Wolford, D.S., Johnson, A.L., 1996. Golden anniversary of the AISI specification. In: 13th International Specialty Conference on Cold-formed Steel Structures (St. Louis, MO).

Zhang, X.G., 1996. Market Case Study – Styltech Framing, Arcelor, Presentation to IISI, 200410. In: Corrosion and Electrochemistry of Zinc. Plenum Publishing Corporation, New York, NY.

Zhang, X.G., 2005. Corrosion of zinc and its alloys. In: Metals Handbook. ASM International, Materials Park, OH.

Part One

Codes, specifications and design methods for cold-formed steel structures

Recent code development and design standards for cold-formed steel structures

2

C. Yu[1], H. Chen[2]
[1]University of North Texas, Denton, TX, United States; [2]American Iron and Steel Institute, Washington, DC, United States

2.1 Introduction

American Iron and Steel Institute (AISI) design and construction standards are developed and maintained by the AISI Committee on Specifications for the Design of Cold-Formed Steel Structural Members and the AISI Committee on Framing Standards. Both committees are operated following "Essential requirements: Due process requirements for American National Standards" established by the American National Standards Institute (2012). Those standards developed by the committees go through ANSI-conducted public review and approval processes to become American National Standards (ANS). The specification and standards developed by AISI are intended for adoption and use in the United States, Canada, and Mexico. In 2012 all AISI standards published in 2007 were either reaffirmed as the ANS or published as a new edition. A complete list of AISI standards that were updated in 2012 is provided in Appendix A.

During 2014–15 the contents of cold-formed steel (CFS) framing standards (AISI S200 standard series) and AISI S100 were reorganized. As a result, AISI S240, *North American Standard for Cold-Formed Steel Structural Framing*, and AISI S400, *North American Standard for Seismic Design of Cold-Formed Steel Structural Systems*, were published in 2015, replacing AISI S200, S210, S211, S212, S213, and S214. The reorganized AISI S100 was published in 2016.

The following sections provide an overview of major technical changes and additions included in the 2012, 2015, and 2016 editions of the standards.

2.2 Revisions of AISI S100, *North American specification for the design of cold-formed steel*

2.2.1 Technical changes and additions in AISI S100-12

AISI S100 is a standard that addresses individual CFS member and connection design, and is applicable in the United States, Canada, and Mexico. After the publication of the

Recent Trends in Cold-Formed Steel Construction. http://dx.doi.org/10.1016/B978-0-08-100160-8.00002-5

2007 edition, Supplement nos 1 and 2 were published in 2009 and 2010, respectively. An overview paper for the supplements was published (Chen et al., 2010). The following paragraphs summarize the major technical changes made in the 2012 edition compared to the 2007 edition. The reference numbers of chapters, sections, and tables in this text correspond to those in AISI S100-12.

(a) Section A2: material. This section lists American Society for Testing Materials (ASTM) standards applicable to CFS applications, establishes the criteria for using other steels in CFS design, and provides ductility requirements for different CFS applications. In the 2012 edition the list of applicable steels was grouped by their minimum elongation requirements over a 2 in (50-mm) gauge length: a specified minimum elongation of 10% or greater, 3% to less than 10%, and less than 3%. The permitted uses and restrictions are then specified for each group. For example, steels with a specified minimum elongation of 10% or greater can be used without restriction as long as the steels meet the requirements (specified in Section A2.3.1). Steels with elongation from 3% to 10% can be used with reduced yield stress and tensile strength (specified in Section A2.3.2). Steels with a specified minimum of less than 3% may be used in multiple web configurations such as roofing, siding, and floor decking provided the adjustments specified in Section A2.3.3 are met. Additionally, ASTM A1063/A1063M, "Standard Specification for Steel Sheet, Twin Roll Cast, Zinc-Coated (Galvanized) by the Hot-Dip Process," was added in 2012.

(b) Section B1.3: corner radius-to-thickness ratios. This is a newly added section to address cold-formed cross-sections with a larger corner radius. Research (Zeinoddini and Schafer, 2010) indicated that the provisions in Chapter B may be insufficiently conservative in predicting the effective width if the corner radius-to-thickness ratio is larger than 10. To take into consideration the corner radius effect, a rational engineering analysis method such as the direct strength method may be employed. A prescriptive method applicable for $10 < R/t \leq 20$ is also provided in Commentary Section B1.3, and is shown below.

To determine the effective width of the element between the left and right radius R_1 and R_2, a reduced plate-buckling coefficient k_R can be determined:

$$k_\mathrm{R} = kR_{R1}R_{R2} \qquad [2.1]$$

where

k = plate-buckling coefficient determined in accordance with S100-12 Sections B2—B5, as applicable.

$$R_{R1} = 1.08 - (R_1/t)/50 \leq 1.0 \qquad [2.2]$$

$$R_{R2} = 1.08 - (R_2/t)/50 \leq 1.0 \qquad [2.3]$$

where

R_1, R_2 = inside bend radius (see Fig. 2.1)

t = thickness of element (see Fig. 2.1).

(c) Section B2.5: uniformly compressed elements restrained by intermittent connections. This section is used to determine the effective widths for elements such as decks that are restrained by intermittent connections. In 2012 provisions for determining the effective width of e, as illustrated in Fig. 2.2, were added.

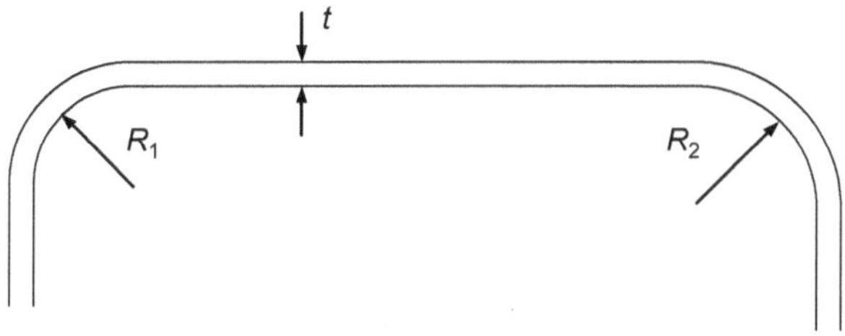

Figure 2.1 Corner radius.

(d) Section C3.4.1: web crippling strength (resistance) of webs without holes. Based on experimental findings (Yu, 2009a,b) that the values provided in AISI S100 Table C3.4.1−2 for interior two-flange loading or reactions may be insufficiently conservative for small-depth C-sections, a limit of an out-to-out web depth greater than or equal to 4.5 in (110 mm) was added.

(e) Section C3.6: combined bending and torsional loading. This section takes into consideration the torsional effect for singly or doubly symmetric section members subjected to bending and torsional loading by applying a reduction factor, R, to the nominal flexural strength determined based on initial yield. The reduction factor, R, was revised in 2012, as shown in Eq. [2.4]:

$$R = \frac{f_{\text{bending\_max}}}{f_{\text{bending}} + f_{\text{torsion}}} \leq 1 \qquad [2.4]$$

where

$f_{\text{bending\_max}}$ = bending stress at extreme fiber, taken on the same side of the neutral axis as f_{bending}

f_{bending} = bending stress at location in cross-section where combined bending and torsion stress is maximum

f_{torsion} = torsional warping stress at location in cross-section where combined bending and torsion stress effect is maximum.

Eq. [2.4] enables one to accommodate situations where the maximum stress due to combined bending and torsional warping occurs at the tip of a flange stiffener, and at web−flange or flange−lip junctions.

(f) Section D3.3: bracing of axially loaded compression members. This section contains provisions for determining the force required to brace a concentrically loaded compression member. The required bracing force is calculated based on the axial load in the column, and is also permitted to be determined using a frame analysis that takes into consideration second-order effects (ie, considering both P-δ and P-Δ effects). Since the brace force is

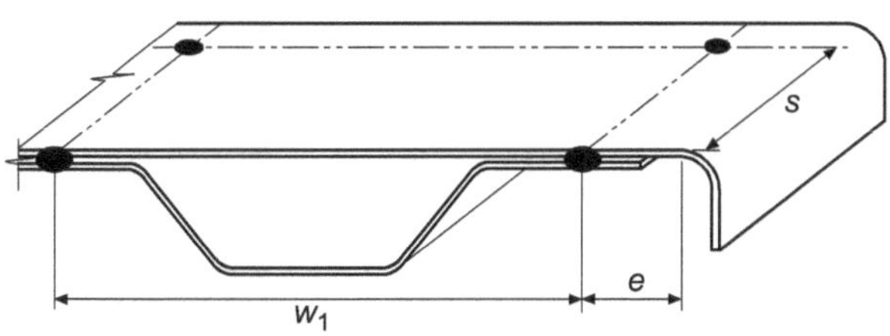

Figure 2.2 Effective width at edge stiffener.

based on the axial load in the column, the commentary cautions users that if the axial load in the column is increased, the corresponding brace member needs to be checked to make sure it is still adequate to brace the column with increased axial load.

(g) Section D6.1.1: flexural members having one flange through fastened to deck and sheathing. This section determines the flexural strength of a beam with its tension flange braced by deck or sheathing and its compression flange unbraced. Since the provisions were developed based on tests, application was limited by the tested systems. Based on a review of the conducted tests, some conditions were revised to accommodate member sizes commonly used in the industry: the member depth upper limit was revised to 12 in (305 mm); the upper limit of the depth/flange width is increased to 5.5 in; flange width to $\geq$2.125 in (54.0 mm); and ratio of tensile strength to design yield stress to $\geq$1.08.

(h) Section E3: bolted connections. Yu and Xu (2010) tested steel-to-steel bolted connections without washers on oversized and short-slotted holes. Based on the test data, they developed new equations for bearing factor C and new values for modification factor m_f. The hole dimensions investigated by Yu and Xu (2010) were consistent with those defined in Table E3a. Provisions were added to determine the strength of bolted connections for slotted or oversized holes without a washer installed. The slotted or oversized hole sizes are defined in AISI Table E3a. The bearing strength, P_n, without consideration of bolt hole deformation, can be determined from:

$$P_n = Cm_f dtF_u \qquad [2.5]$$

where C = coefficient determined in accordance with AISI S100 Table E3.3.1-1; m_f = modification factor for type of bearing connection determined according to AISI S100 Table E3.3.1-2; d = nominal bolt diameter; t = uncoated sheet thickness; and F_u = tensile strength of sheet as defined in Section A2.1 or A2.2.

The provisions are also applicable for oversized or short-slotted holes within the lap of lapped or nested Z-members which satisfy the following conditions:

(1) 1/2 in (12.7 mm) diameter bolts only, with or without washers or backup plates
(2) maximum slot size is 9/16 × 7/8 in (14.3 × 22.2 mm), slotted vertically
(3) maximum oversize hole is 5/8 in (15.9 mm) diameter
(4) minimum member thickness is 0.060 in (1.52 mm) nominal

Table E3a Maximum size of bolt holes

Nominal bolt diameter (d in)	Standard hole diameter (d_h in)	Oversized hole diameter (d_h in)	Short-slotted hole dimensions (in)	Long-slotted hole dimensions (in)	Alternative short-slotted hole[a] dimensions (in)
<1/2	d + 1/32	d + 1/16	(d + 1/32) by (d + 1/4)	(d + 1/32) by (2 1/2 d)	
≥1/2	d + 1/16	d + 1/8	(d + 1/16) by (d + 1/4)	(d + 1/16) by (2 1/2 d)	9/16 by 7/8

[a]The alternative short-slotted hole is only applicable for d = 1/2 in.

Table E3.3.1-1 Bearing factor, *C*

Thickness of connected part, t, in (mm)	Connections with standard holes: Ratio of fastener diameter to member thickness, d/t	Connections with standard holes: C	Connections with oversized or short-slotted holes: Ratio of fastener diameter to member thickness, d/t	Connections with oversized or short-slotted holes: C
$0.024 \le t < 0.1875$ ($0.61 \le t < 4.76$)	$d/t < 10$	3.0	$d/t < 7$	3.0
	$10 \le d/t \le 22$	$4 - 0.1(d/t)$	$7 \le d/t \le 18$	$1 + 14/(d/t)$
	$d/t > 22$	1.8	$d/t > 18$	1.8

Note: Oversized or short-slotted holes within the lap of lapped or nested Z-members as defined in Section E3 are permitted to be considered as standard holes.

Table E3.3.1-2 Modification factor, m_f, for type of bearing connection

Type of bearing connection	m_f
Single shear and outside sheets of double shear connection using standard holes with washers under both bolt head and nut	1.00
Single shear and outside sheets of double shear connection using standard holes without washers under both bolt head and nut, or with only one washer	0.75
Single shear and outside sheets of double shear connection using oversized or short-slotted holes parallel to the applied load without washers under both bolt head and nut, or with only one washer	0.70
Single shear and outside sheets of double shear connection using short-slotted holes perpendicular to the applied load without washers under both bolt head and nut, or with only one washer	0.55
Inside sheet of double shear connection using standard holes with or without washers	1.33
Inside sheet of double shear connection using oversized or short-slotted holes parallel to the applied load with or without washers	1.10
Inside sheet of double shear connection using short-slotted holes perpendicular to the applied load with or without washers	0.90

Note: Oversized or short-slotted holes within the lap of lapped or nested Z-members as defined in Section E3 are permitted to be considered as standard holes.

(5) maximum member yield stress is 60 ksi (410 MPa, and 4220 kg/cm^2)
(6) minimum lap length measured from center of frame to end of lap is 1.5 times the member depth.

(i) Section E3.4: shear and tension in bolts. The nominal tensile and shear strengths of bolts were updated so the values are consistent with those in the American Institute of Steel Construction (AISC) specification (AISC, 2010).

(j) Section E4.5: combined shear and tension (of screws). This section includes three combined shear and tension checks for screw connections: combined shear and pull-over, combined shear and pull-out, and combined shear and tension, where combined shear and pull-out and combined shear and tension are newly added.

For combined shear and pull-out:

$$\frac{Q}{P_{ns}} + \frac{T}{P_{not}} \leq \frac{1.15}{\Omega} \quad \text{For ASD} \tag{2.6}$$

and

$$\frac{Q_u}{P_{ns}} + \frac{T_u}{P_{not}} \leq 1.15\ \phi \quad \text{For LRFD} \tag{2.7}$$

where Q, Q_u, and T, T_u = shear and tension forces, respectively, determined in accordance with allowable strength design (ASD) or load resistance factor design (LRFD) load combinations; P_{ns} = nominal shear strength of sheet per fastener $= 4.2\left(t_2^3 d\right)^{1/2} F_{u2}$; P_{not} = nominal pull-out strength of sheet per fastener $= 0.85 t_c d F_{u2}$; Ω = safety factor = 2.55; ϕ = resistance factor = 0.60; t_2 = thickness of member not in contact with crew head or washer; F_{u2} = yield stress of t_2; d = nominal diameter of screw; and t_c = lesser of depth of penetration and thickness of t_2. Eqs. [2.6] and [2.7] are applicable with the following requirements satisfied:

(1) 0.0297 in (0.754 mm) $\leq t_2 \leq$ 0.0724 in (1.84 mm)
(2) No. 8, 10, 12, or 14 self-drilling screws with or without washers
(3) $F_{u2} \leq$ 121 ksi (834 MPa or 8510 kg/cm^2)
(4) $1.0 \leq F_u/F_y \leq 1.62$.

For screw combined shear and tension:

$$\frac{T}{P_{ts}} + \frac{V}{P_{ss}} \leq \frac{1.3}{\Omega} \tag{2.8}$$

$$\frac{T_u}{P_{ts}} + \frac{V_u}{P_{ss}} \leq 1.3\ \phi \tag{2.9}$$

where P_{ts} = nominal tension strength of screw as reported by manufacturer or determined by independent laboratory testing; P_{ss} = nominal shear strength of screw as reported by manufacturer or determined by independent laboratory testing; Ω = safety factor = 3.0; and ϕ = resistance factor = 0.5.

(k) Section F1.1: load and resistance factor design and limit states design, and Section D4: CFS light-frame construction. With the development of AISI S220, *North American Cold-Formed Steel Framing Standard—Nonstructural Members*, provisions related to nonstructural members are moved from AISI S100 to AISI S220. In Section F1.1 the value of target reliability index, β_o, for nonstructural interior partition wall studs was removed. Similarly, Section D4 was revised by removing the provisions related to nonstructural members.

(l) Appendix 1: design of CFS structural members using direct strength methods. Appendix 1 provides a rational engineering analysis approach in determining strengths of CFS members. The following new provisions were added.

  (l-1) Reserve capacity. New provisions were added to take into consideration the inelastic reserve capacity of CFS members when subjected to local, distortional, and/or global buckling.

  (l-2) Members with holes. New provisions were added to determine the flexural and compressive strength of members with holes in the web along the member length. The method considers the hole effect when determining local, distortional, and global buckling strengths of the member, and limits the member strength to the yielding strength on the net section. The commentary discusses how to use either a prescriptive method or numerical approaches, such as the finite strip method, to obtain the local, distortional, and global buckling strengths while including hole effects.

  (l-3) Shear strength. Provisions were added to predict the shear strength of CFS members using the direct strength method. The method enables users to predict shear strength for members with stiffeners.

2.2.2 *Reorganized AISI S100-16*

The 2016 edition of AISI S100 was reorganized by adopting the chapter structure of the hot-rolled steel specification in AISC 360, *Specification for Structural Steel Buildings*. Table 2.1 shows how AISI-S100-07 (also AISI S100-12) was revamped to be parallel to AISC 360. As indicated in Table 2.1, the paralleling is readily doable for AISC Chapters A–H and J. However, for AISC Chapters I, K–M, and the appendices, some variation was necessary to include completely the unique provisions for CFS design.

In terms of content updates, AISI S100-16 integrated second-order system analysis for structural system stability. The direct strength method is moved from an appendix to the main body of the specification and appears in a parallel fashion to the effective width method. This allows engineers to choose which method best serves their design and enables them to enter a new design domain opened up by the direct strength method.

2.3 Revisions of AISI cold-formed steel framing standards

2.3.1 *Technical changes in S200-12,* North American Cold-Formed Steel Framing—General Provisions, *and S201-12,* North American Cold-Formed Steel Framing—Product Data

These two standards were reorganized in a code synchronization effort to eliminate duplications and redundancy, as well as to clear any ambiguities among AISI and ASTM standards and building codes. Specific areas considered include material thickness, physical dimensions and tolerance, mechanical properties, coating corrosion resistance, and labeling requirements. Another major reorganization was due to the development of a new AISI S220, *North American Cold-Formed Steel Framing*. The design

Table 2.1 Mapping of AISI S100-07 to new AISI S100-16

AISI–S100–07	AISC–360–10	AISI–S100–XX Strawman
A. General provisions B. Elements	A. General provisions B. Design requirements C. Design for stability	A. General provisions (→A) B. Design requirements (→A) C. Design for stability C1. System (new B_1,B_2+→App. 2) C2. Bracing (new +→D3)
C. Members C1. Properties C2. Tension C3. Flexural members C4. Concentrically loaded compression members C5. Combined axial load and bending	D. Design of members for tension E. Design of members for compression F. Design of members for flexure G. Design of members for shear H. Design of members for combined forces and torsion	D. Members in tension (→C2) E. Members in compression (→C4) F. Members in flexure (→C3) G. Members in shear & web Cr. (→C3) H. Members under combined forces (→C5,C3)
D. Structural assemblies and systems D1. Built-up sections D2. Mixed systems D3. Lateral and stability bracing D4. Cold-formed steel light-frame construction D5. Floor, roof, or wall steel diaphragm construction D6. Metal roof and wall systems	I. Design of composite members	I. Assemblies and systems I1. Built-up sections (→D1) I2. Steel deck diaphragms (→D5) I3. Mixed material assemblies (→D2) I4. Light steel framing (→D4) I5. Rack systems (ref. RMI) I6. Metal building secondary systems (→D6)
E.Connections and joints	J. Design of connections	J. Connections and joints (→E)
F. Tests for special cases G. Design of cold-formed steel structural members for cyclic loading (fatigue) App. 1 Design of cold-formed steel structural members using the direct strength method App. 2 Second-order analysis App. A Provisions applicable to the United States and Mexico App. B Provisions applicable to Canada	K. Design of HSS and box members connections L. Design for serviceability M. Fabrication and erection N. Quality control and quality assurance App. 1 Design by inelastic analysis App. 2 Design for ponding App. 3 Design for fatigue App. 4 Structural design for fire... App. 5 Evaluation of existing structures App. 6 Stability bracing for columns & beams App. 7 Alt. methods of design for stability App. 8 Approx. second-order analysis	K. Available strength for special cases 1.1 Rational analysis (→A) 1.2 Test standards (ref. only) 1.3 Reliability via testing(→F) L. Design for serviceability (I_{eff}) M. Design for fatigue (→G) App. 1 Effective width of elements (→B) App. 2 Elastic buckling of members (new) App. A Provisions applicable to the United States and Mexico (→App.A) App. B....Applicable to Canada (→App.B)
Blue chapter/section numbers are reflected in third column terms of their new location.	Grey sections of AISC are not covered or intended to be covered in AISI	→A indicates all or part of AISI–S100–07 would be in the new section.

RMI, Rack Manufacturers Institute; *HSS*, Hollow Structural Section.
Based on Schafer, B., Chen, H., Manley, B.E., Larson, J.W., 2015. Enable cold-formed steel system design through new AISI standards, 2015 Structures Congress, Portland, OR.

provisions related to nonstructural members were moved from AISI S200 and AISI S201 to AISI S220. Consequently, AISI S200 and AISI S201 are written for structural members, while AISI S220 is specifically for nonstructural members (a review paper on AISI S220, LaBoube et al. (2012), was published separately).

2.3.2 *Technical changes in AISI S214-12,* North American Cold-Formed Steel Framing Standard—Truss Design

The major change in this standard relates to the provisions of truss responsibilities. Those provisions were extracted from AISI S202, *Code of Standard Practice for Cold-Formed Steel Structural Framing*, and added to AISI S214.

2.3.3 *Technical changes in Supplement 1 to AISI S211-07,* North American Cold-Formed Steel Framing Standard—Wall Stud Design

Supplement 1 to AISI S211 includes updates of the referenced standards and deletions of the provisions related to nonstructural members.

2.3.4 *Technical changes in Supplements 2 and 3 to AISI S230-07,* North American Cold-Formed Steel Framing Standard—One- or Two-Family Dwellings

Supplement 2 to AISI S230 deleted the reference to wind exposure A, which is no longer used in ASCE 7; and changed SDC D0, D1, and D2 to D_0, D_1, and D_2. Supplement 3 to S230 ensures that S230-07 with Supplement 2 is in compliance with ASCE 7−10 (2010). An equivalent wind load table is added which converts ASCE 7 basic wind speeds to AISI S230 basic wind speeds (Table A1-3).

Table A1-3 Conversion of ASCE 7 basic wind speeds to AISI S230 basic wind speeds (mph)[a]

ASCE 7 basic wind speed	110	115	126	139	152	164	177	190
AISI S230 basic wind speed	85	90	100	110	120	130	140	150

For SI: 1 mph = 1.61 km/h = 0.447 m/s.
[a]ASCE 7 permits linear interpolation between the contours of the basic wind speed maps.

Some connection requirements are revised accordingly.

2.3.5 *AISI S220-15,* North American Cold-Formed Steel Framing Standard—Nonstructural members

This standard addresses the design, installation, and testing analysis of cold-formed nonstructural members, and its first edition was published in 2011. The nonstructural member is defined as "A member in a steel-framed system that is not a part of the gravity load resisting system, lateral force resisting system or building envelope." Examples of nonstructural members include a member in a steel-framed system which is limited to a transverse (out-of-plane) load of not more than 10 lb/ft^2 (0.48 kPa), a superimposed axial load, exclusive of sheathing materials, of not more than 100 lb/ft (1.46 kN/m), or a superimposed load of not more than 200 lbs (0.89 kN). The strength of nonstructural members can be determined by either a noncomposite or a composite design approach. The noncomposite assembly design approach utilizes the design provisions of AISI S100 but with adjusted safety and resistance factors, $\Omega_N = 0.9\ \Omega$ and $\phi_N = 1.1\ \phi$, where Ω, ϕ = safety and resistance factors from relevant section of AISI S100, and Ω_N, ϕ_N = corresponding safety and resistance factors for nonstructural members. The noncomposite assembly design approach can also be done via testing, with the test results evaluated in accordance with AISI S100-16 Section K2 but with a target reliability index, $\beta_0 = 1.6$. The composite assembly design approach is generally accomplished by testing, with the test results evaluated in accordance with AISI S100-16 Section K2. Detailed logistics of design provisions and design examples can be found in LaBoube et al. (2012). In the 2015 edition, requirements for testing bridging connectors in accordance with AISI S915 and testing composite assemblies in accordance with AISI S916 were specified.

2.3.6 *AISI S240-15,* North American Standard for Cold-Formed Steel Structural Framing

In 2015 AISI developed a new unified standard, AISI S240, to address requirements for construction with CFS structural framing that are common to prescriptive and engineered design. The standard integrated a set of previous AISI CFS framing standards into one document.

- AISI S200-12, *North American Standard for Cold-Formed Steel Framing—General Provisions.*
- AISI S210-07 (2012), *North American Standard for Cold-Formed Steel Framing—Floor and Roof.*
- *System Design* (reaffirmed 2012).
- AISI S211-07 (2012), *North American Standard for Cold-Formed Steel Framing—Wall Stud Design* (reaffirmed 2012).
- AISI S212-07 (2012), *North American Standard for Cold-Formed Steel Framing—Header Design* (reaffirmed 2012).
- AISI S213-07w/S1-09 (2012), *North American Standard for Cold-Formed Steel Framing—Lateral Design with Supplement 1* (reaffirmed 2012).
- AISI S214-12, *North American Standard for Cold-Formed Steel Framing—Truss Design.*

AISI S240 supersedes all previous editions of these individual AISI standards except for the specific seismic requirements. Modifications were made in S240 to align the provisions with AISI S400, as follows.

- The applicability of AISI S240 for seismic design was limited to applications where specific seismic detailing is not required.
- Definitions no longer needed in AISI S240 were removed and the remaining definitions were revised, if needed, for consistency with AISI S400.
- Seismic-specific tables for *nominal shear strength* (*resistance*) were deleted.
- Seismic-specific safety factors and resistance factors were deleted.
- Other seismic-specific requirements were removed, as appropriate, and the remaining requirements were generalized for applicability to wind, seismic, or other lateral loads.

Also in this first edition, a new Chapter F on testing was added to allow reference to applicable AISI 900-series test standards. Methods for *truss* tests, formerly in Section E7, were moved to Appendix 2.

2.3.7 *AISI S400-15,* North American Standard for Seismic Design of Cold-Formed Steel Structural Systems

In 2015 AISI published a new seismic standard, AISI S400, to address the design and construction requirements of CFS structural members and connections used in seismic force-resistance systems in buildings and other structural systems. The first edition represents a merging of AISI S110, *Standard of Seismic Design of Cold-Formed Steel Structural Systems—Special Bolted Moment Frame* (2007), and the seismic design portions of AISI S213, *North American Standard for Cold-Formed Steel Framing—Lateral Design* (2007), with Supplement nos. 1—09. Some seismic design requirements stipulated in ANSI/AISC 340-10, *Seismic Provisions for Structural Steel Buildings*, which were suitable for CFS structural systems were also adopted in the new AISI S400. AISI S400 should be applied in conjunction with AISI S100, *North American Specification for the Design of Cold-Formed Steel Structural Members*, and AISI S240, *North American Standard for Cold-Formed Steel Structural Framing*.

2.4 Conclusions

The AISI, through its standards development process, is working to enable engineers to utilize and understand the much broader efficiencies of CFS structural members and systems through updating and developing design specifications and standards. The chapter presents major changes in recent years and new developments for the 2016 edition of CFS design documents in the United States. As the AISI Specification Committee continues to advance the AISI standards, two major initiatives have been established for current development efforts: promoting CFS framing in midrise construction; and enabling performance-based design practices. The AISI standards will continue evolving by adopting the latest research findings and reflecting best practices in industries.

Acknowledgment

The authors appreciate the assistance and material provided by Dr. Ben Schafer of Johns Hopkins University for this chapter.

References

AISC 360, 2010. Specification for Structural Steel Buildings. American Institute of Steel Construction, Chicago, IL.

ASCE 7-10, 2010. Minimum Design Loads For Buildings and Other Structures. American Society of Civil Engineers, Reston, VA.

American National Standards Institute, 2012. Essential Requirements: Due Process Requirements for American National Standards. Washington, DC.

Chen, H., Brockenbrough, R.L., Haws, R., November 2010. An overview of recent changes and additions to AISI standards. In: Proceedings of the Twentieth International Specialty

Conference on Cold-Formed Steel Structures. Missouri University of Science and Technology, Rolla, MO.
LaBoube, A.R., Chen, H., Larson, J., November 2012. New standard AISI S220, North American Cold-formed Steel Framing Standard–Nonstructural Members. In: Proceedings of Twenty-First International Specialty Conference for Cold-Formed Steel Structures. Missouri University of Science and Technology, MO.
Schafer, B., Chen, H., Manley, B.E., Larson, J.W., 2015. Enable cold-formed steel system design through new AISI standards. In: 2015 Structures Congress. Portland, OR.
Yu, C., 2009a. Web Crippling Strength of Cold-Formed Steel NUFRAME Members. Report No. 20090112-01. University of North Texas, Denton, TX.
Yu, C., 2009b. Web Crippling Strength of Cold-Formed Steel NUFRAME Members. Report No. 20090217-01. University of North Texas, Denton, TX.
Yu, C., Xu, K., 2010. Cold-Formed Steel Bolted Connections Using Washers on Oversized and Slotted Holes – Phase 2. Research Report RP10-2. American Iron and Steel Institute, Washington, DC.
Zeinoddini, V., Schafer, B.W., November 2010. Impact of corner radius on cold-formed steel member strength. In: Proceedings of the Twentieth International Specialty Conference on Cold-formed Steel Structures. Missouri University of Science and Technology, Rolla, MO, pp. 1–15.

Appendix A: AISI standard updates

Designation	Title	Published editions
AISI S100	North American Specification for Design of Cold-Formed Steel Structural Members	Updated in 2012 and 2016
AISI S110	Standard for Seismic Design of Cold-Formed Steel Structural Systems—Special Bolted Moment Frames	Reaffirmed in 2012 and integrated into AISI S400-15
AISI S200	North American Cold-Formed Steel Framing—General Provisions	Updated in 2012 and integrated into AISI S240-15
AISI S201	North American Cold-Formed Steel Framing—Product Data	Updated in 2012
AISI S202	North American Cold-Formed Steel Framing—Code of Standard Practice	Updated in 2015
AISI S210	North American Cold-Formed Steel Framing—Floor and Roof System Design	Reaffirmed in 2012 and integrated into AISI S240-15
AISI S211	North American Cold-Formed Steel Framing—Wall Stud Design	Reaffirmed in 2012 and integrated into AISI S240-15

Designation	Title	Published editions
AISI S212	North American Cold-Formed Steel Framing—Header Design	Reaffirmed in 2012 and integrated into AISI S240-15
AISI S213	North American Cold-Formed Steel Framing—Lateral Design	Reaffirmed in 2012 and integrated into AISI S240-15
AISI S214	North American Cold-Formed Steel Framing—Truss Design	Updated in 2012 and integrated into AISI S240-15
AISI S220	North American Cold-Formed Steel Framing—Nonstructural Members	Published in 2011 and updated in 2015
AISI S230	North American Cold-Formed Steel Framing—Prescriptive Method	Reaffirmed in 2012 and updated in 2015
AISI S240	North American Standard for Cold-Formed Steel Structural Framing	Published in 2015
AISI S400	North American Standard for Seismic Design of Cold-Formed Steel Structural Systems	Published in 2015

AISI design procedures and practical examples for cold-formed steel structures

3

W. Zhang[1], *C. Yu*[2]
[1]Tongji University, Shanghai, China; [2]University of North Texas, Denton, TX, United States

3.1 Introduction

In 1949 the American Iron and Steel Institute (AISI) published the first edition of its "Light Gage Steel Design Manual," which was intended to supplement the design specification and facilitate its application to ordinary design problems (Kaehler and Seaburg, 1996). Since then AISI has been developing and updating the manual, which was later renamed the "Cold-Formed Steel Design Manual" (AISI, 1968). Today, the manual provides the latest design information and examples for conformance with S100 *North American Specification for the Design of Cold-Formed Steel Structural Members* (AISI, 2012). The 2013 edition of the "Cold-Formed Steel Design Manual" (AISI, 2013) consisted of two volumes with eight parts.

- *Part I: Dimensions and Properties.* Contains information regarding the availability and properties of steels referenced in the specification; tables of section properties; and formulas and examples of calculations of section properties.
- *Part II: Beam Design.* Contains tables and charts to aid in beam design, and beam design example problems.
- *Part III: Column Design.* Contains tables to aid in column design, and column design example problems.
- *Part IV: Connections.* Contains tables to aid in connection design, and connection example problems.
- *Part V: Supplementary Information.* Contains a table of specification cross-references to the examples provided in the design manual; design procedures of a specification nature which are not included in the specification itself, either because they are infrequently used or are regarded as too complex for routine design; and other information intended to assist users of cold-formed steel (CFS).
- *Part VI: Test Procedures.* Contains a bibliography of other pertinent test methods, and test calibration example problems.
- *Part VII: North American Specification.* Contains the 2012 edition of the *North American Specification for the Design of Cold-Formed Steel Structural Members.*
- *Part VIII: Commentary on the North American Specification.* Contains the 2012 edition of the *Commentary on the North American Specification for the Design of Cold-Formed Steel Structural Members.*

Recent Trends in Cold-Formed Steel Construction. http://dx.doi.org/10.1016/B978-0-08-100160-8.00003-7

Up-to-date design examples were added in the AISI Design Manual (2013) to illustrate new design provisions in AISI S100 (2012).

1. Round and rectangular tubular section member design in Parts II and III.
2. C-section members subjected to combined bending and torsional loading in Part II.
3. Distortional buckling of C-section members in Parts II and III.
4. Sigma-shaped flexural and compression member design by the direct strength method in Parts II and III.
5. Web crippling strength of beam webs with bearing stiffeners in Part II.
6. Frame design with consideration of second-order analysis in Part III.

AISI's "Cold-Formed Steel Design Manual" is focused on member or connection design. For framing or system design, the Cold-Formed Steel Engineers Institute (CFSEI) has published a series of technical notes to address specific topics related to CFS framing for residential and commercial construction. Currently the CFSEI technical notes series consists of 10 categories.

- Durability and corrosion protection.
- Fasteners and connection hardware.
- Component assemblies (trusses and wall panels).
- General topics.
- Floor and joist systems.
- Wall systems.
- Roof and ceiling systems.
- Thermal, fire, and acoustic.
- Lateral systems.
- Other technical documents.

In addition to the technical notes, the CFSEI also periodically releases research reports, design guides, and technical papers to promote the use of CFS in the construction industry.

In recent years CFS framing has become an attractive and competitive structural system for midrise residential and office buildings due to its light weight, fast installation, low waste and noise during construction, and noncombustibility. Among the three sheathing materials for shear walls approved in AISI S213 *North American Standard for Cold-Formed Steel Framing—Lateral Design* (AISI, 2012) and the International Building Code (IBC, 2015), flat steel is the only noncombustible material for Types I and II construction and other building types such as medical care facilities, etc., which require noncombustible materials for their load-bearing structures. Fig. 3.1 shows the failure mechanism of CFS shear walls sheathed by flat steel sheet: the tension field action dominated the load-bearing behavior of these shear walls.

The existing design methods in AISI S213 and the IBC for CFS-sheathed shear walls are solely based on experimental data with limited options in sheathing and framing configurations. To improve the situation, an analytical model and closed-form design equations for steel-sheet-sheathed CFS shear walls were developed by Yanagi and Yu (2014). Named the effective strip method, it offers an alternative

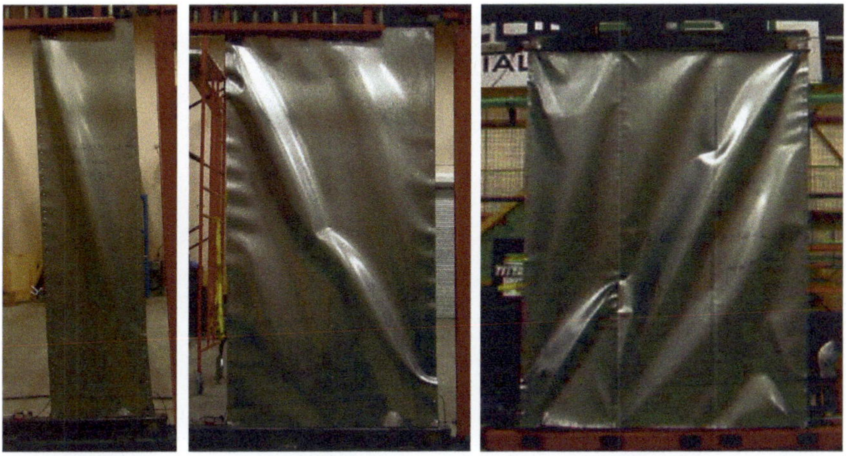

Figure 3.1 Failure mechanism of steel-sheet-sheathed shear walls.
Yanagi, N., Yu, C., 2013. Effective strip model for cold-formed steel shear wall using steel sheet sheathing. In: Proceedings of the 21st International Specialty Conference on Cold-Formed Steel Structures, St. Louis, MO, April 2013.

approach for engineers to predict the normal strength of CFS shear walls. The method was approved by the AISI Committee on Steel Framing and will be included in AISI S240 and AISI S400. A design example for the effective strip method is provided in this chapter.

In any structural system, the connection method and its details are always the critical elements to achieve the desired structural behavior and ensure the minimum safety level defined by building codes and design specifications. The connections in CFS structures are no different to those in any other structural system, and are even more critical as fasteners are most commonly used and sometimes have dual responsibilities: load bearing and energy dissipation. Among the various connections and connectors, a thin-walled CFS clip angle is a common method used in CFS framing. Fig. 3.2 shows applications of CFS clip angles to connect the stud to the track in CFS framing, and attach the curtain wall framing member to the main structure. As illustrated in Fig. 3.2, a CFS clip angle can be subjected to shear, axial, bending, and a combination of those three. However, CFS clip angle design has not been developed in the major design specifications—AISI S100, the Australian/New Zealand Standard AS/NZS 4600 Cold-Formed Steel Structures (AS/NZS 4600, 2005), and Eurocode 3 Design of Steel Structures (2006). To address the knowledge gap in the behavior and strength of CFS clip angles, a research project was conducted by Yu et al. (2015) at the University of North Texas to develop design methods for clip angles subjected to shear, compression, and tension forces. This chapter gives examples of shear and compression designs of CFS clip angles.

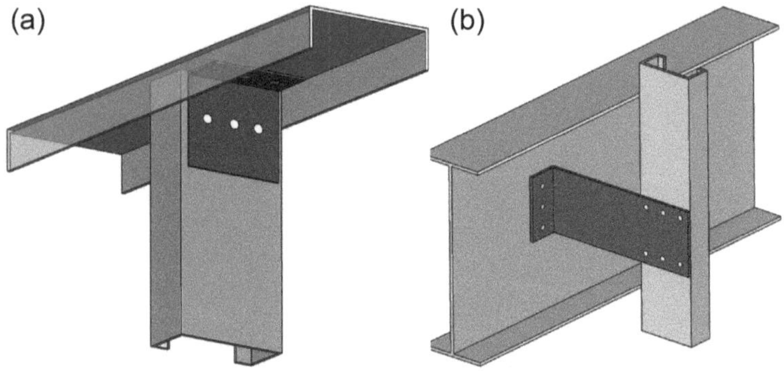

Figure 3.2 Examples of applications of CFS clip angles: (a) CFS framing; (b) curtain wall framing.

3.2 Sheet steel shear wall design—effective strip method

3.2.1 *The effective strip method*

The effective strip method for determining the nominal shear strength (resistance) for Type I shear walls with steel sheet sheathing is based on research by Yanagi and Yu (2014). The method assumes a sheathing strip carries the lateral load via a tension field action, as illustrated in Fig. 3.3. The shear strength of the shear wall is controlled by the tensile strength of the effective sheathing strip, which is determined as the lesser of the fasteners' tensile strength and the yield strength of the effective sheathing strip. The statistical analysis in Yanagi and Yu (2014) yielded a load and resistance factor design (LFRD) resistance factor of 0.79 for the effective strip method. To be consistent with the resistance factors (0.60 for LRFD) specified in AISI S213 (2012) Section E2.3.2, the original design equation in Yanagi and Yu (2014) was adjusted accordingly.

The nominal shear strength (resistance) per unit length for a Type I shear wall with steel sheet sheathing can be determined in accordance with the effective strip method as:

$$V_n = \text{minimum}(1.33P_n \cos\alpha,\ 1.33W_e tF_y \cos\alpha) \quad [3.1]$$

where

P_n = nominal shear strength (resistance) of screw connections within the effective strip width, W_e, on the steel sheet sheathing

$$\alpha = \text{Arctan}(h/w) \quad [3.2]$$

h = shear wall height
w = shear wall length
t = design thickness of steel sheet sheathing
F_y = yield stress of steel sheet sheathing

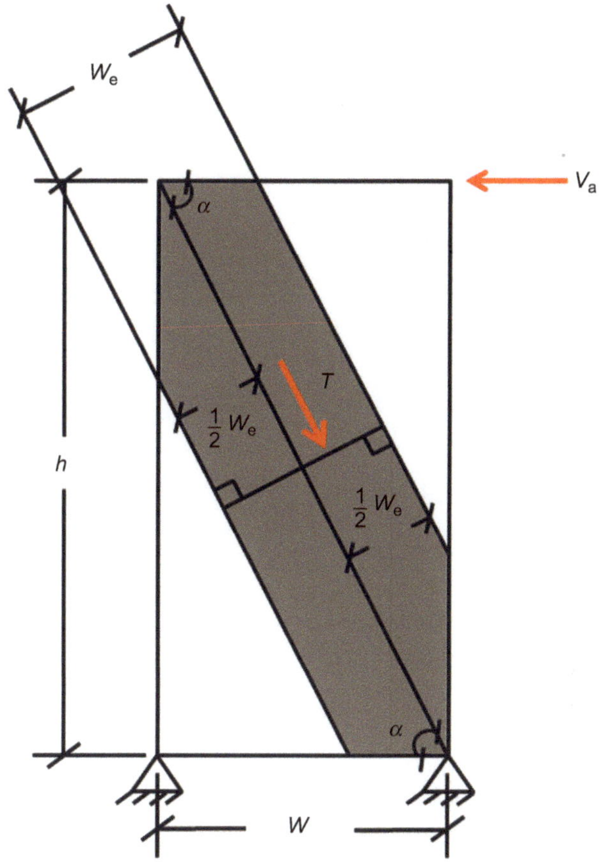

Figure 3.3 Effective strip model for steel sheet sheathing.
Yanagi, N., Yu, C., 2013. Effective strip model for cold-formed steel shear wall using steel sheet sheathing. In: Proceedings of the 21st International Specialty Conference on Cold-Formed Steel Structures, St. Louis, MO, April 2013.

$$W_e = W_{max}, \quad \text{when } \lambda \leq 0.0819 \qquad [3.3]$$

$$= \rho W_{max}, \quad \text{when } \lambda > 0.0819 \qquad [3.4]$$

where

$$W_{max} = w/\sin \alpha \qquad [3.5]$$

$$\rho = \frac{1 - 0.55(\lambda - 0.08)^{0.12}}{\lambda^{0.12}} \qquad [3.6]$$

$$\lambda = 1.736\frac{\alpha_1\alpha_2}{\beta_1\beta_2\beta_3^2 a} \quad [3.7]$$

where

$$\alpha_1 = F_{\text{ush}}/310.3 \quad (\text{for } F_{\text{ush}} \text{ in MPa})$$

$$= F_{\text{ush}}/45 \quad (\text{for } F_{\text{ush}} \text{ in ksi})$$

$$\alpha_2 = F_{\text{uf}}/310.3 \quad (\text{for } F_{\text{uf}} \text{ in MPa})$$

$$= F_{\text{uf}}/45 \quad (\text{for } F_{\text{uf}} \text{ in ksi})$$

$$\beta_1 = t_{\text{sh}}/0.457 \quad (\text{for } t_{\text{sh}} \text{ in mm})$$

$$= t_{\text{sh}}/0.018 \quad (\text{for } t_{\text{sh}} \text{ in inches})$$

$$\beta_2 = t_{\text{f}}/0.457 \quad (\text{for } t_{\text{f}} \text{ in mm})$$

$$= t_{\text{f}}/0.018 \quad (\text{for } t_{\text{f}} \text{ in inches})$$

$$\beta_3 = s/152.4 \quad (\text{for } s \text{ in mm})$$

$$= s/6 \quad (\text{for } s \text{ in inches})$$

F_{ush} = tensile strength of steel sheet sheathing
F_{uf} = minimum tensile strength of framing materials
t_{sh} = design thickness of steel sheet sheathing
t_{f} = minimum design thicknesses of framing members
s = screw spacing on the panel edges
a = wall aspect ratio (h:w).

The effective strip method is valid within the following range of parameters.

1. Designation thickness of stud, track, and stud blocking: 0.838–1.37 mm (0.033–0.054 in.).
2. Designation thickness of steel sheet sheathing: 0.457–0.838 mm (0.018 –0.033 in.).
3. Screw spacing at panel edges: 50.8–152 mm (2–6 in.).
4. Height to length aspect ratio (h:w): 1:1–4:1.
5. Sheathing screw shall be minimum No. 8.
6. Yield stress of steel sheet sheathing shall not be greater than 345 MPa (50 ksi).

3.2.2 Design example

In this section, a design example of 1220 × 2440 mm (4 ft × 8 ft) CFS shear walls using 1.092 mm (43 mil) framing and 0.838 mm (33 mil) sheet steel sheathing is provided. The nominal yield stress is 230 MPa (33 ksi) for the framing and 345 MPa

(50 ksi) for the sheathing. Sheathing-to-framing fasteners at the perimeter are No. 8 self-drilling tapping screws spaced at 76 mm (3 in.).

Step 1: Estimating the effective strip width.

$$F_{ush} = 448.2 \text{ MPa}$$

$$F_{uf} = 310.3 \text{ MPa}$$

$$t_{sh} = 0.879 \text{ mm}$$

$$t_f = 1.146 \text{ mm}$$

$$a = (2440 \text{ mm})/(1220 \text{ mm}) = 2.0$$

$$\alpha = \tan^{-1} a = \tan^{-1}(2.0) = 63.43 \text{ degrees}$$

Maximum effective width of the steel sheet sheathing

$$W_{max} = \frac{W}{\sin\alpha} = \frac{W}{\sin(63.43 \text{ degree})} = \frac{1220 \text{ mm}}{\sin(63.43 \text{ degree})} = 1363 \text{ mm}$$

$$\alpha_1 = F_{ush}/(310.3 \text{ MPa}) = (448.2 \text{ MPa})/(310.3 \text{ MPa}) = 1.444$$

$$\alpha_2 = F_{uf}/(310.3 \text{ MPa}) = (310.3 \text{ MPa})/(310.3 \text{ MPa}) = 1.0$$

$$\beta_1 = t_{sh}/(0.457 \text{ mm}) = (0.879)/(0.457 \text{ mm}) = 1.923$$

$$\beta_2 = t_f/(0.457 \text{ mm}) = (1.146)/(0.457 \text{ mm}) = 2.508$$

$$\beta_3 = s/(152.4 \text{ mm}) = (76 \text{ mm})/(152.4 \text{ mm}) = 0.499$$

$$\lambda = 1.736\frac{\alpha_1\alpha_2}{\beta_1\beta_2\beta_3^2 a} = 1.736 \times \frac{1.444 \times 1.0}{1.923 \times 2.508 \times (0.499)^2 \times 2.0} = 0.601$$

$$\lambda = 0.601 > 0.0819$$

Effective strip width of the steel sheet sheathing

$$W_e = \rho W_{max} = \frac{1 - 0.55(\lambda - 0.08)^{0.12}}{\lambda^{0.12}} W_{max}$$

$$= \frac{1 - 0.55 \times (0.601 - 0.08)^{0.12}}{(0.601)^{0.12}} \times (1363 \text{ mm}) = 712.0 \text{ mm}$$

Step 2: Determining the nominal shear capacity of individual connections. Connection shear limited by tilting and bearing

$$t_1 = 0.879 \text{ mm}$$

$$t_2 = 1.146 \text{ mm}$$

$$d = 4.166 \text{ mm}$$

$$F_{u1} = 448.2 \text{ MPa}$$

$$F_{u2} = 310.3 \text{ MPa}$$

$$t_2/t_1 = (1.146 \text{ mm})/(0.879 \text{ mm}) = 1.304$$

$$1.0 < t_2/t_1 < 2.5$$

For $t_2/t_1 \leq 1.0$,

$$P_{ns} = \min \begin{cases} 4.2(t_2^3 d)^{1/2} F_{u2} = 4.2\left\{(1.146 \text{ mm})^3 \times (4.166 \text{ mm})\right\}^{1/2} \times (310.3 \text{ MPa}) = 3.263 \text{ kN} \\ 2.7 t_1 d F_{u1} = 2.7 \times (0.879 \text{ mm}) \times (4.166 \text{ mm}) \times (448.2 \text{ MPa}) = 4.431 \text{ kN} \\ 2.7 t_2 d F_{u2} = 2.7 \times (1.146 \text{ mm}) \times (4.166 \text{ mm}) \times (310.3 \text{ MPa}) = 4.000 \text{ kN} \end{cases}$$

For $t_2/t_1 \geq 2.5$,

$$P_{ns} = \min \begin{cases} 2.7 t_1 d F_{u1} = 2.7 \times (0.879 \text{ mm}) \times (4.166 \text{ mm}) \times (448.2 \text{ MPa}) = 4.431 \text{ kN} \\ 2.7 t_2 d F_{u2} = 2.7 \times (1.146 \text{ mm}) \times (4.166 \text{ mm}) \times (310.3 \text{ MPa}) = 4.000 \text{ kN} \end{cases}$$

By linear interpolation of the smallest of the above two cases

$$P_{ns} = 3.412 \text{ kN}$$

Connection shear limited by end distance

$$t = 0.879 \text{ mm}$$

$$\alpha = \tan^{-1} a = \tan^{-1}(2) = 63.4 \text{ degrees}$$

$$w_f = \text{flange width of stud} = 41.2 \text{ mm}$$

$e = \frac{w_f}{2 \cos \alpha} = \frac{41.2 \text{ mm}}{2 \times (\cos 63.4 \text{ degrees})} = 46.0$ mm (assume the screws are installed at the center of the flange of the outer stud)

$$F_u = 448.2\ \text{MPa}$$

$$P_{ns} = teF_u = (0.879\ \text{mm}) \times (46.0\ \text{mm}) \times (448.2\ \text{MPa}) = 18.114\ \text{kN}$$

Connection shear limited by shear in screw.

The nominal shear strength of the screw is provided by the manufacturer. It is assumed that a No. 8−18 Phillips truss head screw by HILTI is used. The screw shear strength can be found in a HILTI self-drilling screws report (ESR-2196, 2013).

$$P_{ss} = 5.204\ \text{kN}$$

$$P_{ns,s} = \min\{3.412\ \text{kN}, 18.114\ \text{kN}, 5.204\ \text{kN}\} = 3.412\ \text{kN}$$

Similarly,

$$P_{ns,t} = 3.412\ \text{kN (sheathing to track connection)}$$

$$P_{ns,ts} = 3.412\ \text{kN (sheathing to track and stud connection)}$$

Step 3: Determining the nominal shear strength of the shear wall.

$$t_{sh} = 0.879\ \text{mm}$$

$$F_y = 344.7\ \text{MPa}$$

$$V_n = \text{minimum}\left\{\left(\frac{W_e}{2s\sin\alpha}P_{ns,t} + \frac{W_e}{2s\cos\alpha}P_{ns,s} + P_{ns,ts}\right)\cos\alpha, W_e t F_y \cos\alpha\right\}$$

Limited by connection capacity

$$\left(\frac{W_e}{2s\sin\alpha}P_{ns,t} + \frac{W_e}{2s\cos\alpha}P_{ns,s} + P_{ns,ts}\right)\cos\alpha$$

$$= \left(\frac{712.0\ \text{mm}}{2 \times (76\ \text{mm}) \times \sin(63.43\ \text{degree})} \times (3.412\ \text{kN}) + \frac{712.0\ \text{mm}}{2 \times (76\ \text{mm}) \times \cos(63.43\ \text{degrees})} \times (3.412\ \text{kN}) + 3.412\ \text{kN}\right) \times \cos(63.43\ \text{degree})$$

$$= 25.50\ \text{kN}$$

Limited by sheathing yield capacity

$$W_e t_{sh} F_y \cos\alpha = (712.0\ \text{mm}) \times (0.879\ \text{mm}) \times \left(0.3447\ \text{kN/mm}^2\right) \times \cos(63.43\ \text{degrees}) = 96.49\ \text{kN}$$

$$V_n = \text{minimum}\{25.50\ \text{kN}, 96.49\ \text{kN}\} = 25.50\ \text{kN}$$

Nominal shear strength of the shear wall

$$V_n = 25.50\ \text{kN}(20.90\ \text{kN/m})$$

3.3 Shear design of load-bearing clip angle connectors

3.3.1 Shear strength without consideration of clip angle deformation

The nominal shear strength (resistance), V_n, of the cantilevered leg of a clip angle is calculated as follows:

$$V_n = 0.17\lambda^{-0.8} F_y B t \le 0.35 F_y B t \quad [3.8]$$

where

$$\lambda = \sqrt{\frac{F_y}{F_{cr}}} \quad [3.9]$$

$$F_{cr} = \frac{k\pi^2 E}{12(1-\mu^2)} \left(\frac{t}{B}\right)^2 \quad [3.10]$$

$$k = 2.569 \left(\frac{L}{B}\right)^{-2.202} \quad [3.11]$$

B = width of cantilevered leg measured parallel to the applied shear force
L = flat length of cantilevered leg measured from the center of the first line of fasteners to the bend line.

These equations are valid within the following range of parameters and boundary conditions.

Clip angle design thickness: 0.838–2.464 mm (0.033–0.097 in.).
Clip angle design yield strength: 228–345 MPa (33–50 ksi.).
L/B ratio: 0.18–1.40.
The fastener pattern allows full engagement of the cantilevered leg in bearing the shear load.

3.3.2 Shear strength with consideration of clip angle deformation

When deformation of the clip angle is a design consideration, the nominal shear strength (resistance) V_n of the cantilevered leg of a clip angle is calculated in accordance with this section. In addition, the available strength must not exceed the nominal shear strength obtained in accordance with Section 3.3.1.

$$V_n = 9000\,\alpha\left(\frac{Bt}{L}\right) \qquad [3.12]$$

where

$\alpha = 0.175$ N/mm for SI units
$= 1$ lb/in. for US customary units.

The same parametric ranges listed in Section 3.3.1 apply to the above equations.

3.3.3 Design example

This example illustrates the use of the design methods proposed in this chapter. The overall dimensions of the cantilevered leg of a clip angle are 121.9 × 68.6 mm, as shown in Fig. 3.4. The design thickness is 1.438 mm and the inside corner radius is 2.156 mm. The material yield stress is 227.5 MPa.

1. Nominal shear strength without consideration of deformation.

$$L = 68.6 - 12.7 - 2.156 - 1.438 = 52.306 \text{ mm}$$

$$k = 2.569\left(\frac{L}{B}\right)^{-2.202} = 2.569\left(\frac{52.306}{121.9}\right)^{-2.202} = 16.55$$

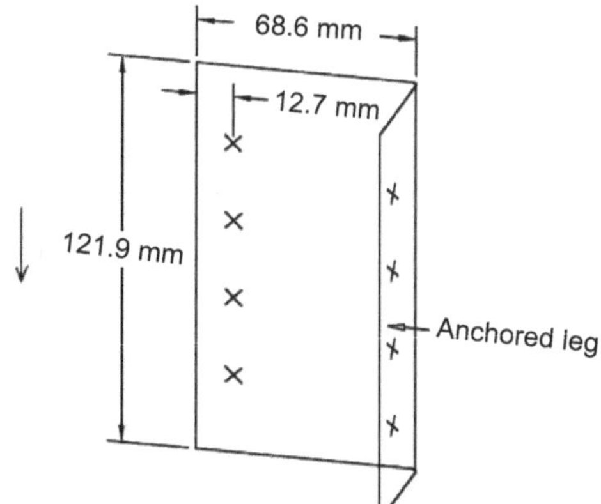

Figure 3.4 Dimensions and details of a clip angle in shear.

$$F_{cr} = \frac{k\pi^2 E}{12(1-\mu^2)}\left(\frac{t}{B}\right)^2 = \frac{16.55\pi^2 203,000}{12(1-0.3^2)}\left(\frac{1.438}{121.9}\right)^2 = 422.55 \text{ MPa}$$

$$\lambda = \sqrt{\frac{F_y}{F_{cr}}} = \sqrt{\frac{227.5}{422.55}} = 0.734$$

$$V_n = 0.17\lambda^{-0.8}F_y Bt = (0.17)\left(0.734^{-0.8}\right)(227.5)(121.9)(1.438)\left(10^{-3}\right) = 8.68 \text{ kN}$$

$$< 0.35F_y Bt = (0.35)(227.5)(121.9)(1.438)\left(10^{-3}\right) = 13.96 \text{ kN}$$

2. Nominal shear strength with consideration of deformation.

$$V_n' = 9000\alpha\left(\frac{Bt}{L}\right) = (9000)(0.175)\left(\frac{121.9 \times 1.438}{52.306}\right)\left(10^{-3}\right) = 5.3 \text{ kN}$$

$$< V_n = 8.68 \text{ kN}$$

3.4 Compression design of load-bearing clip angle connectors

3.4.1 Compression design method

The nominal compression strength (resistance), P_n, of the cantilevered leg of a clip angle is calculated as follows:

$$P_n = F_n A_g \tag{3.13}$$

where

$$A_g = B't \tag{3.14}$$

$$F_n = 0.0028\lambda^{1.44}F_{cr} \leq 0.4F_y \tag{3.15}$$

$$\lambda = \frac{L}{t} \tag{3.16}$$

$$F_{cr} = \frac{k\pi^2 E}{12(1-\mu^2)}\left(\frac{t}{L}\right)^2 \tag{3.17}$$

k = buckling coefficient can be determined by interpolation in Table 3.1 = 0.90 as a conservative value.

Table 3.1 **Theoretical *k* values**

L/B	*k*
0.1	0.993
0.2	0.988
0.3	0.983
0.4	0.978
0.5	0.973
0.6	0.969
0.7	0.964
0.8	0.960
0.9	0.956
1.0	0.952
1.5	0.938
2.0	0.929

B' is taken as the lesser of the actual clip angle width or the Whitmore section width, which is determined using a spread-out angle of 30 degrees along both sides of the connection, beginning at the first row of fasteners in the cantilevered leg of the clip angle.

The above equations are valid within the following range of parameters.

Clip angle design thickness: 0.838−2.997 mm (0.033−0.118 in.).
Clip angle design yield strength: 228−354 MPa (33−50 ksi).
L/B ratio: 0.18−1.40.

3.4.2 Design example

This example illustrates the use of the compression deign method proposed in this chapter. The overall dimensions of the cantilevered leg of a clip angle are 121.9 × 68.6 mm, as shown in Fig. 3.5. The design thickness is 1.438 mm and the inside corner radius is 2.156 mm. The material yield stress is 227.5 MPa.

Determine the flat width of the cantilevered leg in the loading direction:

$$L = 68.6 - 12.7 - 2.156 - 1.438 = 52.306 \text{ mm}$$

$$0.18 < \frac{L}{B} = \frac{52.306}{121.9} = 0.43 < 1.4$$

$$k = 0.977 \quad \text{(Interpolation from Table 1)}$$

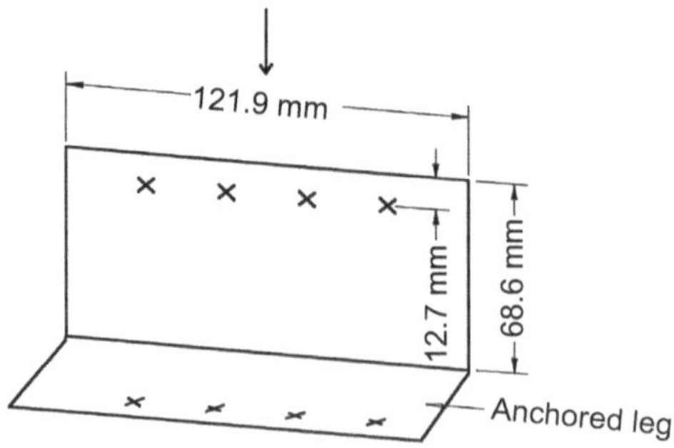

Figure 3.5 Dimensions and details of a clip angle in compression.

$$B' = 121.9 \text{ mm}$$

$$\lambda = \frac{L}{t} = \frac{52.306}{1.438} = 36.37$$

$$F_{cr} = \frac{k\pi^2 E}{12(1-\mu^2)}\left(\frac{t}{L}\right)^2 = \frac{(0.977)(\pi^2)(203,000)}{12(1-0.3^2)}\left(\frac{1.438}{52.306}\right)^2 = 135.48 \text{ MPa}$$

$$F_n = 0.0028\lambda^{1.44}F_{cr} = (0.0028)(36.37^{1.44})(135.48) = 67.07 \text{ MPa}$$

$$\leq 0.4F_y = (0.4)(227.5) = 91.0 \text{ MPa}$$

$$F_n = 67.07 \text{ MPa}$$

$$A_g = B't = (121.9)(1.438) = 175.3 \text{ mm}^2$$

Nominal compression strength:

$$P_n = F_n A_g = (67.07)(175.3)10^{-3} = 11.76 \text{ kN}$$

3.5 Conclusions and future work

In the chapter, design examples to support a newly developed design method for CFS shear walls and clip angle connectors are provided. As the application of CFS becomes more popular in midrise buildings, the demand for a high-strength, high-ductility lateral force-resisting system is getting stronger. The flat sheet sheathed CFS shear wall is one favorable candidate for such a system. From the effective strip method, it can be seen that thicker framing, thicker sheathing, and more screws will lead to higher shear strength. However, how those factors affect the postpeak behavior and energy dissipation performance of the shear walls requires further research.

The clip angle design methods assume line patterns for the screws and successful prevention of screw failures. However, in actual projects the clip angle connectors may be installed using a random pattern for the screw locations, therefore how the screw pattern affects the strength of the clip angle requires further investigation. Some clip angles may use spot welding instead of fasteners as the connection method, so additional research is needed to study welded clip angles in steel framing. The design methods assume fastener failures were prevented, and the clip angles failed in the cantilevered legs. Members' failures were expected. The proposed design equations for shear and compression might be appropriate to use as a reliability index for members when resistance factors and safety factors were determined. However, if one considers the clip angle connector as a critical connection component that should have higher reliability than a structural member, a reliability index for connections should be used. Furthermore, when clip angles are used in secondary structural components of a building, such as curtain walls, a reliability index smaller than that for the primary structural components might be reasonable. A comprehensive study on the reliability of the presented design methods is needed to ensure safe and efficient use of CFS clip angles.

Acknowledgments

The authors would like to recognize funding support from the US National Science Foundation through grant "NSF-CMMI-0955189: Comprehensive Research on Cold-Formed Steel Sheathed Shear Walls, Special Detailing, Design, and Innovation," and the American Iron and Steel Institute through a project titled "Load Bearing Clip Angle Design." Any opinions, findings, conclusions, or recommendations expressed in this publication are those of the authors and do not necessarily reflect the views of the National Science Foundation or AISI.

References

AISI Design Manual, 1968. Cold-formed Steel Design Manual, 1968 Edition. American Iron and Steel Institute, Washington, DC.

AISI Design Manual, 2013. Cold-formed Steel Design Manual, 2013 Edition. American Iron and Steel Institute, Washington, DC.

AISI S100, 2012. North American Specification for Cold-formed Steel Structural Members, 2012 Edition. American Iron and Steel Institute, Washington, DC.

AISI S213, 2012. North American Standard for Cold-formed Steel Framing – Lateral Provisions. American Iron and Steel Institute, Washington, DC.

AS/NZS 4600, 2005. Australian/New Zealand Standard – Cold-formed Steel Structures. Joint Technical Committee BD-082, Sydney-Wellington.

Eurocode 3, 2006. EN 1993-1-3 Design of Steel Structures – Part 1-3: General Rules – Supplementary Rules for Cold-formed Members and Sheeting. CEN Central Secretariat, Brussels, Belgium.

HILTI, 2013. ESR-2196: HILTI Self-drilling and Self-piercing Screws. ICC-ES Report. ICC Evaluation Service, Brea, CA.

IBC, 2015. International Building Code. International Code Council, Washington, DC.

Kaehler, R., Seaburg, P.A., October 17–18, 1996. A new AISI cold-formed steel design manual. In: Proceeding of Thirteenth International Specialty Conference on Cold-formed Steel Structures, St. Louis, Missouri, US.

Yanagi, N., Yu, C., 2014. Effective strip method for the design of cold-formed steel framed shear wall with steel sheet sheathing. ASCE Journal of Structural Engineering 140 (4).

Yu, C., Yousof, M., Mahdavian, M., 2015. Load Bearing Clip Angle Design. Research Report RP15-2. American Iron and Steel Institute, Washington, DC.

Direct strength method—a general approach for the design of cold-formed steel structures

4

D. Camotim, P.B. Dinis, A.D. Martins
Instituto Superior Técnico, Universidade de Lisboa, Lisbon, Portugal

4.1 Introduction

Rather than a "method" for the design of cold-formed steel (CFS) members, a designation that merely reflects its current codification, the Direct Strength Method (DSM) should be viewed as an "approach" (or a "procedure") that can be adopted to develop methodologies for the design of a wide variety of thin-walled structural systems, namely CFS structures. This versatility/universality provides the explanation for both the rapid (almost instantaneous) acceptance of the DSM by the technical/scientific community and the fast-growing number of successful applications of this design approach. The large amount of recent and ongoing research activity on DSM design applications will certainly be amply evidenced and reflected in upcoming versions of design codes concerning thin-walled structures.

4.1.1 *The basic idea and its origin*

CFS members invariably display very slender thin-walled open cross-sections, a feature making them highly susceptible to several instability phenomena, namely local (L), distortional (D), and global (G—flexural or flexural–torsional) buckling. Thus their overall structural response and ultimate strength are affected, to a larger or smaller extent, by these instability phenomena, which explains why they must necessarily be incorporated in CFS specifications (but may often be ignored in the conventional design of members with "compact" cross-sections). In view of this, it is not at all surprising that CFS members were the "birthplace" of the DSM—indeed, they provided the ideal setting for its development and widespread use.

Traditionally, the design of thin-walled steel members against local buckling has used the "Effective Width Method" (EWM), originally proposed by Von Kármán and later calibrated by Winter, which is incorporated (in various forms) in the current versions of nearly all (worldwide) codes addressing the design of such members. In the specific context of CFS structures, this method can be found, for example, in Eurocode 3 (1) Parts 1–3 and 1–5 (EC3-1–3 and EC3-1–5 (CEN, 2006a,b)) and (2) AISI (American Iron and Steel Institute) S100 *North American Specification for the Design of Cold-Formed Steel Structural Members* (AISI, 2012).

Recent Trends in Cold-Formed Steel Construction. http://dx.doi.org/10.1016/B978-0-08-100160-8.00004-9

The advent of increasingly more complex cross-section shapes (eg, those with several lips and/or intermediate stiffeners) and the unveiling of distortional buckling as a potential cause for the failure of CFS thin-walled members with lipped cross-sections paved the way for the development and popularity of the DSM. Indeed, after a few fairly unsuccessful attempts to predict efficiently (safely and accurately) distortional failures by means of design methods based on an "effective cross-section" concept (eg, the methodology prescribed in EC3-1−3 (CEN, 2006a)), it became clear that a more rational approach was needed. Ensuing research work led, after a decade or so, to the DSM, first codified in North America in 2004 (AISI, 2004), and almost simultaneously included in the Australian/New Zealand standard (AS/NZS, 2005). A few years later it was also adopted in Brazil (ABNT, 2010). The DSM provided a unified approach for the design of CFS members under compression (columns) and bending (beams) exhibiting local, distortional, global, and local−global interactive failures. To foster its use, AISI funded the publication of a design guide for the application of the method (Schafer, 2006). The DSM's status and scope have grown fast and steadily in the last few years, as can be seen by comparing the 2012 and 2004 North American specification versions.

The DSM may be viewed as an extension of the use of column curves for global buckling, due to the fact that it is based on the assumption that a member's ultimate strength can be accurately predicted solely on the basis of its elastic buckling and yield stresses—ie, $f_n = F(f_{crL}, f_{crD}, f_{crG}, f_y)$, where f_n is the member's nominal strength, f_{crL}, f_{crD}, f_{crG} are the elastic local, distortional, and global buckling stresses, and f_y is the steel yield stress—Fig. 4.1 provides the basic scheme concerning the application of the DSM. The elastic buckling and yield stresses are plugged into simple "Winter-type" equations that provide a "direct" estimate of the member's ultimate strength—instead of the effective cross-section, which is the cornerstone of the EWM, the DSM is based on accurate knowledge of the member's buckling stress associated with the failure mode under consideration, which means that computational tools to evaluate it are indispensable.

When compared with the EWM, the DSM exhibits four major advantages, all stemming from the fact that the cross-section is viewed as a whole. Indeed, in the DSM (1) wall-restraint effects are automatically taken into account (instead of the EWM wall-by-wall approach), (2) no cross-section classification or effective width calculations are needed, (3) it becomes possible to provide strength estimates for members failing in distortional modes (distortional buckling is regarded as a separate limit state), and (4) can explicitly take into account buckling mode interactions. In addition, the DSM provides a rational and systematic framework for the design of thin-walled

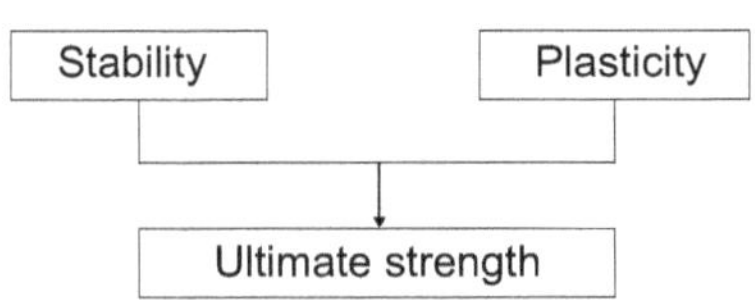

Figure 4.1 Basic scheme for the application of the DSM design approach.

structural systems (not only members) made of various materials (not only CFS)—of course, the development of a given application needs proper calibration and validation (ie, comparison with a fair number of experimental and/or numerical results) (Schafer, 2008).

4.1.2 Historical perspective

The designation "Direct Strength Method" was first mentioned in the pioneering work of Schafer and Peköz (1998a,b), in the context of developing new design approaches for CFS beams. In particular, the authors explored the application of strength curves to provide a direct way to assess (1) distortional resistance of lipped channel and Z-section beams, and (2) the local and distortional resistance of deck sections with multiple longitudinal intermediate stiffeners in the compression flange. The method had its roots in the design method proposed earlier by Hancock et al. (1994) when investigating the distortional resistance of thin-walled sections under compression or bending.[1] These authors (1) adopted the effective section approach, (2) adapted Winter's expression/curve to distortional buckling, and (3) modified its coefficients to obtain more accurate estimates. The end product of this research effort was

$$\frac{b_e}{b} = \left(\frac{f_{crD}}{f_y}\right)^{0.6}\left[1 - 0.25\left(\frac{f_{crD}}{f_y}\right)^{0.6}\right] \qquad [4.1]$$

where b_e is the effective part of the wall width b. The width reduction given by Eq. [4.1] was performed for all cross-section walls, and the method was included in the 1996 edition of the Australian/New Zealand *Standard for Cold-Formed Steel Structures* (AS/NZS, 1996).

A couple of years later, Schafer and Peköz (1998a,b), after going through Hancock et al. (1994), proposed a modification of Eq. [4.1] which consisted of replacing its coefficients (0.6/0.25) by new/reduced ones (0.4/0.15). They showed that their proposal provided, in general, accurate and reliable predictions of the ultimate strength of CFS beams failing in distortional modes, clearly outperforming the design procedure prescribed by the AISI specification at the time. Mainly due to the research carried out by Schafer at Johns Hopkins University (Schafer, 2002a,b, 2006), this first application of the DSM was fairly quickly followed by similar applications for (1) CFS columns failing in local, distortional, and global modes, and (2) CFS beams failing in local and global modes. Such research efforts led to the so-called "first generation of DSM design curves," soon to be codified in North America (AISI, 2004) and Australia and New Zealand (AS/NZS, 2005), which are addressed in detail in Section 4.2.

[1] Since Hancock credited his methodology to earlier work by Trahair on the ultimate strength of columns failing in global (flexural—torsional) modes, Schafer (2008) rightfully (but also rather humbly) argued that "the DSM is basically the extension of older ideas to new instability limit states."

The potential of the DSM approach for the design of thin-walled (not necessarily CFS) members and structural systems was quickly recognized worldwide, and led to a plethora of experimental and numerical investigations aimed at the development, validation, and possible codification of DSM-based design methodologies for a wide variety of structural problems. It may easily be argued that the "frontrunners" of this "race," in the sense that the corresponding design curves had already been codified in North America (AISI, 2012), were the studies concerning CFS (1) beams failing due to shear or combined bending and shear, advocated by Pham and Hancock at the University of Sydney, and (2) perforated columns and beams failing in local, distortional, and global modes, investigated by Moen and Schafer at Johns Hopkins University—these two design approaches are addressed in some detail in Section 4.3. But the "race" is full of "competitors," as is clearly evidenced in Section 4.4, which includes a selection of recent or ongoing studies carried out with the final goal of achieving DSM-based design expressions/curves for various problems. It is worth noting that (1) this selection is by no means exhaustive (the main purpose of Section 4.4 is to illustrate the power and versatility of the DSM approach) and necessarily reflects the authors' research activity and interests; also, due to space limitations and the focus of this book, (2) only applications to CFS members and structural systems are considered. Nevertheless, the reader should be aware that DSM solutions to problems involving other materials are currently being sought.

It should be pointed out that the DSM bears some resemblance (1) to the "general method" prescribed in Part 1−1 of Eurocode 3 (EC3-1−1 (CEN, 2005)) for the design of structural components (members or plane frames/subframes) against lateral and lateral−torsional failures, and (2) to the application of the "Overall Interaction Concept," an ambitious endeavor championed by Boissonnade et al. (2013) that circumvents the need to perform a cross-section classification procedure (currently unavoidable when designing according to Eurocode 3). Finally, one must mention the "Continuous Strength Method" (CSM), devised and developed mainly due to the efforts of Gardner and associates at Imperial College London (Gardner, 2008; Gardner et al., 2013; Afshan and Gardner, 2013), and also not requiring any cross-section classification. Although the CSM and DSM philosophies are different, they share some features and may be viewed as alternative design approaches to some structural problems. The CSM was initially proposed in the context of the design of stainless steel (nonlinear stress−strain relationship) members subject to compression or bending against local failures (Ashraf et al., 2006) and later extended to cover high-strength and carbon steel members (Gardner, 2008). The application of the method hinges on an experimentally derived "base curve" relating the cross-section resistance and deformation capacity, with the latter determining the cross-section ability to evolve into the strain-hardening region, thus sustaining a higher loading—naturally, this design approach makes it possible to take advantage of the added strength due to strain-hardening. An additional benefit of the CSM is the fact that it readily and explicitly provides ductility information.

4.2 First-generation codified DSM design curves

After presenting some practical issues related to the application of the DSM, namely (1) the geometric and material limitations of the so-called "prequalified members" and (2) the accurate calculation of member elastic buckling stresses/loads/moments, this section addresses the development, validation, and application of the first generation of codified DSM design expressions (strength curves). They concern (1) axially compressed members (columns), and (2) members subjected to major- and/or minor-axis bending (beams). Although the chapter focuses on the provisions of the current AISI S100 (AISI, 2012), it should be noted that the curves prescribed by the Australian/New Zealand standard (AS/NZS, 2005) are exactly the same.

4.2.1 Prequalification

AISI S100 (AISI, 2012) allows use of the DSM to determine the nominal axial (P_n) and flexural (M_n) strengths of any CFS columns and beams. However, this specification makes a distinction between "prequalified" and "nonprequalified" columns and beams, in the sense that different safety (Ω) and resistance (ϕ) factors must be employed—naturally, those applying to the prequalified members are less stringent.

For both columns and beams, the prequalified cross-section shapes include lipped channels (plain and web-stiffened) and zeds—complex lips may be considered in the plain lipped channel columns and zed-section beams (Young and Yan, 2004; Schafer et al., 2006). In addition, prequalified columns and beams include (1) hat and rack cross-sections, and (2) hat (flange-stiffened) and trapezoid (web-flange-stiffened) deck cross-sections, respectively. Fig. 4.2 shows the prequalified column and beam cross-section shapes, indicating the wall angles and dimensions whose values or (mostly) ratios are covered by the prequalification. The members must satisfy the geometric and material criteria presented in Tables 1.1.1−1 (columns) and 1.1.1−2 (beams) of AISI S100 Appendix 1 (AISI, 2012), which are those exhibited by the columns and beams considered in the development, calibration, and validation of the DSM design curves/expressions.

4.2.2 Determination of critical buckling loads/moments

Since the accurate determination of the column and beam buckling loads/moments is a key aspect in the application of the DSM, it is essential for designers to be equipped with efficient computational tools to perform member elastic buckling analyses aimed at obtaining local (P_{crL} or M_{crL}), distortional (P_{crD} or M_{crD}), and/or global/Euler (P_{crG} or M_{crG}) critical buckling (bifurcation) loads/moments (or stresses). Once these buckling loads/moments are known, the application of the DSM is quite straightforward. Therefore, it is not surprising that Section 1.1.2.1 of the *Commentary on the North American Specification for the Design of Cold-Formed Steel Structural Members*, published jointly with the standard (AISI, 2012), provides fairly detailed guidelines for the calculation of member critical buckling loads and moments.

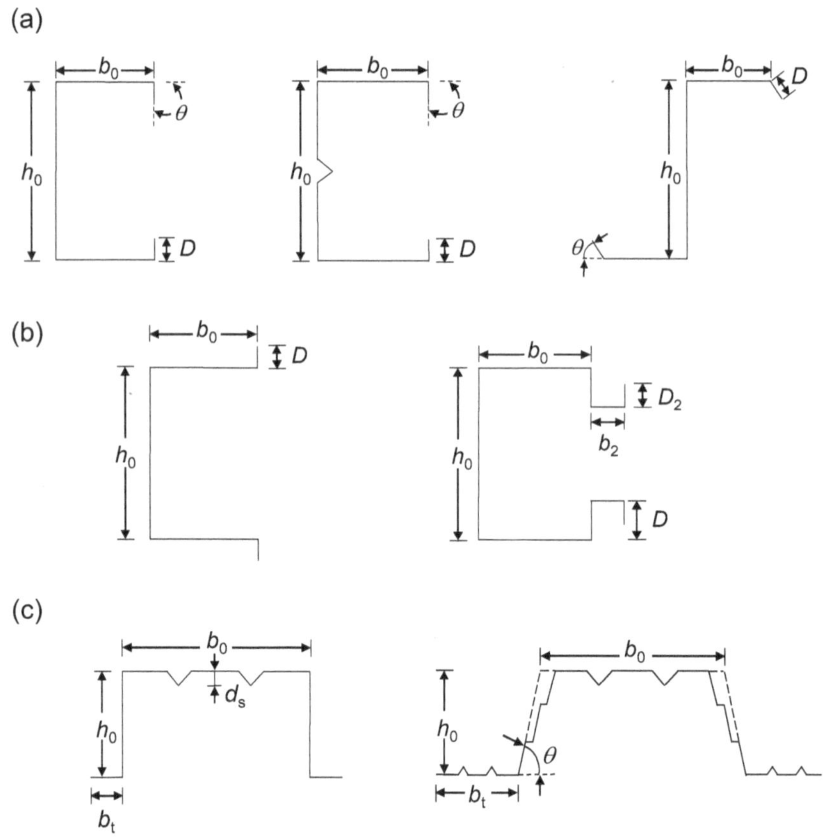

Figure 4.2 Prequalified cross-section shapes: (a) columns and beams; (b) columns; and (c) beams.

Nowadays, the column and beam local, distortional, and global critical buckling loads/moments can be easily calculated by resorting to either shell finite element (SFE), constrained finite strip (cFS), or generalized beam theory (GBT) elastic linear buckling analyses. While the SFE analyses must generally be performed with commercial software packages (ABAQUS (Simulia Inc., 2015) and ANSYS (Swanson Analysis Systems Inc., 2015) are the most commonly used), it is possible to carry out the others in freely available open source codes, namely CUFSM (Schafer et al., 2012; Li et al., 2014) and GBTUL (Bebiano et al., 2014, Bebiano et al., Submitted for publication).[2] Although SFE analyses are more versatile, in the sense that they can be applied to columns and beams with all types of support and loading conditions, the cFS or GBT analyses should be employed whenever possible. Indeed, due to their modal character, they make it much easier to identify the nature of the critical buckling modes (or calculate the critical buckling

[2] The references concern the latest versions of the aforementioned computer programs—note, however, that there are earlier versions available that provide accurate buckling solutions for most problems of practical interest.

load associated with a specific buckling mode nature)—the SFE analyses often entail the need to consider a large number of buckling modes to determine a given elastic buckling load/moment. In particular, it is worth mentioning that, as far as prismatic CFS members are concerned, the GBTUL capabilities match very closely those exhibited by SFE analysis (Bebiano et al., submitted for publication; Camotim and Basaglia, 2013).

Fig. 4.3 illustrates the evaluation, by means of GBT analyses, of the critical buckling stresses ($f_{crL}, f_{crD}, f_{crG}$) of the fixed-ended lipped channel column with the cross-section depicted in Fig. 4.3(a), with web, flange, lip widths, and thickness equal to $b_w = 105$ mm, $b_f = 82$ mm, $b_s = 15$ mm, and $t = 1.0$ mm (the corner radii are disregarded). The analyses were performed considering a GBT cross-section discretization

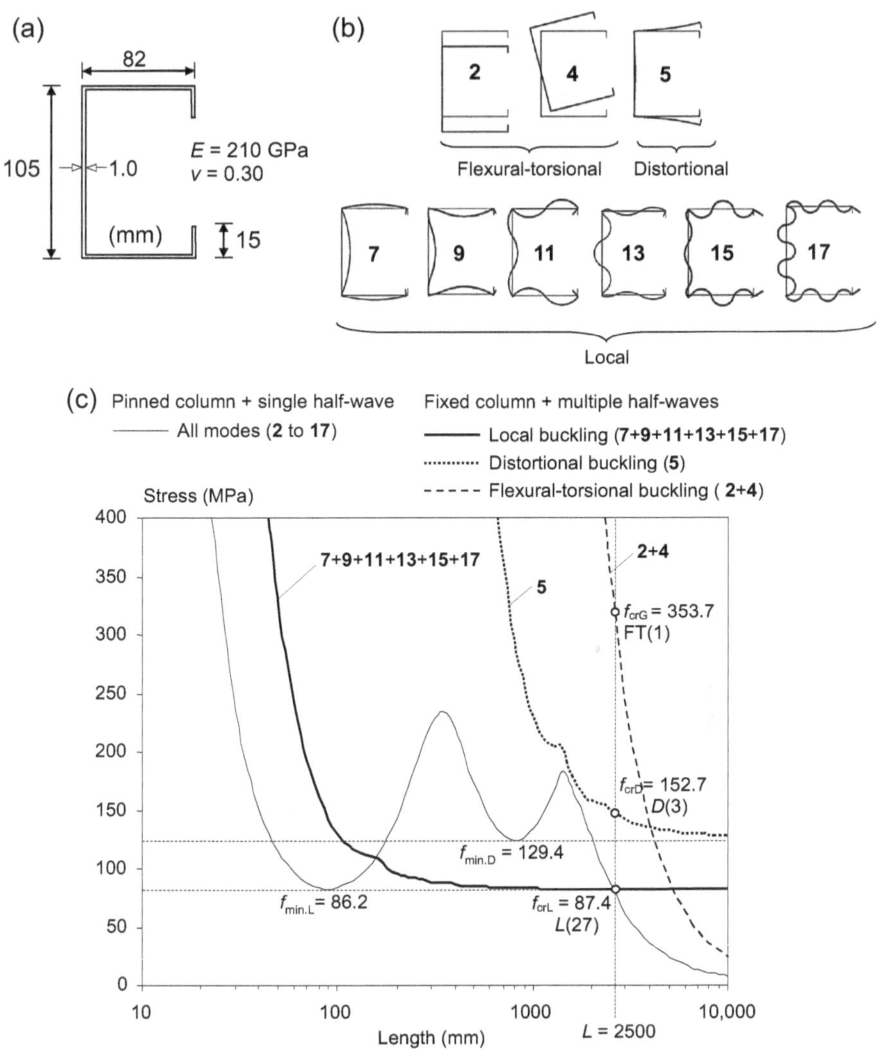

Figure 4.3 Lipped channel column: (a) cross-section geometry; (b) most relevant GBT deformation modes (used to calculate $f_{crL}, f_{crD}, f_{crG}$); and (c) buckling curves $f_{crL}(L), f_{crD}(L), f_{crG}(L)$ (fixed- and pin-ended columns).

involving three intermediate nodes in the web and flanges, which leads to 17 deformations modes. Fig. 4.3(b) displays the in-plane shapes of: (1) all symmetric local modes (7, 9, 11, 13, 15, 17), (2) the symmetric distortional mode 5, and (3) the global modes 2 (major-axis flexure) and 4 (torsion). The thin solid curve depicted in Fig. 4.3(c) is the well-known "column signature curve," corresponding to simply supported columns buckling in single half-wave modes—it may be obtained from semianalytical FS or GBT analyses. The remaining three curves, obtained from numerical GBT analyses, concern fixed-ended column buckling in multiple half-wave modes. The thick solid, dotted, and dashed curves provide, respectively, the variation of f_{nL}, f_{nD}, and f_{nG} with L—for instance, for a column with $L = 2500$ mm, one has $f_{\mathrm{crL}} = 87.4$ MPa, $f_{\mathrm{crD}} = 152.7$ MPa, and $f_{\mathrm{crG}} = 353.7$ MPa (the three white circles in Fig. 4.3(c)), which are the values required to apply the DSM. Note that the two local minima appearing in the column signature curve ($f_{\mathrm{min.L}} = 86.2$ MPa and $f_{\mathrm{min.D}} = 129.4$ MPa) provide lower bounds of the fixed-ended column local and distortional critical buckling stresses.

4.2.3 Columns (compressive strength)

To develop the DSM, Schafer (2002a, 2008) collected a fairly large data bank of experimental failure loads concerning CFS columns with various cross-section shapes (plain and web-stiffened lipped channels, hats, zeds, and racks) and failing in local, distortional, and global modes. Then, after determining analytically and/or numerically (FS analyses) the critical local, distortional, and global buckling loads of all the columns involved, he selected the strength curves/expressions to be included in the DSM. Concerning local failures, the curve selected was similar to that previously developed by Schafer and Peköz (1998a,b, 1999) for beams (see Section 4.2.4)—a Winter-type curve given by

$$\begin{cases} P_{\mathrm{nL}} = P_{\mathrm{y}} & \text{if} \quad \lambda_{\mathrm{L}} \leq 0.776 \\ P_{\mathrm{nL}} = P_{\mathrm{y}} \left(\dfrac{P_{\mathrm{crL}}}{P_{\mathrm{y}}}\right)^{0.4} \left[1 - 0.15\left(\dfrac{P_{\mathrm{crL}}}{P_{\mathrm{y}}}\right)^{0.4}\right] & \text{if} \quad \lambda_{\mathrm{L}} > 0.776 \end{cases} \qquad [4.2]$$

where $\lambda_{\mathrm{L}} = (P_{\mathrm{y}}/P_{\mathrm{crL}})^{0.5}$ is the local slenderness, P_{y} is the squash load, and P_{crL} is the critical local buckling load. For distortional failures, a curve proposed by Kwon and Hancock (1992) was embraced and cast in a DSM format as[3]

$$\begin{cases} P_{\mathrm{nD}} = P_{\mathrm{y}} & \text{if} \quad \lambda_{\mathrm{D}} \leq 0.561 \\ P_{\mathrm{nD}} = P_{\mathrm{y}} \left(\dfrac{P_{\mathrm{crD}}}{P_{\mathrm{y}}}\right)^{0.6} \left[1 - 0.25\left(\dfrac{P_{\mathrm{crD}}}{P_{\mathrm{y}}}\right)^{0.6}\right] & \text{if} \quad \lambda_{\mathrm{D}} > 0.561 \end{cases} \qquad [4.3]$$

[3] It is worth noting that, in this work, Kwon and Hancock investigated the interaction between local and distortional buckling. Since they found no evidence of adverse effects due to this interaction, they proceeded to assess the column distortional strength independently. Later studies, addressed in Section 4.4.1, showed that local–distortional buckling erodes the column failure load.

where $\lambda_D = (P_y/P_{crD})^{0.5}$ is the distortional slenderness and P_{crD} is the critical distortional buckling load. Finally, for global failures, the curve appearing in the 1996 AISI specification was adopted—it reads

$$\begin{cases} P_{nG} = P_y \ 0.658^{\lambda_G^2} & \text{if} \quad \lambda_G \leq 1.5 \\ P_{nG} = P_y \ \dfrac{0.877}{\lambda_G^2} & \text{if} \quad \lambda_G > 1.5 \end{cases} \quad [4.4]$$

where $\lambda_G = (P_y/P_{crG})^{0.5}$ is the global slenderness and P_{crG} is the critical global (flexural, torsional, or flexural–torsional) buckling load. Moreover, to account for the possible occurrence of local–global (L–G) interactive failures, Schafer proposed replacing P_y and $\lambda_L = (P_y/P_{crL})^{0.5}$ by P_{nG} and $\lambda_L = (P_{nG}/P_{crL})^{0.5}$ in Eq. [4.2], leading to

$$\begin{cases} P_{nL} = P_{nG} & \text{if} \quad \lambda_L \leq 0.776 \\ P_{nL} = P_{nG}\left(\dfrac{P_{crL}}{P_{nG}}\right)^{0.4}\left[1 - 0.15\left(\dfrac{P_{crL}}{P_{nG}}\right)^{0.4}\right] & \text{if} \quad \lambda_L > 0.776 \end{cases} \quad [4.5]$$

which becomes Eq. [4.2] when P_{crG} is much higher than P_y (ie, for short or heavily braced columns). Fig. 4.4(a) shows a comparison between the shapes of the three DSM column design curves.

Then, the column nominal axial strength is equal to $P_n = \min(P_{nD}, P_{nG}, P_{nL})$, where P_{nD}, P_{nG}, and P_{nL} are given by Eqs. [4.3]–[4.5], respectively. It is worth noting that, due to the different amounts of postcritical strength reserve associated with local, distortional, and global buckling, the natures of the critical buckling and failure modes are not necessarily the same—thus the need to evaluate P_{nD}, P_{nG}, and P_{nL}.

The merits and reliability of the DSM design approach for columns were also assessed by Schafer, as discussed and summarized in Schafer (2008). It was clearly

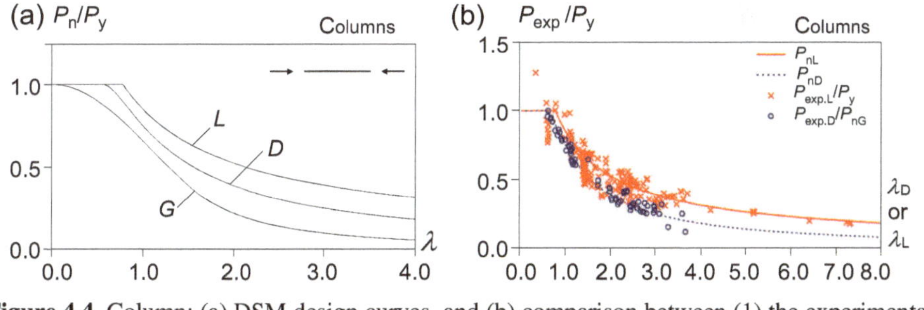

Figure 4.4 Column: (a) DSM design curves, and (b) comparison between (1) the experimental local and distortional failure loads, and (2) the corresponding DSM predictions. *(Schafer, B.W., 2008. Review: the direct strength method of cold-formed steel member design. Journal of Constructional Steel Research 64 (7-8), 766-778.)*

shown that this design methodology provides good estimates for the whole set of 249 experimental failure loads, as attested by test-to-predicted-failure load ratio (P_u/P_n) indicators: (1) average and standard deviation equal to 0.98 and 0.14 (the web-stiffened lipped channel test results are responsible for the below 1.0 average); and (2) Load and Resistance Factor Design (LRFD) resistance factor equal to $\phi_c = 0.85$—exactly the value recommended by the North American specification for CFS compression members.[4] Fig. 4.4(b), adapted from AISI (2012) and Schafer (2008), provides an overall comparison between (1) the experimental failure loads used to develop the DSM local and distortional design curves, and (2) the corresponding DSM predictions.

4.2.4 Beams (bending strength)

As pointed out earlier, the DSM was first mentioned by Schafer and Peköz (1998a,b) in the context of the design of beams. The authors addressed two problems, namely (1) the distortional failure of beams with lipped channel (plain or web-stiffened) and zed cross-sections, and (2) the local and distortional failure of beams with deck sections exhibiting multiple intermediate stiffeners in the compression flange (Schafer, 2008). They collected a large amount of available test data and, on the basis of the experimental failure moments, proposed design curves against distortional and local failures, which were later validated against further experimental (Yu and Schafer, 2003, 2006) and numerical (Yu and Schafer, 2007) failure moments. Concerning distortional failures, the codified curve is exactly the one appearing in Schafer and Peköz (1999) and Hancock et al. (1994) (even if the latter is cast in a different form), namely

$$\begin{cases} M_{nD} = M_y & \text{if} \quad \lambda_D \leq 0.673 \\ M_{nD} = M_y \left(\dfrac{M_{crD}}{M_y}\right)^{0.5} \left[1 - 0.22\left(\dfrac{M_{crD}}{M_y}\right)^{0.5}\right] & \text{if} \quad \lambda_D > 0.673 \end{cases} \qquad [4.6]$$

where $\lambda_D = (M_y/M_{crD})^{0.5}$ is the distortional slenderness, M_y is the cross-section first yield moment ($M_y = S_f f_y$, where S_f is the gross section elastic modulus), and M_{crD} is the critical distortional buckling moment. For local failures, the best curve first presented in Schafer and Peköz (1998a,b) was selected—it reads

$$\begin{cases} M_{nL} = M_y & \text{if} \quad \lambda_L \leq 0.776 \\ M_{nL} = M_y \left(\dfrac{M_{crL}}{M_y}\right)^{0.4} \left[1 - 0.15\left(\dfrac{M_{crL}}{M_y}\right)^{0.4}\right] & \text{if} \quad \lambda_L > 0.776 \end{cases} \qquad [4.7]$$

[4] At this stage it should be pointed out that a numerical investigation by Landesmann and Camotim (2013) suggested that the DSM distortional design curve may be unsafe for columns with "nonfixed-ended" support conditions. Recall that the vast majority of the experimental results used to develop and calibrate this design curve concern fixed-ended columns.

where $\lambda_L = (M_y/M_{crL})^{0.5}$ is the local slenderness and M_{crL} is the critical local buckling moment. For global failures, like in the columns, the curve appearing in the 1996 AISI specification was adopted:

$$\begin{cases} M_{nG} = M_{crG} & \text{if} \quad M_{crG} < 0.56 \ M_y \\ M_{nG} = \dfrac{10}{9} M_y \left(1 - \dfrac{10\, M_y}{36\, M_{crG}}\right) & \text{if} \quad 2.78 \ M_y \geq M_{crG} \geq 0.56 \ M_y \\ M_{nG} = M_y & \text{if} \quad M_{crG} > 2.78 \ M_y \end{cases} \qquad [4.8]$$

where M_{crG} is the critical global (lateral–torsional) buckling moment. Concerning L–G interactive failures, it should be pointed out that no experimental evidence was obtained for beams (unlike for columns). Nevertheless, Schafer proposed the adoption of a procedure similar to the one followed for the columns: replacing M_y and $\lambda_L = (M_y/M_{crL})^{0.5}$ by M_{nG} and $\lambda_L = (M_{nG}/M_{crL})^{0.5}$ in Eq. [4.7], leading to

$$\begin{cases} M_{nL} = M_{nG} & \text{if} \quad \lambda_L \leq 0.776 \\ M_{nL} = M_{nG}\left(\dfrac{M_{crL}}{M_{nG}}\right)^{0.4}\left[1 - 0.15\left(\dfrac{M_{crL}}{M_{nG}}\right)^{0.4}\right] & \text{if} \quad \lambda_L > 0.776 \end{cases} \qquad [4.9]$$

which becomes Eq. [4.7] when M_{crG} is much higher than M_y (ie, for short or heavily braced beams). Fig. 4.5(a) shows a comparison between the shapes of the three DSM beam design curves. Then, similarly to the columns, the beam nominal bending strength is equal to $M_n = \min(M_{nD}, M_{nG}, M_{nL})$, where M_{nD}, M_{nG}, and M_{nL} are given by Eqs. [4.6], [4.8], and [4.9], respectively.

As for columns, the merits and reliability of the DSM design approach for beams were assessed by Schafer, and the output can be found in Schafer (2008). The design methodology provides good estimates for the whole set of 559 experimental failure

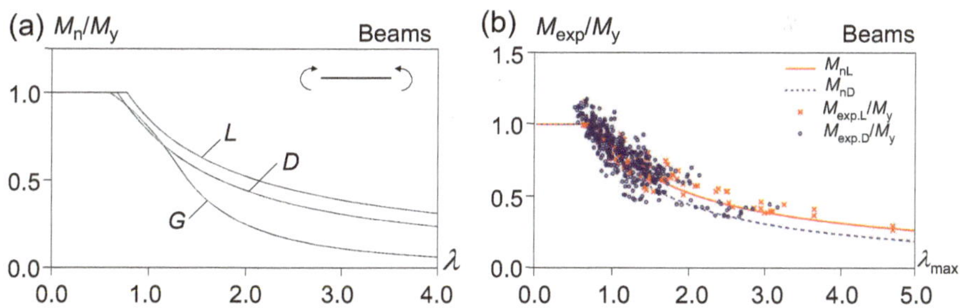

Figure 4.5 Beam: (a) DSM design curves, and (b) comparison between (1) the experimental local and distortional failure moments, and (2) the corresponding DSM predictions. *(Schafer, B.W., 2008. Review: the direct strength method of cold-formed steel member design. Journal of Constructional Steel Research 64 (7-8), 766-778.)*

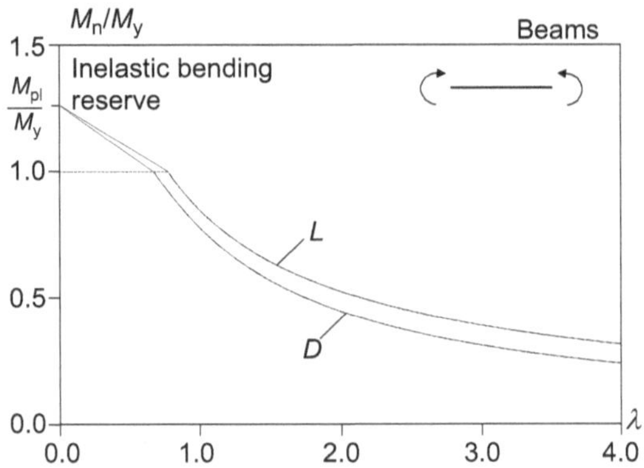

Figure 4.6 Currently codified (AISI, 2012) beam DSM local and distortional design curves incorporating the cross-section inelastic bending strength reserve in the low slenderness range.

loads, as attested by test-to-predicted-failure load ratio (M_u/M_n) indicators: (1) average and standard deviation equal to 1.09 and 0.12; and (2) LRFD resistance factor equal to $\phi_c = 0.92$—higher than the value recommended by the North American specification for CFS bending members ($\phi = 0.90$). Fig. 4.5(b), adapted from AISI (2012) and Schafer (2008), provides an overall comparison between (1) the experimental failure moments used to develop the DSM local and distortional design curves, and (2) the corresponding DSM predictions.[5]

Observation of Eqs. [4.6]−[4.8] readily shows that the cross-section inelastic bending strength reserve is completely ignored in beam DSM design—and this was the case until the 2007 AISI S100 edition. However, an investigation reported by Shifferaw and Schafer (2012), which was originated by the observation of the test results reported in Yu and Schafer (2003, 2006), provided clear evidence that this strength reserve is available and therefore should be taken into account. These findings were incorporated in the latest AISI S100 edition (AISI, 2012), through the replacement of the local, distortional, and global "yield plateaus" by linear approximations whose expressions differ depending on (1) the failure mode nature (local, distortional, global), and (2) whether the cross-section has first yield in compression or in tension (the first case also covers symmetric cross-sections).[6] For illustrative purposes, Fig. 4.6 shows the currently codified beam DSM local and distortional design curves, which take into account the cross-section inelastic bending strength reserve in the low slenderness range.

[5] In Fig. 4.5(b), $\lambda_{max} = (M_y/M_{cr})^{0.5}$, where M_{cr} is the beam critical/lowest (local or distortional) buckling moment. The use of this "mixed slenderness" was due to difficulties in distinguishing between local and distortional failures (due to the bracing and support conditions employed in the beam tests) (Schafer, 2008). These difficulties led Schafer to perform tests on beams designed to exhibit clear local (Yu and Schafer, 2003) and distortional (Yu and Schafer, 2006) failures.

[6] This means that the ensuing beam design curves may also be viewed as "second generation," even if the changes are restricted to the low slenderness range.

Before closing this section, there are a few aspects concerning the codified beam DSM design curves that are worth mentioning.

1. Although most of the experimental results considered in the development of the beam DSM design curves involved lipped channel and zed-section beams subjected to major-axis bending, it was assumed that these curves remain valid for beams under minor-axis bending, which can only exhibit local or distortional (buckling) failures (lateral–torsional buckling can never occur).[7] This assumption is partially backed by the quality of the DSM predictions of experimental failure moments of hat section and deck section beams bent about the minor axis.
2. Since the DSM design curves were basically developed for beams under uniform bending (the vast majority of the available experimental results were obtained from four-point bending tests), they provide lower bound estimates for beams acted by nonuniform bending moment diagrams. To obtain more accurate predictions, it is indispensable to take into account the bending moment diagram shape. Although this issue (a) has satisfactorily solved for lateral–torsional failures long ago (through the value of M_{crG}), and (b) is only marginally relevant for local failures (very short half-wave lengths), it has not been adequately tackled for distortional and L–G interactive failures. The only studies on this topic were reported by Yu and Schafer (2007) and Bebiano et al. (2007), who showed that incorporating exact critical buckling moments (instead of uniform ones) in the DSM expressions leads to slightly less conservative ultimate strength estimates, particularly for beams failing in distortional modes. It is clear that further investigation is needed on this topic.
3. A numerical investigation on lipped channel beams under major-axis bending (Landesmann and Camotim, 2015a) unveiled that the current DSM distortional design curve often provides unsafe failure moment predictions in the high slenderness range—the amount of overestimation depends on the beam cross-section geometry and end support conditions, particularly those concerning the warping restraint. This finding is not excessively surprising, since Fig. 4.5(b) clearly shows that the vast majority of the failure moment data employed to develop Eq. [4.6] is associated with $\lambda_D \leq 1.5$ and, within this slenderness range, these expressions yield good-quality estimates. Moreover, the four-point bending tests providing most of the experimental data concern uniformly bent beam spans with "warping continuity" at their end cross-sections, which are support conditions falling in between "free warping" and "prevented warping" (although closer to the latter).

4.3 Second-generation codified DSM design curves

This section addresses the more recently codified DSM design curves, only included in the current version of the North American specification (AISI, 2012) and not yet appearing in any other code. They concern (1) the shear strength of short-to-intermediate (eg, braced) beams, (2) the combined bending-shear strength of those same beams, and (3) the extension of the first-generation design curves to cover columns and beams with holes (perforated).

[7] It seems logical to presume that this design curve is also valid for skew bending, even if this is never explicitly mentioned.

4.3.1 Shear strength (beams)

Concerning the determination of shear strength, beam prequalification involves only the web, and AISI S100 (AISI, 2012) makes a distinction between "flat webs" and "webs with (multiple) small intermediate (longitudinal) stiffeners." Fig. 4.7 shows prequalified flat and stiffened webs, indicating the dimensions whose values or (mostly) ratios are covered by the prequalification. The beam webs must satisfy the geometric criteria presented in Table 1.1.1−3 of AISI S100 Appendix 1 (AISI, 2012), which are those exhibited by the beams considered to develop, calibrate, and validate the DSM design curves/expressions (Pham and Hancock, 2012a).

When tension field action is negligible, namely in beams whose webs have no (or largely spaced) transverse stiffeners, the DSM design expressions are merely those appearing in the AISI S100 main specification (AISI, 2012), but cast in the "DSM format":

$$\begin{cases} V_n = V_y & \text{if} \quad \lambda_v \le 0.815 \\ V_n = 0.815\ \sqrt{V_{cr}\ V_y} & \text{if} \quad 0.815 < \lambda_v \le 1.227 \\ V_n = V_{cr} & \text{if} \quad \lambda_v > 1.227 \end{cases} \qquad [4.10]$$

where $\lambda_v = (V_y/V_{cr})^{0.5}$ is the slenderness associated with shear buckling, V_y is the cross-section (web) yield shear force ($V_y = 0.6\,A_w f_y$, where A_w is the nominal web area and $0.6 f_y \approx f_y/\sqrt{3}$ is the shear yield stress), and V_{cr} is the cross-section elastic shear buckling force of either the whole cross-section or its web (including the intermediate stiffeners, if present), calculated as indicated in Pham et al. (2014). In beams whose webs have closely spaced transverse stiffeners (spacing not exceeding twice the web depth), tension field action can be taken into account Pham and Hancock (2012b) and Pham et al. (2014) proposed a design curve similar to Eqs. [4.2] and [4.7], which was subsequently shown to estimate quite well the experimental results reported in Pham and Hancock (2012a) and LaBoube and Yu (1978a). The curve reads

$$\begin{cases} V_n = V_y & \text{if} \quad \lambda_v \le 0.776 \\ V_n = V_y \left(\dfrac{V_{cr}}{V_y}\right)^{0.4} \left[1 - 0.15\left(\dfrac{V_{cr}}{V_y}\right)^{0.4}\right] & \text{if} \quad \lambda_v > 0.776 \end{cases} \qquad [4.11]$$

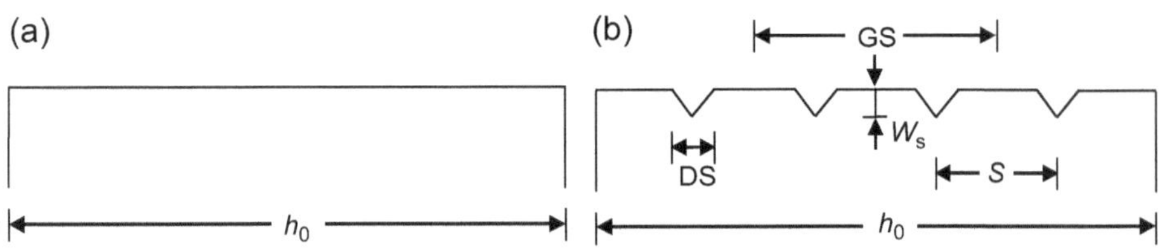

Figure 4.7 Prequalified (a) flat and (b) stiffened webs of beams subjected to shear.

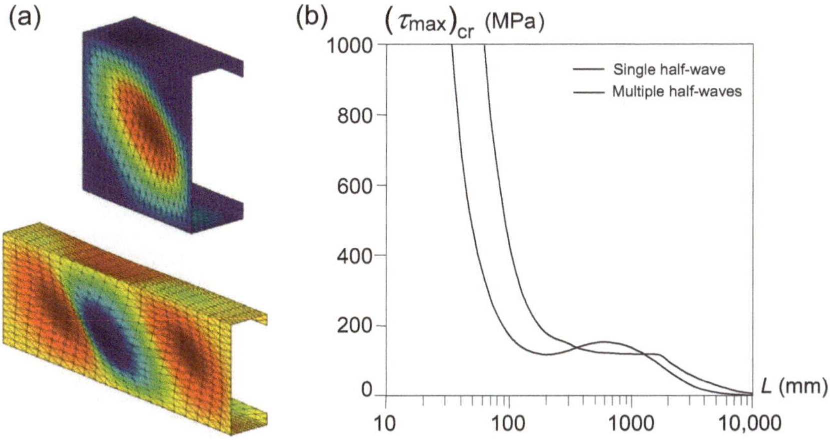

Figure 4.8 (a) Single and three half-wave local buckling modes of a lipped channel under shear and (b) curves $(\tau_{max})_{cr}$ versus L concerning the single and multiple half-wave shear buckling of beams (Pham et al., 2014).

and corresponds to beams/webs buckling and failing in local modes (see Fig. 4.8(a)).[8] Concerning the values of V_{cr}, AISI S100 (AISI, 2012) stipulates that it can be obtained by means of an elastic buckling analysis of either the whole beam segment/cross-section or just its web (including possible intermediate stiffeners) subjected to pure shear, and suggests the use of semianalytical (Hancock and Pham, 2013) or spline (Pham and Hancock, 2012c) FS analysis. However, such calculations can also be performed using other methods, namely GBT (but not with code GBTUL). Fig. 4.8(b), taken from Pham et al. (2014), illustrates curves providing the variation of the maximum critical shear buckling stress with the beam length (logarithmic scale) for single and multiple half-wave buckling.

The merits and the reliability of the DSM design approach for CFS sections in shear were assessed and discussed by Pham and Hancock (2012a), using failure load data concerning tests performed at the University of Sydney (36 tests involving plain lipped channel and SupaCee section beams) and the University of Missouri−Rolla (35 tests involving plain lipped channel beams). The estimates provided by the DSM approach were quite good, as attested by V_u/V_n indicators. The averages and standard deviations of this ratio are equal to (1) 1.29/0.30 (Sydney) and 1.05/0.12 (Missouri−Rolla), corresponding to LRFD resistance factors $\phi = 0.92$ and $\phi = 0.89$, respectively, if tension field action is ignored; and to (2) 1.02/0.07 (Sydney) and 0.84/0.13 (Missouri−Rolla), corresponding to LRFD resistance factors $\phi = 1.06$ and $\phi = 0.71$, respectively, if tension field action is taken into account. Pham and Hancock (2012a) concluded that a resistance factor of $\phi = 0.90$ is appropriate for both cases.

[8] Although beams subjected to pure shear may buckle in distortional modes (Pham et al., 2014), the corresponding failures have not been investigated so far and are therefore not covered by Eqs. [4.10] and [4.11].

4.3.2 *Combined bending-shear strength (beams)*

As mentioned in AISI (2012), the DSM equations to estimate the combined bending-shear strength are based on the interaction equation appearing in the main specification (Section 3.3.2—LFRD), given by

$$\left(\frac{\overline{M}}{\phi_b\,M_{nx0}}\right)^2+\left(\frac{\overline{V}}{\phi_v\,V_n}\right)^2\le 1.0 \qquad [4.12]$$

for beams whose webs have no (or largely spaced) transverse stiffeners (no tension field action), where (1) $\overline{M}$ and $\overline{V}$ are the acting bending moment and shear force, (2) M_{nx0} and V_n are the individual/"pure" bending (about the centroidal x axis) and shear strengths, and (3) ϕ_b and ϕ_v are the corresponding LFRD resistance factors. In beams whose webs have closely spaced transverse stiffeners (tension field action), whenever $\overline{M}$ and $\overline{V}$ are such that $\overline{M}/(\phi_b\,M_b)>0.5$ and $\overline{V}/(\phi_v\,V_n)>0.7$, they must satisfy, instead of Eq. [4.12], the alternative condition

$$0.6\,\frac{\overline{M}}{\phi_b\,M_{nx0}}+\frac{\overline{V}}{\phi_v\,V_n}\le 1.3 \qquad [4.13]$$

which was first proposed in LaBoube and Yu (1978b) and is less stringent than Eq. [4.12]— naturally, $\overline{M}/(\phi_b\,M_b)$ and $\overline{V}/(\phi_v\,V_n)$ cannot exceed 1.0. Fig. 4.9 shows the interaction diagrams defined by Eqs. [4.12] and [4.13], which are valid for beams with or without adequate web transverse stiffening (tension field action), respectively.

In Eqs. [4.12] and [4.13] (plus its limits), the nominal bending and shear strength (M_{nx0} and V_n) are obtained from the DSM design expressions/curves addressed in

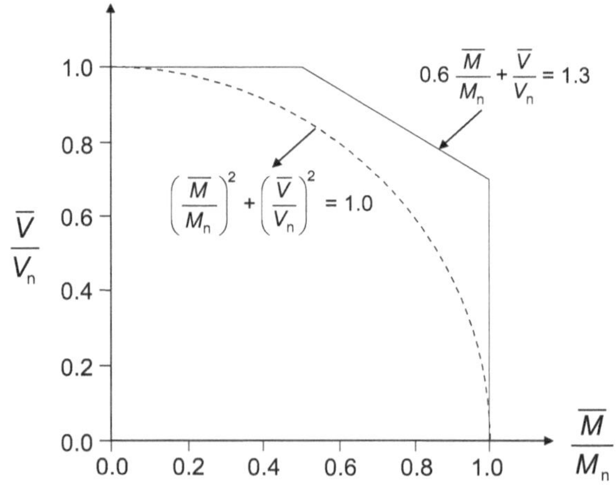

Figure 4.9 Interaction diagrams for beams with or without web transverse stiffeners (tension field action).

Sections 4.2.4 and 4.3.1, respectively (Pham and Hancock, 2012a). Following the validation procedure carried out by Pham and Hancock (2012a), who used experimental results obtained at the University of Sydney concerning (1) lipped channel and zed-section lapped purlins (Pham and Hancock, 2009), and (2) plain lipped and SupaCee channel beams, the following guidelines for the choice of the M_{nx0} and V_n values to be included in the interaction equations were established and codified (AISI, 2012):

1. For beams without adequate web transverse stiffening (no tension field action), V_n is obtained from Eq. [4.10] and M_{nx0} is the nominal flexural strength/resistance associated with local buckling (M_{nL}), which is calculated by means of Eq. [4.7].
2. For beams with adequate web transverse stiffening (tension field action may be taken into account), V_n is obtained from Eq. [4.11] and M_{nx0} is the lesser of the nominal flexural strengths/resistances associated with local (M_{nL}) and distortional (M_{nD}) buckling—the latter calculated by means of Eq. [4.6].

4.3.3 Members with holes (axial and bending strength)

Although CFS columns and beams are often manufactured with holes to accommodate the passage of utilities (eg, electrical, plumbing, or heating conduits), up to a few years ago the available design methods for such members were limited to specific hole sizes, shapes, and configurations (spacing) (Moen and Schafer, 2009). Fig. 4.10 shows a few illustrative examples of perforated members. This situation fostered a considerable amount of research devoted to extending the available DSM design approaches to columns and beams with holes (Sputo and Tovar, 2005; Tovar and Sputo, 2005; Moen and Schafer, 2009a,b, 2011). The main difficulties associated with this work stemmed from the fact that, depending on the hole shape, size, and spacing, different interactive buckling modes may be triggered

Figure 4.10 Illustrative examples (a–d) of structural systems formed by perforated members. (a) Steel Stud Manufacturers Association (SSMA)-Product Technical Information; (b and c) Steel Framing Industry Association (SFIA)-Case studies: Convent Hill and City Green, Milwaukee, Wisconsin; (d) Supreme Steel Framing System Association (SSFSA), Supreme Framing System Product Catalog.

and play a relevant role in the member behavior (Schafer, 2008). As far as the DSM is concerned, the key issues were (1) the accurate determination of the critical buckling loads (columns) and moments (beams), (2) the possible need to take into account the effect of the holes in the cross-section resistance, and (3) the collection and/or determination of a large enough set of experimental and/or numerical failure loads and moments, indispensable to calibrate and/or validate the novel DSM design curves. The relevance of the first two issues is attested by the fact that all the current AISI S100 Appendix 1 (AISI, 2012) provisions concerning application of the DSM to columns and beams with holes always state that the nominal axial or bending strengths/resistances associated with local, distortional, global, and L–G interactive failures shall be determined according to the design expressions/curves presented in Sections 4.2.3 (columns) and 4.2.4 (beams), except that (1) the corresponding critical buckling load or moment must be determined including the influence of the hole(s), and, for local and distortional failures, (2) some condition concerning the net cross-section at the hole location must be satisfied. It is still worth mentioning that AISI (2012) does not specify any limitation concerning hole size, shape, and spacing for the application of the DSM to perforated members—thus it is not surprising that the prequalification criteria for perforated and nonperforated columns and beams are exactly the same (see Section 4.2.1).

The accurate determination of critical buckling loads/moments of members with holes is by no means straightforward, since there are no available GBT or FS codes to perform this task.[9] Resorting to SFE analysis is always possible, but very time-consuming and clearly not feasible for routine engineering design. This major hurdle was gradually overcome by Moen and Schafer, who began by investigating the influence of the presence of holes on the elastic buckling behavior of plates subjected to compression or bending (Moen and Schafer, 2009a), and used the results obtained to develop simplified approximate methods to estimate local, distortional, and global buckling loads and moments of CFS columns and beams with holes (Moen and Schafer, 2009b).[10] For instance, the authors concluded that the distortional buckling load/moment of a column/beam with holes in the web is fairly well estimated by that of a (fictitious) column/beam without holes and with a reduced web thickness of $t_{web.hole} = t_{web}\,(1\text{-}L_{hole}/L_{crD})^{1/3}$, where L_{hole} is the hole length and L_{crD} is the distortional half-wave length of the original member without holes (ie, with the gross cross-section). The commentary in AISI (2012) provides quite detailed guidance on the determination of the elastic buckling loads/moments in columns/beams with holes, mainly using FS analysis. Fig. 4.11(a)–(c) shows the configurations of local, distortional, and global (flexural–torsional) buckling mode shapes of lipped channel columns with two fairly spaced slotted holes, obtained from SFE analyses (Moen and Schafer, 2011).

[9] Although Eccher et al. (2008) developed an isoparametric FS method to perform elastic buckling analysis of perforated members, there is no available computer code that can be readily and easily used by the design office community.

[10] Note that Smith and Moen (2014) recently extended this work to cover pallet rack columns exhibiting perforation patterns (instead of evenly spaced discrete holes).

Figure 4.11 (a) Local, (b) distortional, and (c) global (flexural–torsional) buckling mode shapes of lipped channel columns with two fairly spaced slotted holes. Moen, C.D., Schafer, B.W., 2009b. Elastic buckling of cold-formed steel columns and beams with holes. Engineering Structures 31 (12), 2812-2824.

The basis for the application of the DSM to CFS columns with holes can be found in Moen and Schafer (2011), where the codified design curves are addressed.[11] Besides the P_{crL}, P_{crD}, and P_{crG} values, the differences with respect to the expressions presented in Section 4.2.3 for columns without holes stem from the need to make changes in the inelastic regime, since it was shown that the existing design expressions/curves are adequate when failure is governed by elastic stability effects. These changes deal only with the design curves associated with local (or L–G interactive) and distortional failures. Concerning the former, it is required that P_{nL} satisfies the additional condition

$$P_{nL} \leq P_{ynet} = A_{net} f_y \qquad [4.14]$$

where P_{ynet} is the strength/resistance of the column net cross-section at the location of the hole, which has an area equal to A_{net}. For distortional failures, P_{nD} must satisfy additional conditions, applicable in the low-to-moderate distortional slenderness range and discretely dependent on the value of λ_D (two ranges)—such conditions are

$$\begin{cases} P_{nD} = P_{ynet} & \text{if} \quad \lambda_D \leq \lambda_{d1} \\ P_{nD} = P_{ynet} - \left(\dfrac{P_{ynet} - P_{d2}}{\lambda_{d2} - \lambda_{d1}}\right)(\lambda_D - \lambda_{d1}) & \text{if} \quad \lambda_{d1} < \lambda_D \leq \lambda_{d2} \end{cases} \qquad [4.15]$$

where $\lambda_D = (P_y/P_{crD})^{0.5}$, with P_y based on the gross cross-section area A_g ($P_y = A_g f_y$), $\lambda_{d1} = 0.561(P_{ynet}/P_y)$, $\lambda_{d2} = 0.561\ [14(P_y/P_{ynet})^{0.4} - 13]$, and $P_{d2} = P_y\ [1 - 0.25\ (1/\lambda_{d2})^{1.2}]\ (1/\lambda_{d2})^{1.2}$. Fig. 4.12(a,b) shows the ensuing DSM distortional design curves for two P_{ynet}/P_y values, namely 0.8 and 0.6 (Moen and Schafer, 2011).

The merits and reliability of the DSM design approach for CFS columns with holes were assessed and discussed in Moen and Schafer (2009)—failure load data consisting of 78 experimental (some reported in Moen and Schafer (2008)) and 385 numerical results, involving lipped channel columns with web holes, were considered. It was shown that the codified DSM design curves (option four in Moen and Schafer (2009)) provide quite good estimates for the whole set of failure loads. Indeed, the P_u/P_n values

[11] At this stage it should be mentioned that Yao (2012) reported a very significant amount of work, carried out at the University of Sydney, on the application of the DSM to perforated columns. However, to the authors' best knowledge (and surprise), the output of this research effort was never properly disseminated (eg, by means of international journal articles).

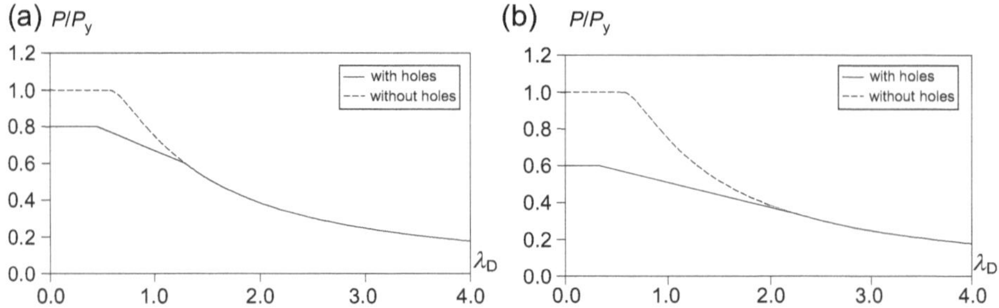

Figure 4.12 DSM distortional design curves for columns with holes: P_{ynet}/P_y equal to (a) 0.80 and (b) 0.60.

associated with experimental and numerical global, distortional, and local or L–G interactive failures exhibited (1) averages 1.04–1.19, standard deviations 0.14–0.19, and LRFD resistance factors (target reliability index of $\beta_0 = 2.5$) $\phi_c = 0.79$–0.85 for the experimental failure loads; and (2) averages 1.04–1.16, standard deviations 0.06–0.11, and LRFD resistance factors (also for $\beta_0 = 2.5$) $\phi_c = 0.87$–0.91 for the numerical failure loads. Note that all ϕ_c values are either better than or quite close to the $\phi_c = 0.85$ recommended by AISI S100 (AISI, 2012).

The basis of application of the DSM to CFS beams with holes is reported in Moen and Schafer (2009, 2010), which also address the codified design curves. As for the columns, the expressions presented in Section 4.2.4 (for beams without holes) that do not take into account the cross-section inelastic bending strength reserve (ie, Eqs. [4.6], [4.8], and [4.9])[12] just need to be complemented with additional conditions concerning the inelastic regime—of course, the calculation of the M_{crL}, M_{crD}, and M_{crG} values must adequately reflect the presence of the holes. Moreover, these conditions are similar to their column counterparts. Indeed, it is stated that M_{nL} and M_{nD} must also satisfy

$$M_{nL} \leq M_{ynet} = S_{fnet}\ f_y \qquad [4.16]$$

$$\begin{cases} M_{nD} = M_{ynet} & \text{if} \quad \lambda_D \leq \lambda_{d1} \\ M_{nD} = \min \begin{cases} M_{ynet} - \left(\dfrac{M_{ynet} - M_{d2}}{\lambda_{d2} - \lambda_{d1}}\right)(\lambda_D - \lambda_{d1}) \\ M_y \left(\dfrac{M_{crD}}{M_y}\right)^{0.5} \left[1 - 0.22\left(\dfrac{M_{crD}}{M_y}\right)^{0.5}\right] \end{cases} & \text{if} \quad \lambda_{d1} < \lambda_D \leq \lambda_{d2} \end{cases} \qquad [4.17]$$

[12] Probably because the research efforts (1) to account for the cross-section inelastic bending strength reserve in the DSM design and (2) to extend its domain of application to columns and beams with holes took place more or less at the same time, their joint incorporation in AISI S100 (AISI, 2012) was not addressed.

respectively, where M_{ynet} is the strength/resistance of the beam net cross-section at the location of the hole, which has an elastic modulus equal to S_{fnet}, $\lambda_D = (M_y/M_{crD})^{0.5}$, with M_y based on the gross cross-section modulus S_f ($M_y = S_f f_y$), $\lambda_{d1} = 0.673(M_{ynet}/M_y)^3$, $\lambda_{d2} = 0.673\ [1.7(M_y/M_{ynet})^{2.7} - 0.7]$, and $M_{d2} = M_y\ [1 - 0.22\ (1/\lambda_{d2})]\ (1/\lambda_{d2})$.[13] The ensuing DSM distortional design curves are qualitatively similar to those depicted in Fig. 4.12(a,b) and, naturally, depend on the M_{ynet}/M_y moment ratio (Moen and Schafer, 2009, 2010).

As for the columns, the merits and reliability of the DSM design approach for CFS beams with holes were assessed and discussed in Moen and Schafer (2009). The failure moment data considered consisted of 144 experimental (some reported in Moen et al. (2013)) and 204 numerical results involving lipped channel beams with web holes. It was shown once more that the codified DSM design curves (again option four in Moen and Schafer (2009)) provide good estimates for the whole set of failure moments. Indeed, the M_u/M_n values associated with experimental and numerical distortional and local or L−G interactive failures (global failures are not addressed) exhibited (1) averages 0.87 and 0.88, standard deviations 0.14 and 0.12, and LRFD resistance factors (for $\beta_0 = 2.5$) $\phi_b = 0.81$ and $\phi_b = 0.85$, for the experimental failure moments; and (2) averages 1.06 and 1.07, standard deviations 0.13 and 0.09, and LRFD resistance factors (also for $\beta_0 = 2.5$) $\phi_b = 0.86$ and $\phi_b = 0.89$, for the numerical failure moments—with one exception, all these ϕ_b values are fairly close to the $\phi_b = 0.90$ recommended by AISI S100 (AISI, 2012) .

4.4 Recent advances and future developments in DSM-based cold-formed steel design

As mentioned in Section 4.1.2, there is a significant amount of recent and/or ongoing research on the development, calibration, validation, and (possible) codification of DSM-based design approaches for a wide variety of structural problems. Indeed, just browsing though recent issues of the most prestigious international journals covering topics related to the design of cold-formed (carbon[14]) steel members and structural systems will give an idea of how many DSM applications are being sought around the globe to handle specific failures—all sharing the (more or less elusive) final goal of achieving codification. A selection of these applications is addressed next, inevitably reflecting the authors' research activity and interests but also illustrating the DSM's power and versatility. Due to space limitations, only brief overviews of selected DSM applications are provided, with a few key references that will help the interested reader acquire and grasp the most relevant information.

[13] It is worth noting that Eq. [4.17] states explicitly that the second expression cannot exceed its counterpart for beams without holes (Eq. [4.6]), employed with M_{crD} calculated taking into account the holes. No similar explicit statement appears in AISI S100 (AISI, 2012) for columns with holes failing in distortional modes.

[14] No application of the DSM to the design of cold-formed stainless steel members is dealt with in this chapter.

4.4.1 Interactive failures involving distortional buckling

Since the currently codified column and beam DSM design curves cover only local, distortional, global, and L–G interactive failures, there is ongoing research work on the understanding and prediction of the ultimate strength erosion caused by the remaining interaction phenomena, namely those involving distortional buckling: local–distortional (L–D), distortional–global (D–G), and local–distortional–global (L–D–G) interaction. It is fair to say that the vast majority of this research has been devoted to columns, although investigation on beams has been initiated (eg, Martins et al. (2015a) and Nandini and Kalyanaraman (2010), on L–D and L–D–G interaction, respectively). Concerning L–D interaction, by far the most studied among the above three coupling phenomena and for which there exists clear experimental evidence of ultimate strength erosion (eg, Kwon and Hancock (1992), Yang and Hancock (2004), Kwon et al. (2009), Young et al. (2013), and Dinis et al. (2014a)), Schafer (2002a) suggested a procedure similar to that employed to handle L–G interactive failures: replace P_y by either (1) P_{nL} in Eq. [4.3], leading to the NDL approach (P_{nDL}), or (2) P_{nD} in Eq. [4.2], leading to the NLD approach (P_{nLD}). The latter was shown to provide safe estimates of the experimental failure loads reported in Yang and Hancock (2004) and Kwon et al. (2009).

On the basis of a numerical investigation involving lipped channel columns with quite close (less than 10% apart) local and distortional critical buckling loads, Silvestre et al. (2009, 2012) (1) showed that P_{nD} yields accurate failure load estimates for $\lambda_D < 1.5$ but provides many unsafe predictions for slender columns, and that P_{nDL} and P_{nLD} invariably and similarly underestimate slender column failure loads, and (2) proposed a novel/modified NLD approach (termed here NL-D). It adopts P_{nD} for $\lambda_D \leq 1.5$ and, for slender columns, (a) defines a modified local strength P^*_{nL}, dependent on the critical half-wave length ratio L_{crD}/L_{crL} (obtained from the simply supported column "signature curve"), and (b) provides column ultimate strength estimates ($P_{nL\text{-}D}$) by replacing P_{nL} with P^*_{nL} in the NDL equation. P^*_{nL} is given by

$$\begin{cases} P^*_{nL} = P_y & \text{if} \quad \dfrac{L_{crD}}{L_{crL}} \leq 4 \\[2ex] P^*_{nL} = P_y + \left(1 - 0.25\dfrac{L_{crD}}{L_{crL}}\right)(P_y - P_{nL}) & \text{if} \quad 4 < \dfrac{L_{crD}}{L_{crL}} < 8 \\[2ex] P^*_{nL} = P_{nL} & \text{if} \quad \dfrac{L_{crD}}{L_{crL}} \geq 8 \end{cases} \qquad [4.18]$$

This approach was subsequently shown to be also valid for hat-section, zed-section, and rack-section columns under similar circumstances (ie, affected by strong L–D interaction) (Dinis and Camotim, 2015a). Very recently, Martins et al. (2015b) shed new light on the L–D interaction by clearly identifying and investigating two sources for this coupling phenomenon, namely (1) the closeness between the local and distortional critical buckling loads (causing "true interaction"), and, in the absence of such closeness, (2) the high noncritical slenderness (causing "secondary bifurcation interaction," which is particularly relevant when local buckling is critical). Moreover, these authors showed that "true L—D interaction" effects may cause a relevant failure load erosion when $0.8 \leq P_{crD}/P_{crL} \leq 1.4$. Taking into account recent and ongoing work (Yap and Hancock, 2011; Dinis et al., 2014b; Martins et al., 2015c, 2016) on stiffened lipped channel columns, it seems fair to say that the codification of DSM design expressions to handle column L—D interactive failures will be possible in the near future, covering lipped channel (plain, web-stiffened, and web/flange-stiffened), zed-section, hat-section, and rack-section columns.

Some research work has recently been done to investigate the L—D—G interaction in fixed-ended plain lipped channel columns with fairly close local, distortional, and global critical buckling loads, namely test campaigns carried out in Rio de Janeiro (Santos et al., 2012, 2014a) and Hong Kong (Young et al., 2015). The latter, in particular, provided very clear experimental evidence of ultimate strength erosion caused by the simultaneous occurrence of local, distortional, and global (flexural—torsional) deformations. On the basis of the experimental failure loads obtained, as well as the numerical results of limited parametric studies, it appears that the currently codified DSM design curve associated with L—G interactive failures (Eq. [4.5]) provides efficient (safe and accurate) estimates of the columns undergoing the triple interaction under scrutiny (Dinis et al., 2011, 2012a, 2015; Santos et al., 2014b). Although this conclusion is preliminary and needs further experimental and numerical confirmation, by considering columns with other cross-section shapes, it appears that no new DSM design curve is necessary to handle column L—D—G interactive failures. Finally, one should mention that very little research activity exists on use of the DSM to predict D—G interactive failures—to the authors' best knowledge, the only international journal article available on this topic is Rossi et al. (2010) and deals with cold-formed stainless steel columns.

4.4.2 Equal-leg angle columns

Despite their geometrical simplicity, angles exhibit a very complex structural behavior, which is responsible for the fact that, in AISI S100 (AISI, 2012), short-to-intermediate equal-leg angle columns (1) are not yet prequalified for DSM design and (2) are excluded from the application of the LFRD resistance factor $\phi = 0.85$, valid for all other CFS compression members (Ganesan and Moen, 2012). This is due to the fact that such columns buckle in flexural—torsional modes associated with a critical load plateau, unlike their longer counterparts, which exhibit a "trivial" minor-axis flexural buckling (Dinis et al., 2010). Because the flexural—torsional buckling deformations in equal-leg angle columns are very similar to local deformations, these columns have been said, erroneously, to fail in "L—G interactive modes," which explains why their design is often based on local strength concepts. Indeed, until very recently the most successful attempts to develop a DSM approach to design equal-leg angle columns, namely those reported by (1) Young (2004) for fixed-ended columns, (2) Rasmussen (2006) for pin-ended columns, and (3) Silvestre et al. (2013) for both fixed- and pin-ended columns, involved the use of either the currently codified DSM local design curve or a slightly modified (empirically) version of this same curve. However, this situation was altered by the findings of Dinis et al. (2012b) and Mesacasa et al. (2014), who provided clear numerical evidence that the failure of angle columns stems from the interaction between two global modes (major-axis flexural–torsional and minor-axis flexural), a kind of unique coupling behavioral feature that has very little (if anything) to do with local deformations. Based on these findings, Dinis and Camotim (2015b) proposed a novel rational DSM-based design approach for short-to-intermediate equal-leg angle columns that (1) is valid for both pin-ended and fixed-ended support conditions, (2) uses genuine flexural–torsional strength curves (instead of the local one), and (3) yields quite reliable predictions of the available experimental and numerical failure loads ($\phi_c = 0.85$ becomes applicable). Pending validation against experimental failure loads concerning slender pin-ended columns (a test program is currently under way to fill this gap) and a few simplifications aimed at improving the user-friendliness, this design approach may be deemed ready for codification.

4.4.3 Columns and beams under elevated temperatures

Although CFS constructions are particularly vulnerable to fire conditions,[15] the behavior and strength of CFS members under elevated temperatures are a research topic which only started to attract attention from the scientific community well into this century. Just to illustrate this assertion, note that Eurocode 3 Part 1–2 (EC3-1–2) merely states that "the methods given (for hot-rolled and welded profiles) are also applicable to cold-formed steel members and sheeting within the scope of EN 1993-1–3" (CEN, 2006a), and no mention of the words "fire" or "temperature" can be found. Nevertheless, the literature already contains a fair number of studies dealing with experimental tests, numerical simulations, and design considerations concerning the application of the DSM to columns and beams subjected to elevated temperatures. The major difficulty is handling the temperature dependence of the steel constitutive law, felt through both (1) the reductions of the Young's modulus and nominal yield stress, and (2) the qualitative change of the stress–strain curve, which becomes gradually more nonlinear (instead of bilinear with a well-defined yield plateau) as the temperature rises. Typically, numerical simulations concerning CFS members at elevated temperatures (1) adopt SFE models and (2) are only very rarely (and then quite loosely) combined with thermal analyses aimed at determining the temperature evolution and/or distribution—instead, they just assess the impact of a uniform elevated temperature distribution on the member's ultimate strength (Abreu et al., 2014).

[15] This is because they experience a very fast temperature rise (and ensuing stiffness and strength degradation), due to a combination of high thermal conductivity with a large "section factor," defined as the ratio between the cross-section perimeter exposed to fire and the area.

As is logical to expect, the available work on the application of the DSM to design CFS members under elevated temperature concerns mostly columns under preestablished uniform temperatures. Moreover, the methodology adopted consists, essentially, in inserting buckling and plastic loads calculated with modified/reduced steel material properties (Young's modulus and yield stress—the models to estimate the temperature-dependent reduction of these parameters are invariably experimentally based and may vary considerably) into the codified DSM strength curves, which were originally developed for carbon steel columns at ambient temperature. This means that the effect of the steel stress−strain curve nonlinearity at elevated temperatures is disregarded. The first attempt can be traced back to 2007, when Chen and Young (2007) reported an investigation on the application of the DSM to estimate failure loads of lipped channel columns failing in local, distortional, and global modes under elevated temperatures. Heva and Mahendran (2008) and Ranawaka and Mahendran (2009) carried out similar studies for columns failing in local and distortional modes, respectively. Shahbazian and Wang (2011a,b, 2012) reported a very thorough investigation, carried out in the context of plain and web-stiffened lipped channel columns, on the applicability of the DSM to estimate the ultimate strength of columns failing in global, local (including L−G interactive), and distortional modes. The temperature-dependent CFS constitutive law model prescribed by EC3-1−2 (CEN, 2006) was adopted and the columns analyzed were subjected to uniform and nonuniform cross-section temperature distributions—in the latter case, the nonuniformity was restricted to a linear variation in the web. These authors concluded that the currently codified DSM design curves, used with critical buckling and plastic loads calculated on the basis of the reduced material properties, (1) are adequate to predict the failure loads of columns under uniform elevated temperature, and (2) need to be modified when temperature distribution is nonuniform (new local, global, and distortional design curves were proposed). For columns failing in local and distortional modes under uniform temperature, this conclusion is at odds with earlier findings from other authors (Heva and Mahendran, 2008; Ranawaka and Mahendran, 2009), who provided evidence that the currently codified DSM design curves become unsafe for higher elevated temperatures, particularly if the column is slender. This assertion was recently corroborated by Landesmann and Camotim (2015b), who carried out an extensive parametric study (1) involving columns with lipped channel, hat, zed, and rack cross-sections failing in distortional modes for temperatures up to $T = 600°C$, and (2) aimed at assessing how the quality of DSM-based failure load predictions was influenced by the particular temperature-dependent steel constitutive law model adopted (six models were considered). Besides concluding that the quality of the predictions is practically independent from the constitutive law model adopted, these authors also showed that the current DSM distortional design curve only estimates reasonably well the failure loads of columns with low-to-intermediate slenderness ($\lambda_D < 1.5$) and subjected to temperatures up to $T = 300°C$. Outside this range, the predictions are clearly unsafe, with the amount of overestimation increasing with both T and λ_D—this is most likely linked to the growing nonlinearity of the steel stress−strain curve. In summary, it may be said that the application of the DSM to columns under elevated temperatures is definitely feasible, but there are still many unresolved issues that need to be investigated before codification can be considered (even for uniform temperature distributions).

Compared to columns, CFS beams under elevated temperature have been much less studied. It is worth noting the work of (1) Kankanamge and Mahendran (2012) and Landesmann and Camotim (2015b) on simply supported lipped channel beams subjected to uniform bending at uniform elevated temperatures and failing in global (lateral–torsional) and distortional modes, respectively, and (2) Lu et al. (2010) on single- and two-span zed-section purlins acted upon by uniformly distributed transverse loads under different fire scenarios. Only the second of these works deals with application of the DSM to predict beam failure moments. One common conclusion from these three investigations is the dependence of the failure moment on the beam support conditions. It is clear that application of the DSM to the design of beams under elevated temperatures is still in its "infancy."

4.4.4 Web crippling

Web crippling is the term commonly attributed to the failure of CFS beams subjected to concentrated transverse forces (applied point loads or support reactions), and involves the collapse, due to a combination of localized buckling and plasticity, of the web region (panel) in the close vicinity of the force point of application. This same phenomenon is termed "patch loading" in the context of welded steel beams. The susceptibility of CFS beams to web crippling stems from their thin webs, which lead to the development of high transverse normal stresses under the action of a concentrated transverse force—such susceptibility is mostly influenced by the (1) web height and thickness, (2) web–flange internal corner radius and angle, and (3) characteristics of the load application. Concerning this last aspect, Hetrakul and Yu (1978) should be specially mentioned, since they were the first to propose the classification of web-crippling scenarios that is adopted nowadays. This classification is based on whether the concentrated transverse forces causing the stress concentration (1) act in a cross-section located inside the beam span (I) or near its extremities (E), and (2) are applied in just one flange (OF—load "absorbed" by the web) or in the two flanges (TF—load "transmitted" by the web). The combination of these two traits leads to four scenarios, namely IOF (internal one flange), ITF (internal two flange), EOF (end one flange), and ETF (end two flange) web crippling, which are schematically represented in Fig. 4.13.

The provisions for design against web crippling included in the current specifications/codes for CFS structures consist basically of expressions involving (1) geometric parameters (those mentioned in the previous paragraph), (2) steel material properties, and (3) a smaller or larger number of coefficients whose values are empirically determined on the basis of test results. It may be argued that the key contribution to these design approaches must be credited to Winter and Pian (1946) and Hetrakul and Yu (1978), who established/proposed the equations that are in use still today—over the years, these equations underwent just minor adjustments and some recalibrations of their coefficients, which reflected the growing amount of

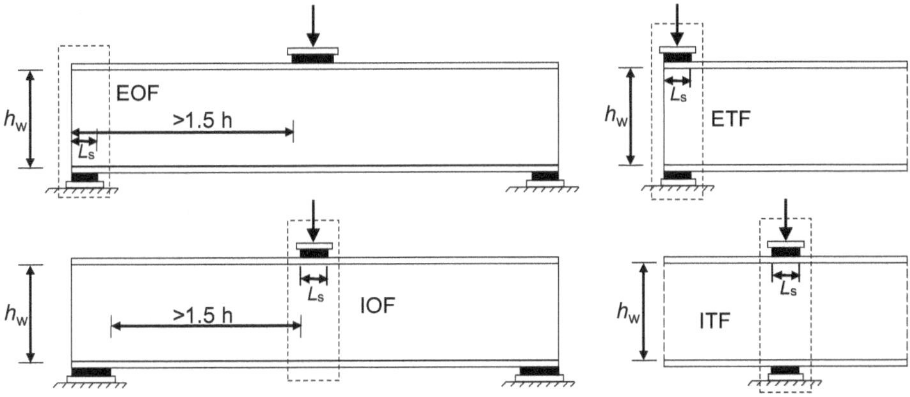

Figure 4.13 Web-crippling scenarios according to the classification proposed by Hetrakul and Yu (1978).

experimental data available. For instance, while (1) EC3-1−3 (CEN, 2006a) contains several equations, each with many coefficients, (2) the AISI S100 (AISI, 2012) includes only one equation and several sets of coefficients. This is because, due to the fully empirical nature of the design procedures, it was necessary to develop specific equations (format and/or coefficient values) for the various web-crippling failures of practical interest. In particular, this approach ignores the concept of "slenderness," which is at the root of the DSM and provides the "backbone" of the most rational design methodologies for steel structures: the ratio between the critical buckling and plastic strengths, indicating the relative importance of instability and plasticity on the collapse under scrutiny. This was recognized by Duarte and Silvestre (2013), who showed it is feasible to use the slenderness concept to estimate ultimate strengths associated with web-crippling failures, thus paving the way for the development of DSM-based design approaches for web crippling. Of course, the major hurdle to overcome is finding tools to calculate the critical buckling and plastic strength values with adequate accuracy.[16]

Mahendran and his coworkers (Keerthan et al., 2014; Gunalan and Mahendran, 2015; Janarthanan et al., 2015) presented applications of the DSM to the web-crippling design of (1) hollow flange channel (LiteSteel) beams under ETF and ITF loading, (2) unlipped channel beams also under ETF and ITF loading, and (3) unlipped channel beams under ITF and IOF loading—unlike the first two studies, the third one concerns beams with restrained flange rotations (ie, fastened-flange support conditions). All these investigations report carefully planned and executed test campaigns and, on the basis of the obtained experimental results, propose/develop design expressions/curves to predict the beam failure loads associated

[16] Obviously, a virtually impossible task prior to the advent and widespread dissemination of powerful numerical tools.

with the loading cases under consideration. These expressions are cast in the "Winter-type" format adopted by the DSM, and their coefficients are calibrated through comparison with the experimental failure load data.[17] Concerning the calculation of the critical buckling and plastic strength/load values required to determine the slenderness, these authors employ analytical expressions providing (1) plate-buckling stresses (with coefficients calibrated by means of SFE analyses) and (2) equivalent yield strengths based on a 45 degrees load distribution to the web middle from the bearing plate edges, respectively. Even if these calculations are clearly approximate, the fact that they enter the calibration procedure ensures the prediction quality.

A different approach to achieve a DSM-based web-crippling design was adopted by Natário (2015) and Natário et al. (2016), who proposed design expressions to predict web-crippling failure loads for the ETF and ITF loading cases, similar to those appearing in Keerthan et al. (2014), Gunalan and Mahendran (2015), and Janarthanan et al. (2015) (ie, cast in "Winter-type" format), but (1) put more effort into calculation of the critical buckling and plastic loads, thus leading to better accuracy, and (2) carried out calibration procedures against larger sets of experimental failure load data, available in the literature for ETF (128 tests) and ITF (130 tests) loading and involving various beam cross-section shapes and support conditions, thus ensuring greater generality, and (3) provided prequalification limits. The critical buckling loads are calculated by means of GBT analyses, based on the formulation presented in Natário et al. (2012),[18] and able to handle the various support conditions considered. The plastic loads are obtained from (approximate) expressions based on the rigid-plastic analysis of yield-line collapse mechanisms identified by means of experimental observation or SFE analyses.

In summary, it may be said that application of the DSM to web-crippling design is already quite successful and has a bright future as far as codification is concerned, particularly because of the lack of quality and generality of the failure load estimates provided by the existing code provisions.

4.4.5 Members under compression and bending (beam-columns)

According to the current specifications for the design of CFS structures, the estimation of the ultimate strength of members subjected to compression and bending (uni- or biaxial) cannot be obtained by means of a genuine DSM-based design approach, in the sense of involving the calculation of critical buckling and plastic load-moment combinations (the essence of the DSM). Nevertheless, it is possible to incorporate the currently codified DSM column and beam nominal strength values, presented in Sections 4.2.3 and 4.2.4, into the linear interaction equations traditionally

[17] Some of these works also include proposals for other (not DSM-based) expressions for web-crippling design.

[18] An user-friendly computer code (GBTWEB) was developed for this purpose (Natário et al., 2015)—it will be made freely available soon.

prescribed by the specifications. In the particular case of AISI S100 (AISI, 2012), these equations read[19]

$$\frac{\Omega_c P}{P_n} + \frac{\Omega_b C_{mx} M_x}{M_{nx} \alpha_x} + \frac{\Omega_b C_{my} M_y}{M_{ny} \alpha_y} \leq 1.0 \qquad [4.19]$$

$$\frac{\Omega_c P}{P_{n0}} + \frac{\Omega_b M_x}{M_{nx}} + \frac{\Omega_b M_y}{M_{ny}} \leq 1.0 \qquad [4.20]$$

and the participation of the DSM is felt through the values of P_n (or P_{n0}), M_{nx}, and/or M_{ny}.

Although (1) the foundations of a genuine DSM-based approach for the design of CFS beam-columns were laid over a decade ago (Schafer, 2002b) and (2) such an approach was successfully applied to the particular case of equal-leg angle members subjected to eccentric axial compression (Rasmussen, 2006), it was not until fairly recently that significant progress toward the application of the DSM to beam-columns was reported (Torabian et al., 2013, 2014a,b). The scope of ongoing work is broad, as it (1) involves experimental studies (Torabian et al., 2015a,b) and numerical simulations, and (2) aims at establishing, assessing the reliability, and achieving codification of a third-generation DSM design approach. This approach (1) is applicable to CFS members subjected to general loadings (any combination of axial compression or tension, major-axis bending, and minor-axis bending) and (2) integrates, under a single/unified formulation, features that are now addressed individually by various available codified first- and second-generation DSM design curves, namely the presence of holes (columns and beams) and consideration of the inelastic bending strength reserve (beams). The current column and beam design curves appear as particular cases of this unified formulation.

The proposed unified DSM formulation requires the calculation of the loading parameter values (ie, applied loading levels[20]) associated with (1) local, distortional, and global critical buckling, and (2) cross-section first yield (or plastic strength, if the inelastic bending strength is taken into account)—they are termed α_{crL}, α_{crD}, α_{crG}, and α_y. The critical buckling loading parameter values can be readily calculated by means of available computer programs, such as CUFSM (Schafer et al., 2012; Li et al., 2014) or GBTUL (Bebiano et al., 2014, Bebiano et al., submitted for publication), considering the stress distribution caused by the particular combination of axial force and (principal or skew) bending moment (assumed to increase proportional to the single parameter α). The determination of α_y is (1) fairly straightforward (based on simple strength of materials concepts)

[19] Equations (C5.2.1-1) and (C5.2.1-2) of the current AISI S100 (AISI, 2012). The meanings of the various quantities appearing in these two linear interaction equations, which are not given here, can be readily obtained from the specification.

[20] Recall that, in general, the loading combines axial compression or tension, major-axis bending, and minor-axis bending. Thus each loading parameter value is always associated with a trio of axial force and bending moment values.

when the cross-section strength corresponds to first yield, but more involved when the inelastic strength reserve is taken into account, particularly because of the lack of cross-section double symmetry—it requires possessing a numerical tool to obtain the cross-section plastic strength for any given combination of P, M_x, and M_y.

4.4.6 Structural systems

Once the application of the DSM to members is well established and covers all aspects of practical relevance, the next logical step is to consider the development of a DSM formulation to design CFS structural systems such as continuous beams, trusses, or frames. While such a formulation looks promising, since it will become possible to determine directly the structural system load-carrying capacity (without having to consider its members individually), for each structural system and loading under consideration it requires (1) efficient and user-friendly computational tools to calculate critical and plastic loading parameters, and (2) established design curves/expressions. Recent exploratory studies dealt with the application of a DSM design approach, based on the concept of "dominant buckling mode nature" (of a structural system) and using the currently codified beam and column strength curves, to (1) two- and three-span lipped channel continuous beams (Basaglia and Camotim, 2013) and (2) simple frames formed by members with various cross-section shapes (Camotim and Basaglia, 2014). The results obtained were encouraging and it seems that this design route may have a promising future, particularly after work on the application of the DSM to beam-columns is completed, since most frame members fall in the beam-column category. However, solid experimental and numerical research findings are needed before the technical/design community is convinced that this avenue can be pursued with a realistic aim of achieving codification (the current situation may be termed a "dream stage").

4.5 Concluding remarks

This chapter began by (1) presenting the philosophy, concepts, and procedures underlying the use of the DSM to design CFS members and structural systems, and (2) providing a historical account of its origin and main developments, focusing on the most important milestones. An overview of the current status of the DSM was reported, addressing its available applications and giving special emphasis to those that are already codified—a distinction was made between (1) the original applications to the compression and bending strength of CFS members and (2) the latest codifications, namely consideration of the inelastic bending strength reserve, the shear and combined shear-bending beam strengths, and applications to perforated columns and beams.[21] Finally, the chapter closed with brief accounts of a selection of recent

[21] Focusing on the provisions of the current (2012) North American specification (AISI, 2012), the only one that contains both the "first generation" and the "second generation" DSM design curves.

or ongoing studies concerning application of the DSM to various problems, illustrating the power and versatility of this approach to the design of CFS members and structural systems.

Acknowledgments

First of all, the authors would like to thank Professor Cheng Yu, editor of this book, for the invitation to elaborate this chapter—it was both a great honor and an even greater pleasure to contribute to this volume. In addition, the third author gratefully acknowledges the financial support of FCT (Fundação para a Ciência e a Tecnologia, Portugal), through doctoral scholarship SFRH/BD/87746/2012.

References

ABNT (Brazilian Standards Association), 2010. Brazilian Standard on Design of Cold-formed Steel Structures (NBR 14762:2010), Rio de Janeiro. (Portuguese).

AISI (American Iron and Steel Institute), 2004. North American Specification (NAS) for the Design of Cold-formed Steel Structural Members (AISI-S100−04) + Respective Commentary. Washington (DC).

AISI (American Iron and Steel Institute), 2012. North American Specification (NAS) for the Design of Cold-formed Steel Structural Members (AISI-S100−12) + Respective Commentary. Washington (DC).

AS/NZS (Standards of Australia and Standards of New Zealand), 1996. Cold-formed Steel Structures. AS/NZS 4600, first ed., Sydney-Wellington.

AS/NZS (Standards of Australia and Standards of New Zealand), 2005. Cold-formed Steel Structures. AS/NZS 4600, second ed., Sydney-Wellington.

Abreu, J.C.B., Vieira, L.M.C., Abu-Hamd, M.H., Schafer, B.W., 2014. Review: development of performance-based fire design for cold-formed steel. Fire Science Reviews 3 (1), 1−15.

Afshan, S., Gardner, L., 2013. The continuous strength method for structural stainless steel design. Thin-Walled Structures 68(July), 42−49.

Ashraf, M., Gardner, L., Nethercot, D.A., 2006. Compression strength of stainless steel cross-sections. Journal of Construction Steel Research 62 (1−2), 105−115.

Basaglia, C., Camotim, D., 2013. Buckling, postbuckling, strength and DSM design of cold-formed steel continuous lipped channel beams. Journal of Structural Engineering (ASCE) 139 (5), 657−668.

Bebiano, R., Dinis, P.B., Silvestre, N., Camotim, D., 2007. On the application of the direct strength method to cold-formed steel beams subjected to non-uniform bending. In: Liew, J.Y.R., Choo, Y.S. (Eds.), Proceedings of 5th International Conference on Advances in Steel Structures (ICASS 2007 − Singapore, 5−7/12), III. Research Publishing, Singapore, pp. 322−327.

Bebiano, R., Gonçalves, R., Camotim, D., 2014. GBTul − Buckling and Vibration of Thin-walled Members. University of Lisbon. Available at: http://www.civil.ist.utl.pt/gbt/ (version 2.0).

Bebiano, R., Camotim, D., Gonçalves, R., 2015. GBTul 2.0 − A Freeware Computer Code for the Buckling and Vibration Analysis of Thin-walled Members (submitted for publication).

Boissonnade, N., Nseir, J., Saloumi, E., 2013. The overall interaction concept: an alternative approach to the stability and resistance of steel sections and members. In: USB Proceedings of Structural Stability Research Council Annual Stability Conference (St. Louis, 16–20/4), pp. 48–66.

Camotim, D., Basaglia, C., 2013. Buckling analysis of thin-walled structures using generalised beam theory (GBT): state-of-the-art report. Steel Construction 6 (2), 117–131.

Camotim, D., Basaglia, C., 2014. On the behaviour, failure and direct strength design of thin-walled steel structural systems. Thin-Walled Structures 81 (August), 50–66.

CEN (Comité Européen de Normalisation), 2005. Eurocode 3: Design of Steel Structures – Part 1–1: General Rules. EN 1993-1-1:2005, Brussels, Belgium.

CEN (Comité Européen de Normalisation), 2006a. Eurocode 3: Design of Steel Structures – Part 1–3: General Rules; Supplementary Rules for Cold-formed Members and Sheeting. EN 1993-1-3:2006, Brussels, Belgium.

CEN (Comité Européen de Normalisation), 2006b. Eurocode 3: Design of Steel Structures – Part 1–5: Plated Structural Elements. EN 1993-1-5:2006, Brussels, Belgium.

CEN (Comité Européen de Normalisation), 2006c. Eurocode 3: Design of Steel Structures – Part 1–2: General Rules; Structural Fire Design. EN 1993-1-2:2005, Brussels, Belgium.

Chen, J., Young, B., 2007. Cold-formed steel lipped channel columns at elevated temperatures. Engineering Structures 29 (10), 2445–2456.

Dinis, P.B., Camotim, D., 2015a. Cold-formed steel columns undergoing local-distortional coupling: behaviour and direct strength prediction against interactive failure. Computers & Structures 147(January), 181–208.

Dinis, P.B., Camotim, D., 2015b. A novel DSM-based approach for the rational design of fixed-ended and pin-ended short-to-intermediate thin-walled angle columns. Thin-Walled Structures 87(February), 158–182.

Dinis, P.B., Young, B., Camotim, D., 2014a. Local-distortional interaction in cold-formed steel rack-section columns. Thin-Walled Structures 81(August), 185–194.

Dinis, P.B., Young, B., Camotim, D., 2014b. Strength, interactive failure and design of web-stiffened lipped channel columns exhibiting distortional buckling. Thin-Walled Structures 81(August), 195–209.

Dinis, P.B., Camotim, D., Batista, E., Santos, E., 2011. Local/distortional/global mode coupling in fixed lipped channel columns: behaviour and strength. Advanced Steel Construction 7 (4), 113–130.

Dinis, P.B., Camotim, D., Young, B., 2015. L-D-G interaction in CFS lipped channel columns – part II: numerical simulations and design considerations. In: Kruis, J., Tsompanakis, Y., Topping, B. (Eds.), Proceedings of 15th International Conference on Civil, Structural and Environmental Engineering Computing (CC 2015 – Prague, 1–4/9) paper 36. (full paper in USB Key Drive Proceedings).

Dinis, P.B., Camotim, D., Silvestre, N., 2010. On the local and global buckling behaviour of angle, T-section and cruciform thin-walled columns and beams. Thin-Walled Structures 48 (10–11), 786–797.

Dinis, P.B., Batista, E.M., Camotim, D., Santos, E.S., 2012a. Local-distortional-global interaction in lipped channel columns: experimental results, numerical simulations and design considerations. Thin-Walled Structures 61(December), 2–13.

Dinis, P.B., Camotim, D., Silvestre, N., 2012b. On the mechanics of angle column instability. Thin-Walled Structures 52(March), 80–89.

Duarte, A.P.C., Silvestre, N., 2013. A new slenderness-based approach for the web crippling design of plain channel steel beams. International Journal of Steel Structures 13 (3), 421–434.

Eccher, G., Rasmussen, K.J.R., Zandonini, R., 2008. Elastic buckling analysis of perforated thin-walled structures by the isoparametric spline finite strip method. Thin-Walled Structures 46 (2), 165–191.

Ganesan, K., Moen, C.D., 2012. LRFD resistance factor for cold-formed steel compression members. Journal of Constructional Steel Research 72 (May), 261–266.

Gardner, L., 2008. The continuous strength method. Proceedings of the Institution of Civil Engineers-Structures and Buildings 161 (3), 127–133.

Gardner, L., Macorini, L., Kucukle, M., 2013. The continuous strength method for steel and composite design. Proceedings of the Institution of Civil Engineers-Structures and Buildings 166 (8), 434–443.

Gunalan, S., Mahendran, M., 2015. Web crippling tests of cold-formed steel channels under two flange load cases. Journal of Constructional Steel Research 110(July), 1–15.

Hancock, G.J., Kwon, Y.B., Bernard, E.S., 1994. Strength design curves for thin-walled sections undergoing distortional buckling. Journal of Construction Steel Research 31 (2–3), 169–186.

Hancock, G.J., Pham, C.H., 2013. Shear buckling of channel sections with simply supported ends using the semi-analytical finite strip method. Thin-Walled Structures 71(October), 72–80.

Hetrakul, N., Yu, W.-W., 1978. Structural Behaviour of Beam Webs Subjected to Web Crippling and a Combination of Web Crippling and Bending. Final Report, Civil Engineering Study 78-4. University of Missouri-Rolla, Missouri.

Heva, Y.B., Mahendran, M., 2008. Local buckling tests of cold-formed steel compression members at elevated temperatures. In: Mahendran, M. (Ed.), Proceedings of 5th International Conference on Thin-walled Structures – Recent Innovations and Developments (ICTWS'08 – Brisbane, 18–20/6), pp. 745–752.

Janarthanan, B., Mahendran, M., Gunalan, S., 2015. Bearing capacity of cold-formed unlipped channels with restrained flanges under EOF and IOF load cases. Steel Construction 8 (3), 146–154.

Kankanamge, N.D., Mahendran, M., 2012. Behaviour and design of cold-formed steel beams subject to lateral-torsional buckling at elevated temperatures. Thin-Walled Structures 61(December), 213–228.

Keerthan, P., Mahendran, M., Steau, E., 2014. Experimental study of web crippling behaviour of hollow flange channel beams under two flange load cases. Thin-Walled Structures 85(December), 207–219.

Kwon, Y.B., Hancock, G.J., 1992. Tests of cold-formed channels with local and distortional buckling. Journal of Structural Engineering (ASCE) 118 (7), 1786–1803.

Kwon, Y.B., Kim, B.S., Hancock, G.J., 2009. Compression tests of high strength cold-formed steel channels with buckling interaction. Journal of Constructional Steel Research 65 (2), 278–289.

LaBoube, R.A., Yu, W.-W., 1978a. Structural Behaviour of Beam Webs Subjected Primarily to Shear. Final Report, Engineering Study 78-2. University of Missouri-Rolla (June).

LaBoube, R.A., Yu, W.-W., 1978b. Structural Behaviour of Beam Webs Subjected to a Combination of Bending and Shear. Final Report, Engineering Study 78-3. University of Missouri-Rolla (June).

Landesmann, A., Camotim, D., 2013. On the direct strength method (DSM) design of cold-formed steel columns against distortional failure. Thin-Walled Structures 67(June), 168–187.

Landesmann, A., Camotim, D., 2015a. Distortional failure and DSM design of cold-formed steel lipped channel beam under elevated temperatures. Thin-Walled Structures 98A(January), 75–93.

Landesmann, A., Camotim, D., 2015b. DSM to predict distortional failures in cold-formed steel columns exposed to fire: effect of the constitutive law temperature-dependence. Computers & Structures 147(15 January), 47–67.

Li, Z., Batista Abreu, J.C., Leng, J., Ádány, S., Schafer, B.W., 2014. Review: constrained finite strip method developments and applications in cold-formed steel design. Thin-Walled Structures 81(August), 2–18.

Lu, W., Mäkeläinen, P., Outinen, J., 2010. Numerical investigation of cold-formed steel purlins in fire. Journal of Structural Mechanics 43 (1), 12–24.

Martins, A.D., Dinis, P.B., Camotim, D., Providência, P., 2015a. Local-distortional interaction in pin-ended lipped channel beams: behaviour, strength and DSM design. In: Camotim, D., et al. (Eds.), CD-ROM Proceedings of 8th International Conference on Advances in Steel Structures (ICASS'2015 – Lisboa, 21–24/7) paper 15.

Martins, A.D., Dinis, P.B., Camotim, D., Providência, P., 2015b. On the relevance of local-distortional interaction effects in the behaviour and design of cold-formed steel columns. Computers & Structures 160(November), 57–89.

Martins, A.D., Dinis, P.B., Camotim, D., 2016. On the influence of localdistortional interaction in the behavior and design of web-stiffened lipped channel columns. Thin-Walled Structures 101 (April), 181–204.

Martins, A.D., Dinis, P.B., Camotim, D., Providência, P., 2015c. Influence of local-distortional interaction in the behaviour and DSM design of stiffened lipped channel columns. In: Camotim, D., et al. (Eds.), CD-ROM Proceedings of 8th International Conference on Advances in Steel Structures (ICASS'2015 – Lisboa, 21–24/7) paper 37.

Mesacasa Jr., E., Dinis, P.B., Camotim, D., Malite, M., 2014. Mode interaction and imperfection-sensitivity in thin-walled equal-leg angle columns. Thin-Walled Structures 81(August), 138–149.

Moen, C.D., Schafer, B.W., 2008. Experiments on cold-formed steel columns with holes. Thin-Walled Structures 46 (10), 1164–1182.

Moen, C.D., Schafer, B.W., 2009. Direct Strength Design of Cold-formed Steel Members with Perforations, Research Report. American Iron and Steel Institute (AISI), Washington, DC.

Moen, C.D., Schafer, B.W., 2009a. Elastic buckling of thin plates with holes in compression or bending. Thin-Walled Structures 47 (12), 1597–1607.

Moen, C.D., Schafer, B.W., 2009b. Elastic buckling of cold-formed steel columns and beams with holes. Engineering Structures 31 (12), 2812–2824.

Moen, C.D., Schafer, B.W., 2010. Extending direct strength design to cold-formed steel beams with holes. In: LaBoube, R., Yu, W.-W. (Eds.), Proceedings of 20th International Specialty Conference on Cold-formed Steel Structures (St. Louis, 3–4/11), pp. 171–183.

Moen, C.D., Schafer, B.W., 2011. Direct strength method for design of cold-formed steel columns with holes. Journal of Structural Engineering (ASCE) 137 (5), 559–570.

Moen, C.D., Schudlich, A., von der Hayden, A., 2013. Experiments on cold-formed steel C-section joists with unstiffened web holes. Journal of Structural Engineering (ASCE) 139 (5), 695–704.

Nandini, P., Kalyanaraman, V., 2010. Strength of cold-formed lipped channel beams under interaction of local, distortional and lateral torsional buckling. Thin-Walled Structures 48 (10–11), 872–877.

Natário, P., 2015. Localized Failure of Thin-walled Steel Members Subjected to Concentrated Loads: Analysis, Behaviour and Design. IST, University of Lisbon (Ph.D. thesis in Civil Engineering).

Natário, P., Silvestre, N., Camotim, D., 2012. Localized web buckling analysis of beams subjected to concentrated loads using GBT. Thin-Walled Structures 61 (December), 27–41.

Natário, P., Silvestre, N., Camotim, D., 2016. Direct strength prediction of web crippling failure of beams under ETF loading. Thin-Walled Structures.

Natário, P., Silvestre, N., Camotim, D., 2015. GBTWEB – GBT-based Code for Web Buckling Analysis of Members under Localised Loads, to be made freely available soon.

Pham, S.H., Pham, C.H., Hancock, G.J., 2014. Direct strength method of design for shear including sections with longitudinal web stiffeners. Thin-Walled Structures 81(August), 19–28.

Pham, C.H., Hancock, G.J., 2009. Direct strength design of cold-formed purlins. Journal of Structural Engineering (ASCE) 135 (3), 229–238.

Pham, C.H., Hancock, G.J., 2012a. Direct strength design of cold-formed C-sections for shear and combined actions. Journal of Structural Engineering (ASCE) 138 (6), 759–768.

Pham, C.H., Hancock, G.J., 2012b. Tension field action for cold-formed sections in shear. Journal of Constructional Steel Research 72(May), 168–178.

Pham, C.H., Hancock, G.J., 2012c. Elastic buckling of cold-formed channel sections in shear. Thin-Walled Structures 61(December), 22–26.

Pham, S.H., Pham, C.H., Hancock, G.J., 2014. Direct strength method of design for shear including sections with longitudinal web stiffeners. Thin-Walled Structures 81 (August), 19–28.

Ranawaka, T., Mahendran, M., 2009. Distortional buckling tests of cold-formed steel compression members at elevated temperatures. Journal of Constructional Steel Research 65 (2), 249–259.

Rasmussen, K.J.R., 2006. Design of slender angle section beam-columns by the direct strength method. Journal of Structural Engineering (ASCE) 132 (2), 204–211.

Rossi, B., Jaspart, J.-P., Rasmussen, K.J.R., 2010. Combined distortional and overall flexural-torsional buckling of cold-formed stainless steel sections: design. Journal of Structural Engineering (ASCE) 136 (4), 361–369.

Santos, E.S., Batista, E.M., Camotim, D., 2012. Experimental investigation concerning lipped channel columns undergoing local-distortional-global buckling mode interaction. Thin-Walled Structures 54 (May), 19–34.

Santos, E.S., Batista, E.M., Camotim, D., 2014a. Cold-formed steel columns under L-D-G interaction: experimental investigation. Steel Construction 7 (3), 193–198.

Santos, E.S., Dinis, P.B., Batista, E., Camotim, D., 2014b. Cold-formed steel lipped channel columns undergoing local distortional-global interaction: experimental and numerical investigation. In: LaBoube, R., Yu, W.-W. (Eds.), Proceedings of 22nd International Specialty Conference on Cold-formed Steel Structures (St. Louis, 5–6/11), pp. 193–211.

Schafer, B.W., 2006. Direct Strength Method Design Guide. American Iron and Steel Institute (AISI), Washington, DC.

Schafer, B.W., 2008. Review: the direct strength method of cold-formed steel member design. Journal of Constructional Steel Research 64 (7–8), 766–778.

Schafer, B.W., Peköz, T., 1998a. Direct strength prediction of cold-formed steel members using numerical elastic buckling solutions. In: Yu, W.-W., LaBoube, R. (Eds.), Proceedings of 14th International Specialty Conference on Cold-formed Steel Structures (St. Louis, 15–16/10), pp. 69–76.

Schafer, B.W., Peköz, T., 1998b. Direct strength prediction of cold-formed steel members using numerical elastic buckling solutions. In: Shanmugam, N., Liew, J.Y.R., Thevendran, V. (Eds.), Thin-walled Structures: Research and Development (ICTWS'98 – Singapore, 2–4/12). Elsevier, pp. 137–144.

Schafer, B.W., 2002a. Local, distortional and Euler buckling in thin-walled columns. Journal of Structural Engineering (ASCE) 128 (3), 289–299.

Schafer, B.W., 2002b. Progress on the direct strength method. In: LaBoube, R., Yu, W.-W. (Eds.), Proceedings of 16th International Specialty Conference on Cold-formed Steel Structures (Orlando, 16–18/10), pp. 647–662.

Schafer, B.W., Sarawit, A., Peköz, T., 2006. Complex edge stiffeners for thin-walled members. Journal of Structural Engineering (ASCE) 132 (2), 212–226.

Schafer, B.W., et al., 2012. CUFSM – Elastic Buckling Analysis of Thin-walled Members with General End Boundary Conditions. Johns Hopkins University. Available at: http://www.ce.jhu.edu/bschafer/ (version 4.05).

Schafer, B.W., Peköz, T., 1999. Laterally braced cold-formed steel flexural members with edge stiffened flanges. Journal of Structural Engineering (ASCE) 125 (2), 118–127.

Shifferaw, Y., Schafer, B.W., 2012. Inelastic bending capacity of cold-formed steel members. Journal of Structural Engineering (ASCE) 138 (4), 468–480.

Shahbazian, A., Wang, Y.C., 2011a. Calculating the global buckling resistance of thin-walled steel members with uniform and non-uniform elevated temperatures under axial compression. Thin-Walled Structures 49 (11), 1415–1428.

Shahbazian, A., Wang, Y.C., 2011b. Application of the direct strength method to local buckling resistance of thin-walled steel members with non-uniform elevated temperatures under axial compression. Thin-Walled Structures 49 (12), 1573–1583.

Shahbazian, A., Wang, Y.C., 2012. Direct strength method for calculating distortional buckling capacity of cold-formed thin-walled steel columns with uniform and non-uniform elevated temperatures. Thin-Walled Structures 53 (April), 188–199.

Silvestre, N., Camotim, D., Dinis, P.B., 2009. Direct strength prediction of lipped channel columns experiencing local-plate/distortional interaction. Advanced Steel Construction 5 (1), 49–71.

Silvestre, N., Camotim, D., Dinis, P.B., 2012. Post-buckling behaviour and direct strength design of lipped channel columns experiencing local-distortional interaction. Journal of Constructional Steel Research 73 (June), 12–30.

Silvestre, N., Dinis, P.B., Camotim, D., 2013. Developments on the design of cold-formed steel angles. Journal of Structural Engineering (ASCE) 139 (5), 680–694.

Simulia Inc., 2015. Abaqus Standard (Version 6.14).

Smith, F.H., Moen, C.D., 2014. Finite strip elastic buckling solutions for thin-walled metal columns with perforation patterns. Thin-Walled Structures 79 (June), 187–201.

Sputo, T., Tovar, J., 2005. Application of direct strength method to axially loaded perforated cold-formed steel studs: longwave buckling. Thin-Walled Structures 43 (12), 1852–1881.

Swanson Analysis Systems Inc., 2015. Ansys Reference Manual (Version 16.2).

Torabian, S., Zheng, B., Schafer, B.W., 2013. Direct Strength Prediction of Cold-formed Steel Beam-columns. Research Interim Report (Year 2) submitted to the American Iron and Steel Institute (AISI), Available at: http://www.ce.jhu.edu/bschafer/dsmbeamcol/.

Torabian, S., Zheng, B., Schafer, B.W., 2014a. Experimental study and modeling of cold-formed steel lipped channel stub beam-columns. In: USB Key Drive Proceedings of Structural Stability Research Council Annual Stability Conference (Toronto, 25–28/3), pp. 366–387.

Torabian, S., Zheng, B., Schafer, B.W., 2014b. Development of a new beam-column design method for cold-formed steel lipped channel members. In: LaBoube, R., Yu, W.-W. (Eds.), Proceedings of 22nd International Specialty Conference on Cold-formed Steel Structures (St. Louis, 5–6/11), pp. 359–375.

Torabian, S., Zheng, B., Schafer, B.W., 2015a. Experimental response of cold-formed steel lipped channel beam-columns. Thin-Walled Structures 89(April), 152–168.

Torabian, S., Fratamico, D.C., Schafer, B.W., 2015b. Experiments on cold-formed steel zee-shaped stub beam-columns. In: Website Proceedings of Structural Stability Research Council Annual Stability Conference (Nashville, 24–27/3).

Tovar, J., Sputo, T., 2005. Application of direct strength method to axially loaded perforated cold-formed steel studs: distortional and local buckliang. Thin-Walled Structures 43 (12), 1882–1912.

Winter, G., Pian, R.H.J., 1946. Crushing Strength of Thin Steel Webs. Engineering Experiment Station, Bulletin 35. Cornell University, New York.

Yang, D., Hancock, G.J., 2004. Compression tests of high strength steel channel columns with interaction between local and distortional buckling. Journal of Structural Engineering (ASCE) 130 (12), 1954–1963.

Yao, Z., 2012. Nonlinear Analysis and Design of Perforated Thin-walled Steel Columns. School of Civil and Mining Engineering, University of Sydney, Australia (Ph.D. thesis in Civil Engineering).

Yap, D.C.Y., Hancock, G.J., 2011. Experimental study of high strength cold-formed stiffened-web C-sections in compression. Journal of Structural Engineering (ASCE) 137 (2), 162–172.

Young, B., 2004. Tests and design of fixed-ended cold-formed steel plain angle columns. Journal of Structural Engineering (ASCE) 130 (12), 1931–1940.

Young, B., Yan, J., 2004. Design of cold-formed steel channel columns with complex edge stiffeners by direct strength method. Journal of Structural Engineering (ASCE) 130 (11), 1756–1763.

Young, B., Silvestre, N., Camotim, D., 2013. Cold-formed steel lipped channel columns influenced by local distortional interaction: strength and DSM design. Journal of Structural Engineering (ASCE) 139 (6), 1059–1074.

Young, B., Dinis, P.B., Camotim, D., 2015. L-D-G interaction in CFS lipped channel columns – part I: experimental investigation. In: Kruis, J., Tsompanakis, Y., Topping, B. (Eds.), Proceedings of 15th International Conference on Civil, Structural and Environmental Engineering Computing (CC 2015 – Prague, 1–4/9) paper 35. full paper in USB Key Drive Proceedings.

Yu, C., Schafer, B.W., 2003. Local buckling tests on cold-formed steel beams. Journal of Structural Engineering (ASCE) 129 (12), 1596–1606.

Yu, C., Schafer, B.W., 2006. Distortional buckling tests on cold-formed steel beams. Journal of Structural Engineering (ASCE) 132 (4), 515–528.

Yu, C., Schafer, B.W., 2007. Simulation of cold-formed steel beams in local and distortional buckling with applications to the direct strength method. Journal of Constructional Steel Research 63 (5), 581–590.

Part Two

Computational analysis of cold-formed steel structures

Advanced computational tools for elastic buckling analysis of cold-formed steel structures

5

Z. Li
SUNY Polytechnic Institute, Utica, NY, United States

5.1 Introduction

Cold-formed steel (CFS) is usually manufactured by rolling or pressing thin gauges of steel sheet at room temperature into desired cross-section members, which have wide use in buildings, bridges, storage racks, various types of equipment, and other applications. Due to the slender nature of the cross-section, their behavior is inevitably complex, with several parallel buckling phenomena influencing performance and limit states. These buckling phenomena are usually categorized as global (eg, flexural or lateral–torsional buckling), distortional, and local (plate) buckling modes. These buckling phenomena are not limited to compression and flexural members: buckling of CFS sections in shear may also occur in either local or global modes.

Though elastic buckling information for CFS members is not a direct predictor of capacity or collapse behavior on its own, both the mode and the load (moment) are important proxies for the actual behavior. In current design codes, such as AISI S100, New Zealand/Australia, and European Union, the design formulae are calibrated through the calculation of elastic critical buckling loads (or moments) to predict the ultimate strength, thus the ability to calculate the associated elastic buckling loads (or moments) is of great importance. Moreover, the buckling mode shapes are commonly employed into nonlinear collapse modeling as initial geometric imperfections.

To examine the elastic buckling behavior of CFS members, analytical models (or closed-form solutions) and numerical methods are available. Although numerical methods using computational modeling are the focus in this chapter, analytical models with their simplicity merit a brief summary of their recent advances. The emphasis, though, is on numerical solution methods, concentrating on computational tools for elastic buckling analysis of CFS members.

5.2 Closed-form solutions and analytical models for elastic buckling analysis

5.2.1 Closed-form solutions

Closed-form solutions are sought through explicit solutions of the governing differential equations. However, due to the complex nature of these equations, explicit

Recent Trends in Cold-Formed Steel Construction. http://dx.doi.org/10.1016/B978-0-08-100160-8.00005-0

solutions have been found for relatively few practical situations. Thus closed-form solutions reside mainly in the analysis of global buckling (ie, flexural "Euler" buckling and flexural–torsional buckling) for beams and columns, and local buckling for plate elements (or assemblies).

Explicit critical loads of global buckling can be provided from classic beam theory, such as the well-known Euler buckling formula. Thus global stability may be commonly understood as rigid cross-section deformations. However, it is worth noting here that this notion is practically useful but mechanically inaccurate. In fact, compared to the critical loads of global buckling provided by numerical solutions such as the finite element method (FEM) and finite strip method (FSM), the error is exactly equal to $E - E/(1 - \nu^2)$, the fundamental rigidity difference between beam and plate theory (Ádány, 2006).

Closed-form solutions to local buckling for a single plate supported along its edges can be provided from classic plate theory when it is subjected to direct compression, bending, shear, or a combination of these stresses in its plane. The critical loads take form by seeking a plate-buckling coefficient k, a function of plate geometry and boundary conditions for a given loading. These buckling coefficients k for various loadings, such as uniform compression, compression and bending, and shear, are summarized in the stability design guide for metal structures (Ziemian, 2010). Yu and Schafer (2007) studied how k varies as a function of normalized plate length for various boundaries and loading, in particular the stress gradient, to demonstrate the effects of relatively plates on elastic local buckling stress. However, for critical loads of members, these closed-form solutions ignore the interaction of plate elements of the whole cross-section, and may result in excessively conservative solutions.

5.2.2 Analytical formulae or models

Over the history of understanding distortional buckling, approximate analytical models had been proposed to develop closed-form hand calculation of critical loads, such as by Sharp (1966), Desmond (1977), Lau and Hancock (1987), Davies and Jiang (1998), and Schafer and Peköz (1999). A new analytical model, similar to Lau and Hancock (1987), was proposed by Li (2009) through a different layout of springs, and showed improved results for predicting critical loads of distortional buckling.

Analytical formulae for analyzing distortional buckling were also pursued by researchers. For various loading and boundary conditions of C- and Z-section members with arbitrary sloping single-lip stiffeners, Silvestre and Camotim (2004a,b) proposed a generalized beam theory (GBT) based on fully analytical formulae to assess distortional critical buckling length and stress. The novelty of these formulae is the utilization of the governing differential equations of GBT, which is able to incorporate folded-plate theory for high accuracy.

Empirical formulae for distortional buckling stress were also proposed using available data obtained from the above analytical models, or numerical solutions such as GBT and FSM using genetic programming (Pala, 2008) or neural networks (Pala, 2006). Though lacking a mechanical basis, these empirical formulae did have the advantage of simplicity in predicting distortional buckling loads.

5.3 Numerical solution methods for elastic buckling analysis

To examine the elastic buckling behavior of CFS members, the main numerical solution methods are the FEM, the FSM, GBT, and more recently the constrained FSM (cFSM).

5.3.1 Finite element method

Due to the thin nature of a CFS cross-section, the application of FEM in stability analysis of CFS members is commonly built from the shell elements so the cross-section deformation can be captured. Given the many advances in computation, shell FEM has become popular in analyzing CFS structures for elastic (eigen) buckling analysis. Its unique applicability to handle complex geometry and boundary conditions makes it a natural choice in many situations. For instance, extensive studies on elastic buckling analysis of CFS plates/members with holes done by Moen and Schafer (2009a,b) were modeled with shell FEM. Moreover, its applicability in nonlinear collapse analysis by taking into consideration material and geometrical nonlinearities proves to be even more powerful for stability analysis of CFS, but this is beyond the scope of this chapter.

5.3.1.1 Shell FEM formulation

Shell FEM requires a user to generate a full three-dimension shell model, then run an eigen buckling analysis to find buckling loads (eigenvalues) and associated buckled mode shapes. Although the formulation of the stiffness matrices may use different plate theories (eg, Kirchhoff–Love theory versus Mindlin–Reissner theory), a similar eigenvalue problem associated with elastic buckling on a perfect structure considering only geometric nonlinearity may be defined as

$$(K_e - \lambda K_g)\Phi = 0 \tag{5.1}$$

where K_e is the conventional elastic stiffness matrix, K_g is the geometric stiffness matrix, λ is the eigenvalue (load factor is compression positive), and Φ is the eigenmode (buckling mode) vector.

5.3.1.2 Modeling sensitivity

Computation modeling of CFS is always an intricate task with a variety of parameters that may influence the analysis. In elastic buckling analysis, three more influential aspects are boundary conditions, element selection, and mesh. Full discussion can be found in a recent journal article (Schafer et al., 2010).

First, boundary conditions necessitate careful treatment in computational modeling of CFS structures. They have an excessive impact on elastic buckling analysis (and on collapse analysis as well); in particular, the issue of longitudinal warping restraint at

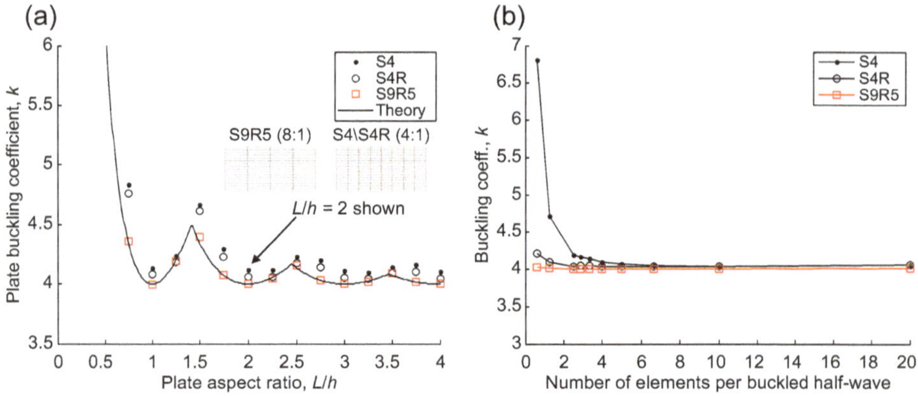

Figure 5.1 Sensitivity of ABAQUS shell elements for local buckling of a simply supported plate in compression: (a) varying physical plate length; (b) varying number of elements along the length.
From Schafer, B.W., Li, Z., Moen, C.D., 2010. Computational modeling of cold-formed steel. Thin-walled Structures 48(10–11), 752–762.

the member ends may commonly be overlooked. In addition, it is worth noting that the real boundary conditions in CFS systems are usually far more complex than those existing in laboratory experiments, and are never as clean as in the computational models.

Second, selection of a shell element appropriate for the modeling at hand is a broad topic. The issues of importance may include order of the element (ie, linear, quadratic, etc.), implementation of Mindlin or Kirchhoff strain assumptions, and integration scheme.

Third, depending on the choice of shell elements, the impact of mesh sensitivity varies. For instance, local buckling results for a simply supported plate in pure compression are illustrated in Fig. 5.1 (Schafer et al., 2010) using different elements—S4, S4S, and S9R5—within the well-known general-purpose finite element method package ABAQUS (Simulia, 2012). As shown in Fig. 5.1(a), the quadratic element (S9R5) allows as little as one element per buckling half-wave (Fig. 5.1(b)) to provide excellent and robust solutions in comparing with (Kirchhoff) thin plate theory. When fewer than five elements are used per buckled half-wave, the linear element (S4) locks, and while the S4R, which includes a reduced integration scheme to lessen the impact of shear locking, performs reasonably well, it is still outperformed by the quadratic S9R5 even when the number of nodes is the same.

5.3.1.3 Limitations

Given the complex buckling modes of CFS members (ie, local, global, distortional), correct identification and classification of these buckling modes and accurate calculation of the associated elastic critical loads (or moments) are crucial in predicting the ultimate strength of a CFS member. While shell FEM can handle arbitrary cross-sections, boundary conditions, and loads, it is limited by the difficulties of modeling and identifying the characteristic buckling modes (local distortional, global), because shell FEM itself provides no means of modal identification, and instead requires a laborious and subjective procedure employing visual investigation to classify the buckling modes in elastic buckling analysis.

5.3.2 Finite strip method

The first paper on the FSM was published by Cheung in the late 1960s. Since then, many developments have been made toward structural analysis with high efficiency. The FSM is a variant of the FEM. For its application to thin-walled structures, instead of a finite element discretization (Fig. 5.2(a)), a thin-walled cross-section is discretized into a series of longitudinal strips (or elements) (Fig. 5.2(b)). Shape functions in the transverse and longitudinal directions are selected to represent the displacement field. Then, similar to FEM, an eigenvalue problem for elastic buckling analysis, similar to Eq. [5.1] (but with less degree of freedoms), can be formulated using different plate theories. Depending on the types of longitudinal shape functions employed, it can be further grouped into semianalytical and spline FSM, discussed below.

5.3.2.1 Semianalytical finite strip method

The semianalytical (conventional) FSM utilizes a set of trigonometric functions to represent the longitudinal field for the strip (or element), while the shape function in the transverse direction is the same as a classical beam finite element. It is the most widely used approach to examining instabilities in a CFS member under longitudinal stress (axial, bending, and/or warping torsion). In particular, the application of semianalytical FSM to members with simply supported ends using a sine function (see Eq. [5.2]) results in an efficient solution and the "signature curve" of the stability of a member in terms of buckling half-wavelength versus buckling load, as implemented in the open source program CUFSM (Schafer and Ádány, 2006) and THIN-WALL (C. for Advanced Structural Engineering, 2006). A typical signature curve for a

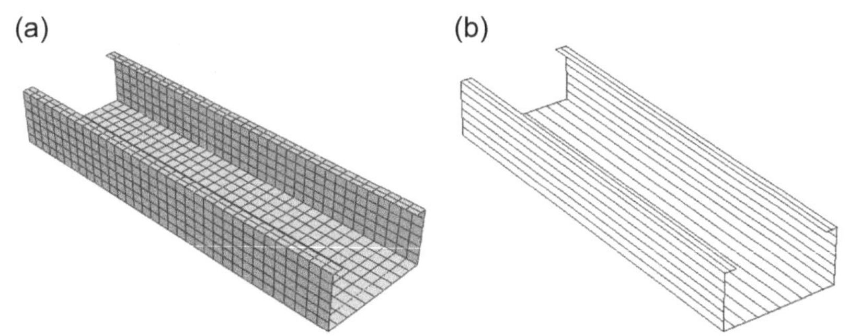

Figure 5.2 Shell FEM versus FSM mesh. (a) Shell FEM mesh (eg, ABAQUS); (b) finite strip mesh.

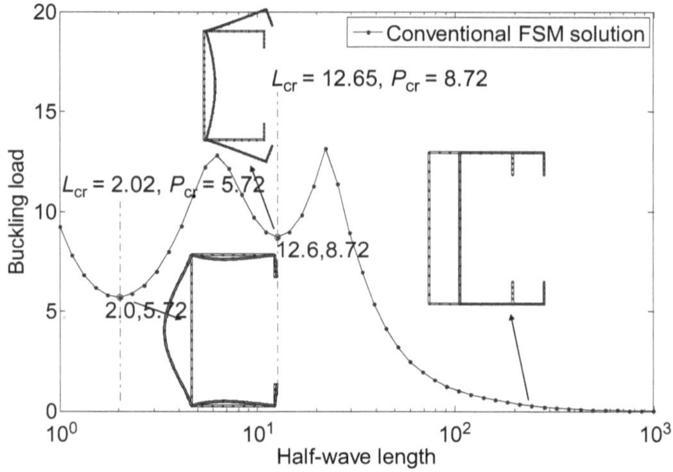

Figure 5.3 A typical signature curve of an FSM solution.
From Li, Z., Schafer. B.W., 2010a. Application of the finite strip method in cold-formed steel member design. Journal of Constructional Steel Research 66(8–9), 971–980.

250S137-33 (Steel Stud Manufacturers Association (SSMA) nomenclature, Product Technical Information, 2001) stud section with a straight-line model under axial compression is shown using CUFSM in Fig. 5.3. Minima in the signature curve shown in Fig. 5.3 indicate the lowest critical load at which a particular buckling mode occurs. Traditionally, local buckling ($P_{cr\ell}$) occurs at the first minimum, distortional buckling (P_{crd}) at the second, and global buckling (P_{cre}) is identified in the descending branch of the curve at a longer half-wavelength equal to the global unbraced length of the member. Note that for a physical length, the identified lowest buckling mode will repeat itself along the physical length at or near the minimal half-wavelength.

New extensions, particularly for CUFSM, are the expansion to other boundary conditions following the approximate longitudinal shape functions, namely:

$$\text{S–S: simple–simple} \quad Y_{[m]} = \sin(m\pi y/a) \qquad [5.2]$$

$$\text{C–C: clamped–clamped} \quad Y_{[m]} = \sin(m\pi y/a)\sin(\pi y/a) \qquad [5.3]$$

$$\text{S–C: simple–clamped} \quad Y_{[m]} = \sin[(m+1)\pi y/a] + (m+1/m)\sin(m\pi y/a) \qquad [5.4]$$

$$\text{C–F: clamped–free} \quad Y_{[m]} = 1 - \cos[(m-1/2)\pi y/a] \qquad [5.5]$$

$$\text{C–G: clamped–guided} \quad Y_{[m]} = \sin[(m-1/2)\pi y/a]\sin(\pi y/2/a) \qquad [5.6]$$

where $[m]$ is the longitudinal term associated with half-wavelength, y represents the coordinate in longitudinal direction, and a is the length of the member. See Li and Schafer (2010) for more details.

Milašinović (2011, 2012) developed a harmonic coupled FSM by including the Green–Lagrange strain terms in the formulations, thus bending and membrane are coupled in the geometric stiffness to give more accurate buckling behavior prediction, especially in large-deflection and elastic postbuckling analysis. Other researchers, such as Bui (2012), sought new implementation using the shallow shell theory of Marguerre instead of the plate theories of Kirchhoff and Mindlin—most existing semianalytical finite strips are based on the plate theories of Kirchhoff and Mindlin.

5.3.2.2 Spline finite strip method

Different from semianalytical FSM, spline FSM uses local spline functions to represent the longitudinal field. In particular, the longitudinal shape functions in the longitudinal directions is a summation of $(m + 3)$ local B^3 splines such that

$$Y(y) = \sum_{i=-1}^{m+1} \alpha_i \phi_i \quad [5.7]$$

where α_i is the displacement parameter and, depending on the longitudinal section length h, ϕ_i the section knot $y = y_i$ is defined as:

$$\phi_i = \frac{1}{6h^3}\begin{cases} (y - y_{i-2})^3 & y_{i-2} \le y \le y_{i-1} \\ h^3 + 3h^2(y - y_{i-1}) + 3h(y - y_{i-1})^2 - 3(y - y_{i-1})^3 & y_{i-1} \le y \le y_i \\ h^3 + 3h^2(y_{i+1} - y) + 3h(y_{i+1} - y)^2 - 3(y_{i+1} - y)^3 & y_i \le y \le y_{i+1} \\ (y_{i+2} - y)^3 & y_{i+1} \le y \le y_{i+2} \\ 0 & \text{else} \end{cases} \quad [5.8]$$

Spline FSM has advantages in modeling more complex types of loading and boundary conditions than those specified above for semianalytical FSM. In particular, spline FSM can conveniently handle members with discrete restraints or supports along the longitudinal directions. In addition, buckling in shear can be easily accounted for by spline FSM. Pham and Hancock (2009, 2012) employed spline FSM to study the elastic buckling of thin-walled plain channel sections and thin-walled channel sections with intermediate web stiffeners in pure shear. Moreover, new developments by Eccher et al. (2008a,b) proposed an isoparametric spline FSM that can be applied to elastic buckling analysis of cold-formed members with perforations.

5.3.2.3 Finite strip method in other forms

Researchers are developing more efficient and accurate FSM solutions by employing new transverse and longitudinal shape functions. Hancock and Pham (2015) modified semianalytical FSM by expanding the transverse shape field to account for the localized loading on thin-walled members. They also developed a complex FSM by using a complex longitudinal shape function based on a trigonometric function similar to semianalytical FSM to provide more efficient buckling analysis, in particular for shear (Hancock and Pham, 2013). Bui and Rondal (2008) explored a new semianalytical FSM based on the Mindlin–Reissner plate theory with the addition of a drilling rotation degree of freedom.

5.3.3 Generalized beam theory

GBT is an enriched beam theory that incorporates the global motions of the cross-section and distortional and local deformations, and is applicable to prismatic members. GBT can provide elastic buckling analysis for CFS members under a variety of loading and boundary conditions. Beside the general solutions in terms of elastic buckling loads and shapes that other methods (eg, FEM and FSM) are able to produce, it is one of only two methods (the other is cFSM, discussed later) that can provide separation and identification of the buckling modes. In addition, work by Gonçalves and Camotim (2012) enabled GBT to perform nonlinear analysis, including geometric and material nonlinearity. Related works on GBT can be found in Abambres et al. (2013, 2014a,b), Basaglia et al. (2011), and Silvestre and Camotim (2003).

5.3.3.1 Cross-section analysis

GBT transforms degrees of freedom from nodal to modal, which can be viewed as an extension of Vlasov's classical bar theory (Silvestre and Camotim, 2002). In particular, by integrating the cross-section deformations, GBT has a foundation in folded-plate theory and uses a unique cross-section analysis procedure to unveil the deformation modes (global, distortional, and local) in displacement fields. Interested readers may refer to Camotim et al. (2004) and Silvestre (2005) for a concise overview of the most important concepts and procedures for performing a cross-section analysis.

Through integration of displacement fields along the cross-section midline coordinate in GBT formulation, the cross-section analysis results in four basic mechanical matrices, C, B, D, and X, that are the generalized section properties which depend only on the cross-section geometry. Matrix C is related to the warping stiffness of the cross-section, and matrix B associates with the transverse bending stiffness of the cross-section. Matrix D ties to the twisting stiffness, while matrix X is second-order section properties relating the cross-section deformations to stress distributions. With these fully populated generalized matrices, a sequence of three eigenvalue problems can be solved to extract the fundamental modes of local, distortional, and global buckling.

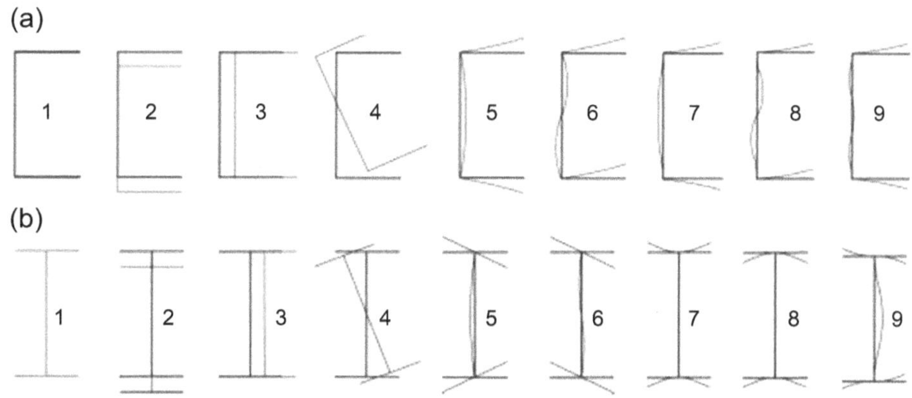

Figure 5.4 In-plane shapes of the nine most relevant deformation modes. (a) U-sections; (b) I-sections.
From Basaglia, C., Camotim, D., Silvestre, N., 2008. Global buckling analysis of plane and space thin-walled frames in the context of GBT. Thin-walled Structures 46(1), 79–101.

1. Distortional and local modes: $(B - \Psi C)\dot{Y} = 0$.
2. Torsion mode: $(\dot{D} - \Psi\dot{C})\ddot{Y} = 0$, where $\dot{C} = \dot{Y}^T C\dot{Y}$ and $\dot{D} = \dot{Y}^T D\dot{Y}$.
3. Bending mode: $(\ddot{X} - \Psi\ddot{C})\dddot{Y} = 0$, where $\ddot{C} = \ddot{Y}^T \dot{C}\ddot{Y}$ and $\ddot{X} = \ddot{Y}^T \dot{X}\ddot{Y}$.

Besides the fundamental modes (some are illustrated in Fig. 5.4), the byproduct of this sequence is that all matrices (*B*, *C*, *D*, and *X*) now possess a modal nature. Matrices *B* and *C* have a diagonal form, and *D* and *X* are partially diagonalized (Ádány et al., 2009).

Gonçalves et al. (2014) also addressed the shear modes in the framework of GBT, and proposed an efficient procedure to extract these modes.

5.3.3.2 Elastic buckling analysis

After performing the cross-section analysis, the linear elastic buckling analysis of a thin-walled member with the appropriate end support conditions can be obtained by solving a governing differential equation (*k* denotes certain mode(s)), expressed in terms of the modal amplitude functions $\phi_k(x)$ as

$$C_{ik}\phi_{k,xxxx} - D_{ik}\phi_{k,xx} + B_{ik}\phi_k + \lambda W_j^0 X_{ijk}\phi_{k,xx} = 0 \qquad [5.9]$$

Differential Eq. [5.9] can be solved analytically using a similar sine longitudinal function to FSM if the end boundary conditions are simply supported. For more general (arbitrary) end support conditions, solutions can only be sought by approximate numerical techniques, such as Galerkin's method, finite differences, or finite elements. Among these techniques, FEM is the most versatile and widely used. GBT-based beam finite elements were first formulated by Silvestre and Camotim (2002a,b). Research by Basaglia et al. (2008, 2009) and Camotim et al. (2010) has led to the use of GBT to analyze the global buckling behavior of plane and space thin-walled frames.

5.3.4 Constrained finite strip method

The cFSM is derived from semianalytical FSM by separating the original displacement field of FSM into four buckling mode categories: global (G), distortional (D), local (L), and shear and transverse (ST) extension. The separation is made possible by enforcing a set of mechanical assumptions, as shown in Table 5.1. Similar deformation modes can also be defined, as shown in Fig. 5.4. More details can be found in Li and Schafer (2010) and Li et al. (2011b).

Using the mechanical criteria, a set of constraint matrices, R_{G}, R_{D}, R_{L}, and R_{ST}, can be formulated for the four mode categories. Thus a key relationship of the original and reduced displacement fields can be expressed as:

$$d = R_{\mathrm{M}} d_{\mathrm{M}} \quad [5.10]$$

where d is the general displacement field, d_{M} is the reduced displacement field that matches buckling mode M, R_{M} is the constraint matrix for buckling mode M, and M refers to the modal space (G, D, L, ST, or any combination thereof).

An extension of cFSM to general end boundary conditions (Li and Schafer, 2013) was implemented in CUFSM (Li and Schafer, 2010). Ádány and Schafer (2014) also enabled cFSM to handle closed cross-sections by generalizing the modes through implementation of the criteria regardless of cross-section topology.

5.3.4.1 Modal decomposition

Modal decomposition is one of the main applications of cFSM, and provides a powerful means for examining any individual or combined modes of interest. Instead of solving the generalized eigenvalue problem in FSM, modal decomposition solves a reduced eigenvalue problem by introducing the constraint matrix to the original eigenvalue problem of FSM. The constrained (or reduced) eigenvalue problem corresponding to mode(s) M is expressed as

$$R_{\mathrm{M}}^{\mathrm{T}} K_{\mathrm{e}} R_{\mathrm{M}} \Phi_{\mathrm{M}} = \Lambda_{\mathrm{M}} R_{\mathrm{M}}^{T} K_{\mathrm{g}} R_{\mathrm{M}} \Phi_{\mathrm{M}} \quad [5.11]$$

where Λ_{M} is a diagonal matrix containing the eigenvalues for the given mode(s) and Φ_{M} is the matrix corresponding buckling modes in its columns. Also, $K_{\mathrm{e,M}} = R_{\mathrm{M}}^{\mathrm{T}} K_{\mathrm{e}} R_{\mathrm{M}}$ and $K_{\mathrm{g,M}} = R_{\mathrm{M}}^{\mathrm{T}} K_{\mathrm{g}} R_{\mathrm{M}}$ may be defined as the elastic and geometric stiffness matrix of the cFSM problem, respectively.

Table 5.1 Mechanical criteria of mode definition

Mechanical criteria	G	D	L	ST
Valsov's hypothesis: $(\gamma_{xy})_{PS} = 0$, $(\varepsilon_x)_{PS} = 0$, v is linear	Yes	Yes	Yes	No
Longitudinal warping: $(\varepsilon_y)_{PS} \neq 0$	Yes	Yes	No	–
Undistorted section: $\kappa_x = 0$	Yes	No	–	–

5.3.4.2 Modal identification

Modal identification is another main application of cFSM. The combined G, D, L, and ST subspaces, as defined in the constraint R matrices, provide a complete alternative basis to the original FSM displacement fields and represent a transformation of the solution from the original nodal degrees of freedom to a basis that separates G, D, L, and ST modes. This basis may be used to evaluate the contribution of the different deformation modes (G, D, L, and ST) to any general deformed shape d after appropriate normalization (ie, scaling). This procedure is known as modal identification. Any displaced shape or buckling mode, d, can be transformed into the basis spanned by the deformation classes through:

$$d = \begin{bmatrix} R_{\mathrm{G}} & R_{\mathrm{D}} & R_{\mathrm{L}} & R_{\mathrm{ST}} \end{bmatrix} \begin{bmatrix} C_{\mathrm{G}} \\ C_{\mathrm{D}} \\ C_{\mathrm{L}} \\ C_{\mathrm{ST}} \end{bmatrix} \qquad [5.12]$$

where C_{G}, C_{D}, C_{L}, and C_{ST} are column vectors that include the contribution coefficients for each vector inside each deformation class.

Then the participation (P_{M}) of space M in a general deformation (d) given the contribution coefficients is calculated using the absolute sum first implemented in Schafer and Ádány (2006)

$$p_{\mathrm{M}} = \|C_{\mathrm{M}}\|_1 \Big/ \sum_{\mathrm{M=G,D,L,ST/O}}^{4} \|C_{\mathrm{M}}\|_1 \qquad [5.13]$$

Fig. 5.5 illustrates modal decomposition and identification for the simply supported end boundary conditions of a C-column. As shown in Li et al. (2011a,b), the columns of R_{M} may be considered as a set of base vectors in this space of mode M and transformation inside space M is also possible, thus the base vectors defined by R_{M} are not unique. The modal decomposition and identification solutions may be affected by different transformations.

5.4 Open source programs and other software packages

5.4.1 CUTWP

CUTWP is free open source software for global buckling analysis coded in MATLAB environments with an easy-to-use single-page interface. It employs classical beam theory and Vlasov warping torsion to provide global member stability solutions (flexural–torsional, lateral–torsional, etc.)—the same mechanics utilized in codes

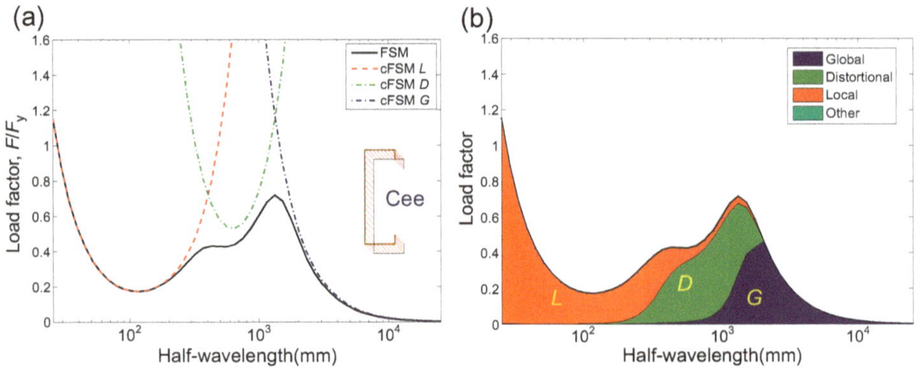

Figure 5.5 Modal decomposition and identification of thin-walled lipped channel member for a simply supported column by cFSM: (a) modal decomposition that separates buckling modes; (b) modal identification that quantitatively categorizes buckling modes.
From Li, Z., Abreu, J.C.B., Leng, J., Ádány, S., Schafer, B.W., 2014. Review: constrained finite strip method developments and applications in cold-formed steel design. Thin-walled Structures 81, 2–18.

and standards. In addition, it allows different global effective length factors (K_xL_x, K_yL_y, K_tL_t) and is able to read CUFSM input files. More information about CUTWP is available at www.ce.jhu.edu/bschafer/cutwp/.

5.4.2 GBTUL

GBTUL, short for GBT at the University of Lisbon, is a free software program that can provide elastic buckling and vibration analyses of prismatic thin-walled members. The kernel incorporates the latest formulations of GBT. GBTUL took advantage of GBT's modal features to provide information and visualization of member deformation modes. With the tool, it is possible to select desired deformation modes in the analysis (a modal decomposition feature). It is also possible to analyze members made of one or several isotropic or orthotropic materials, and under various common support conditions (eg, simple supports, fixed supports, or free ends). In the buckling analyses, the user is able to specify any combination of arbitrary axial force, bending moment, and bimoment diagrams. GBTUL contains a user-friendly and intuitive graphical interface with a small number of commands and providing high-quality two- and three-dimensional outputs. GBTUL can be downloaded at www.civil.ist.utl.pt/gbt/.

5.4.3 CUFSM

CUFSM is a direct implementation of semianalytical FSM to provide elastic buckling analysis for thin-walled cross-section members. New extensions over the years have enabled it to perform analysis on general end boundary conditions. More importantly, cFSM implemented in CUFSM gives it a modal feature similar to GBT, thus modal decomposition to study individual mode or mode class, and mode identification by categorizing the general mode into the four fundamental classes, are available in CUFSM. In the analysis, the user can specify any combination of arbitrary axial force, bending moment, and bimoment. CUFSM provides an easy-to-use interface and has been successfully used by researchers, educators, students, and practicing engineers. While CUFSM is most commonly used for analyzing thin-walled CFS members, it has also been used for a large variety of other materials and applications. CUFSM can be downloaded at www.ce.jhu.edu/bschafer/cufsm/.

5.4.4 Finite element method

The above tools for elastic buckling analysis of CFS are free software that was mainly developed by researchers in the structural stability community, but several general-purpose commercial finite element software packages can be used for elastic buckling analysis. To the author's knowledge, three are widely used for CFS analysis: ABAQUS FEA from Dassault Systemes (www.3ds.com/products-services/simulia/); ANSYS FEA (www.ansys.com/); and ADINA by ADINA R&D, Inc., founded in 1986 by Dr. K. J. Bathe (www.adina.com/).

5.5 Applications in design and development

5.5.1 Design example employing numerical elastic buckling solutions

In CFS member design, the elastic buckling load is a key parameter in predicting the design strength. In current AISI design specifications of CFS members, two design methods can be used: effective width method (also known as the unified method) and direct strength method (DSM). The effective width method performs a reduction of the plates that comprise a cross-section based on the stability of the individual plates and is a widely used design model with long history, but it is intimately tied to classical plate stability and in general requires a laborious iterative procedure. The DSM performs a similar reduction but based on the full cross-section stability (usually from numerical solutions), by extending the global buckling column curve to local and distortional buckling instabilities. In predicting the elastic buckling loads, semianalytical FSM is the most commonly used tool, such as CUFSM implemented in Schafer and Ádány (2006) and Li and Schafer (2010). Other numerical methods for elastic buckling prediction are the cFSM, FEM, and GBT, as discussed above.

For illustration purposes, a 250S137-33 stud section with a straight-line model under axial compression is analyzed using CUFSM for elastic buckling and its signature curve is similar to that shown in Fig. 5.3 (assume the end boundary conditions are simply supported). Elastic local buckling ($P_{cr\ell}$) at the first minimum is 25.4 kN and distortional buckling (P_{crd}) at the second is 38.8 kN. Assume the column is fully restrained for global buckling and the yield stress of this cold-formed member is 344.7 MPa. According to the DSM, the nominal axial strength, P_n, is the minimum of the individual predicted capacities:

$$P_n = \min(P_{ne}, P_{n\ell}, P_{nd}) \qquad [5.14]$$

where the nominal axial strengths, P_{ne} for global buckling (flexural, torsional, or torsional–flexural buckling), $P_{n\ell}$ for local buckling, and P_{nd} for distortional buckling are:

$$P_{ne} = \begin{cases} (0.658\lambda_C^2)P_y & \text{if} \quad \lambda_C \leq 1.5 \\ (0.877\lambda_C^2)P_y & \text{if} \quad \lambda_C > 1.5 \end{cases} \qquad [5.14a]$$

$$P_{n\ell} = \begin{cases} P_{ne} & \text{if} \quad \lambda_\ell \leq 0.776 \\ \left(1 - 0.15(P_{cr\ell}/P_{ne})^{0.4}\right)(P_{cr\ell}/P_{ne})^{0.4}P_{ne} & \text{if} \quad \lambda_\ell \leq 0.776 \end{cases} \qquad [5.14b]$$

$$P_{nd} = \begin{cases} P_y & \text{if} \quad \lambda_d \leq 0.561 \\ \left(1 - 0.25(P_{crd}/P_y)^{0.6}\right)(P_{crd}/P_y)^{0.6}P_y & \text{if} \quad \lambda_d > 0.561 \end{cases} \qquad [5.14c]$$

where $\lambda_C = \sqrt{P_y/P_{cre}}$, $\lambda_\ell = \sqrt{P_{ne}/P_{cr\ell}}$, $\lambda_d = \sqrt{P_y/P_{crd}}$, $P_y = A_g F_y$, A_g is the gross area of the cross-section, and F_y is the yield stress. Note, P_{cre} is minimum of the critical elastic column buckling load in flexural, torsional, or torsional–flexural buckling. This column is fully restrained to global buckling, thus $P_{ne} = P_y$.

The nominal strength of this SSMA 250S137-33 column is therefore 31.6 kN based on DSM calculation.

5.5.2 Automated design integrating computational elastic buckling analysis

As shown above, DSM is the most amenable approach to integrated computational elastic buckling analysis, but challenges in its application exist unless all buckling modes are uniquely defined. In particular, if both local and distortional buckling can be uniquely determined from FSM solutions, such as that shown in Fig. 5.3, the application of DSM has the necessary elastic buckling load inputs. However, if either or both minima is "indistinct" (eg, no corresponding minimum) the FSM solution will pose problems in applying DSM equations. Studies show that this problem is common for SSMA stud sections.

cFSM has the capability to predict the elastic buckling load (or moment) automatically for a given buckling mode (eg, a solution decomposed to include only distortional buckling is possible.) Thus cFSM is a potential solution to the above

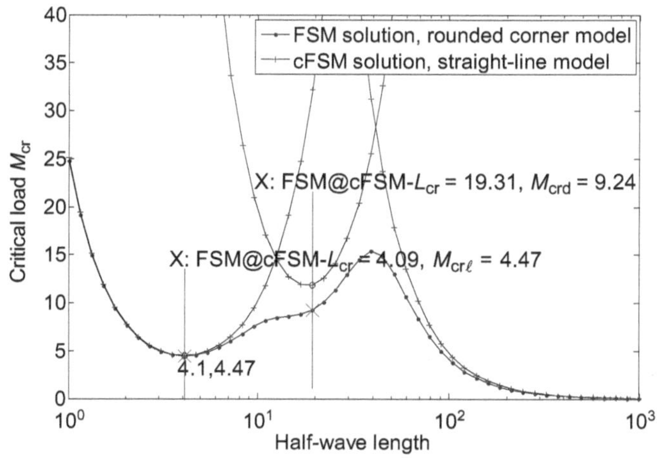

Figure 5.6 Illustration of the recommended solution.
From Li, Z., Schafer, B.W., 2010a. Application of the finite strip method in cold-formed steel member design. Journal of Constructional Steel Research 66(8–9), 971–980.

problem. However, there are two basic issues hampering us from directly using the cFSM solution.

1. DSM's strength expressions are calibrated to the conventional FSM minima instead of pure mode solutions from cFSM.
2. cFSM cannot handle rounded corners and still provide a meaningful separation of local and distortional buckling.

A solution proposed by Li and Schafer (2010a) is that cFSM is only used to aid in the identification of the half-wavelength of interest. The analyst develops a rounded-corner model of the section and runs a conventional FSM model. If the resulting signature curve has unique minima, no further analysis is conducted. If the signature curve does not have unique minima, pure mode cFSM solutions are performed (on a straight-line model only) to identify the appropriate L_{cr}. The P_{cr} (or M_{cr}) at the associated L_{cr} is determined from conventional FSM analysis. A shorthand for this solution method is FSM@cFSM-L_{cr}, which can be illustrated by the example of an SSMA 550S162-43 stud section (Product Technical Information, 2001) with a round-corner model under axial compression, as shown in Fig. 5.6.

5.5.3 New development: modal identification for shell finite element method

As mentioned, while shell FEM is very versatile in modeling CFS members for elastic buckling analysis, it provides no means of uniquely identifying the modes and instead requires a laborious and subjective procedure employing visual investigation to classify the buckling modes.

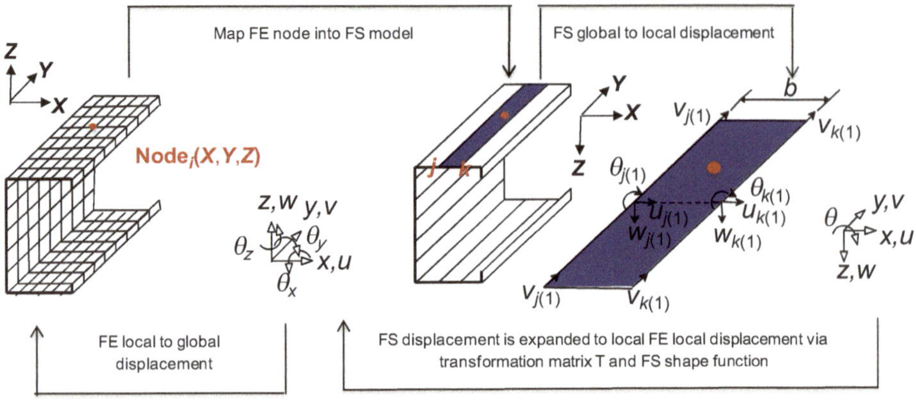

Figure 5.7 Illustration of interpolation of finite strip DOF to FE DOF (finite strip local displacement is illustrated for longitudinal term $m = 1$ of S-S).
From Li, Z., Ádány, S., Schafer, B.W., 2013. Modal identification for shell finite element models of thin-walled members in nonlinear collapse analysis. Thin-walled Structures 67, 15–24.

Recent works by Li et al. (2011a,b, 2013) provided a method for applying modal identification to both elastic and collapse analyses of thin-walled members modeled using shell finite element (FE) analysis. This approach is developed to utilize the current development of the cFSM. The modal identification method takes a general FE displacement vector (general deformed shape or buckling mode) based on a shell element model of a thin-walled member, and accurately assigns quantitative modal participations in terms of the G, D, L, and ST classes. The key step in this approach is to form a similar modal basis for shell FEM as those formed in cFSM for modal identification. Due to the discrepancy of element formulations in shell FEM (6 degree of freedom (DOF), ie, three translational and three rotational displacements) and FSM (4 DOF, ie, three translational and one rotational displacements), the cFSM modal basis serving as the fundamental base is interpreted through a coordinate transformation matrix at the cross-section level and a special set of longitudinal ($Y_{[m]}$, in Eq. [5.2]) shape functions available in FSM (Li and Schafer, 2010) along the length to form a shell FEM modal basis for modal identification. The essential steps of this expansion are illustrated in Fig. 5.7 and the following equation

$$\sum_{m=1}^{q} Y_{[m]} = 1 - \cos\left(\frac{\pi y}{2a}\right) + 1 - \cos\left(\frac{\pi(-y+a)}{2a}\right) + \sum_{m=1}^{q} \sin\left(\frac{m\pi y}{a}\right) \qquad [5.15]$$

With the shell FEM modal basis (R_{FE}) known, any shell FEM displacement vector can be approximated by a linear combination of the vectors in modal basis R_{FE}. A fairly sizable minimization problem is required for assigning the contributions to the fundamental buckling deformation classes, as following:

$$D_{\mathrm{err}} = D_{\mathrm{FE}} - R_{\mathrm{FE}} C_{\mathrm{FE}} \quad \text{and} \quad \min D_{\mathrm{err}}^{\mathrm{T}} D_{\mathrm{err}} \qquad [5.16]$$

where D_{FE} is the FE displacement vector, D_{err} is the residual error, and C_{FE} is the contribution coefficients (for G, D, L, and ST). Participation evaluation in terms of the numerical contribution of G, D, L, and ST and error estimation follow the minimization results.

In fact, modal identification in shell FEM can also be achieved through deformation modes formulated from GBT with a similar minimization problem as above (Cai and Moen, 2015; Nedelcu, 2014; Nedelcu and Cucu, 2014).

Sources of further information and advice

Beside the limitation FEM has in identifying buckling modes, another drawback is that it necessitates expertise and extensive efforts to build an FEM of CFS structures. To aid design and research, in-house tools have been developed by many researchers. For instance, to the author's knowledge, the Thin-Walled Structures research group led by Professor Schafer at Johns Hopkins University has developed a couple of tools to build shell FE models of cold-formed members, such as abaqus_maker and adina_maker to generate the shell FE ABAQUS and ADINA models, respectively. These tools are written in MATLAB codes, and easy to use and modify for necessary changes. Professor Moen's group at Virginia Tech has a similar abaqus_maker but with the capability to add a variety of hole shapes in CFS members. More information can be found at their personal websites (Schafer: www.ce.jhu.edu/bschafer/; Moen: www.moen.cee.vt.edu/category/tools/). In future, a collection of available tools to help designers and researchers build shell FE models would be useful.

Moreover, given the complex buckling modes existing in CFS members (local, global, distortional), correct identification and classification of these modes and accurate calculation of the associated elastic critical loads (or moments) are crucial not only in predicting the ultimate strength of a CFS member but also in understanding the potential coupled instabilities. Future design development taking advantage of the current analysis capability in modal identification and decomposition in cFSM and GBT is a possible direction.

References

Abambres, M., Camotim, D., Silvestre, N., 2013. Physically non-linear GBT analysis of thin-walled members. Computers & Structures 129, 148–165.

Abambres, M., Camotim, D., Silvestre, N., 2014a. GBT-based elastic-plastic post-buckling analysis of stainless steel thin-walled members. Thin-walled Structures 83, 85–102.

Abambres, M., Camotim, D., Silvestre, N., Rasmussen, K.J.R., 2014b. GBT-based structural analysis of elastic-plastic thin-walled members. Computers & Structures 136, 1–23.

Ádány, S., Schafer, B.W., 2014a. Generalized constrained finite strip method for thin-walled members with arbitrary cross-section: primary modes. Thin-walled Structures 84, 150–169.

Ádány, S., Schafer, B.W., 2014b. Generalized constrained finite strip method for thin-walled members with arbitrary cross-section: secondary modes, orthogonality, examples. Thin-walled Structures 84, 123–133.

Ádány, S., Silvestre, N., Schafer, B.W., Camotim, D., 2009. GBT and cFSM: two modal approaches to the buckling analysis of unbranched thin-walled members. Advanced Steel Construction 5 (2), 195–223.

Ádány, S., 2006. Flexural buckling of thin-walled columns: discussion on the definition and calculation, in International Colloquium on Stability and Ductility of Steel Structures. In: International Colloquium on Stability and Ductility of Steel Structures, pp. 249–258.

Basaglia, C., Camotim, D., Silvestre, N., 2008. Global buckling analysis of plane and space thin-walled frames in the context of GBT. Thin-walled Structures 46 (1), 79–101.

Basaglia, C., Camotim, D., Silvestre, N., 2009. GBT-based local, distortional and global buckling analysis of thin-walled steel frames. Thin-walled Structures 47 (11), 1246–1264.

Basaglia, C., Camotim, D., Silvestre, N., 2011. Non-linear GBT formulation for open-section thin-walled members with arbitrary support conditions. Computers & Structures 89 (21–22), 1906–1919.

Bui, H.C., Rondal, J., 2008. Buckling analysis of thin-walled sections by semi-analytical Mindlin Reissner finite strips: a treatment of drilling rotation problem. Thin-walled Structures 46 (6), 646–652.

Bui, H.C., 2012. Semi-analytical finite strip method based on the shallow shell theory in buckling analysis of cold-formed sections. Thin-walled Structures 50 (1), 141–146.

C. for Advanced Structural Engineering, 2006. THIN-wall – A Computer Program for Cross-section Analysis and Finite Strip Buckling Analysis and Direct Strength Design of Thin-walled Structures. Version 2.1. School of Civil Engineering, University of Sydney, Sydney, Australia.

Cai, J., Moen, C.D., 2015. Automated buckling mode identification of thin-walled structures from 3D finite element mode shapes or point clouds. In: Proceedings of the Annual Stability Conference Structural Stability Research Council.

Camotim, D., Silvestre, N., Gonçalves, R., Dinis, P.B., 2004. GBT analysis of thin-walled members: new formulations and applications. In: Thin-walled Structures: Recent Advances and Future Trends in Thin-walled Structures Technology, pp. 137–168.

Camotim, D., Basaglia, C., Silvestre, N., 2010. GBT buckling analysis of thin-walled steel frames: a state-of-the-art report. Thin-walled Structures 48 (10–11), 726–743.

Davies, J., Jiang, C., 1998. Design for distortional buckling. Journal of Constructional Steel Research 46 (1–3), 174–175.

Desmond, T.P., 1977. The Behaviour and Design of Thin-walled Compression Elements with Longitudinal Stiffener. Cornell University, Ithaca (NY).

Eccher, G., Rasmussen, K.J.R., Zandonini, R., 2008a. Elastic buckling analysis of perforated thin-walled structures by the isoparametric spline finite strip method. Thin-walled Structures 46 (2), 165–191.

Eccher, G., Rasmussen, K.J.R., Zandonini, R., 2008b. Linear elastic isoparametric spline finite strip analysis of perforated thin-walled structures. Thin-walled Structures 46 (3), 242–260.

Gonçalves, R., Camotim, D., 2012. Geometrically non-linear generalised beam theory for elastoplastic thin-walled metal members. Thin-walled Structures 51, 121–129.

Gonçalves, R., Bebiano, R., Camotim, D., 2014. On the shear deformation modes in the framework of Generalized Beam Theory. Thin-walled Structures 84, 325–334.

Hancock, G.J., Pham, C.H., 2013. Shear buckling of channel sections with simply supported ends using the Semi-Analytical Finite Strip Method. Thin-walled Structures 71, 72–80.

Hancock, G.J., Pham, C.H., 2015. Buckling analysis of thin-walled sections under localised loading using the Semi-Analytical Finite Strip Method. Thin-walled Structures 86, 35–46.

Lau, S., Hancock, G., 1987. Distortional buckling formulas for channel columns. Journal of Structural Engineering 113 (5), 1063–1078.

Li, Z., Schafer, B.W., 2010a. Application of the finite strip method in cold-formed steel member design. Journal of Constructional Steel Research 66 (8–9), 971–980.

Li, Z., Schafer, B.W., 2010b. Buckling analysis of cold-formed steel members with general boundary conditions using CUFSM: conventional and constrained finite strip methods. In: Proceedings of the 20th Int;l. Spec. Conf. On Cold-formed Steel Structures.

Li, Z., Schafer, B., 2013. Constrained finite strip method for thin-walled members with general end boundary conditions. Journal of Engineering Mechanics 139 (11), 1566–1576.

Li, Z., Hanna, M.T., Ádány, S., Schafer, B.W., 2011a. Impact of basis, orthogonalization, and normalization on the constrained finite strip method for stability solutions of open thin-walled members. Thin-walled Structures 49 (9), 1108–1122.

Li, Z., Joó, A.L., Ádány, S., Schafer, B.W., 2011b. Approximate modal identification in nonlinear collapse analysis of thin-walled members. In: Proceedings of the 2011 Annual Stability Conference.

Li, Z., Ádány, S., Schafer, B.W., 2013. Modal identification for shell finite element models of thin-walled members in nonlinear collapse analysis. Thin-walled Structures 67, 15–24.

Li, Z., Abreu, J.C.B., Leng, J., Ádány, S., Schafer, B.W., 2014. Review: constrained finite strip method developments and applications in cold-formed steel design. Thin-walled Structures 81, 2–18.

Li, L., 2009. Analyses of distortional buckling of cold-formed sigma purlins using EN1993-1-3. Journal of Constructional Steel Research 65 (12), 2099–2102.

Milašinović, D.D., 2011. Geometric non-linear analysis of thin plate structures using the harmonic coupled finite strip method. Thin-walled Structures 49 (2), 280–290.

Milašinović, D.D., 2012. Harmonic coupled finite strip method applied on buckling-mode interaction analysis of composite thin-walled wide-flange columns. Thin-walled Structures 50 (1), 95–105.

Moen, C.D., Schafer, B.W., 2009a. Elastic buckling of thin plates with holes in compression or bending. Thin-walled Structures 47 (12), 1597–1607.

Moen, C.D., Schafer, B.W., 2009b. Elastic buckling of cold-formed steel columns and beams with holes. Engineering Structures 31 (12), 2812–2824.

Nedelcu, M., Cucu, H.L., 2014. Buckling modes identification from FEA of thin-walled members using only GBT cross-sectional deformation modes. Thin-walled Structures 81, 150–158.

Nedelcu, M., 2014. Buckling mode identification of perforated thin-walled members by using {GBT} and shell {FEA}. Thin-walled Structures 82, 67–81.

Pala, M., 2006. A new formulation for distortional buckling stress in cold-formed steel members. Journal of Constructional Steel Research 62 (7), 716–722.

Pala, M., 2008. Genetic programming-based formulation for distortional buckling stress of cold-formed steel members. Journal of Constructional Steel Research 64 (12), 1495–1504.

Pham, C.H., Hancock, G.J., 2009a. Shear buckling of thin-walled channel sections with intermediate web stiffener. In: Proceedings, Sixth International Conference on Advances in Steel Structures, pp. 417–424.

Pham, C.H., Hancock, G.J., 2009b. Shear buckling of thin-walled channel sections. Journal of Constructional Steel Research 65 (3), 578–585.

Pham, C.H., Hancock, G.J., 2012. Elastic buckling of cold-formed channel sections in shear. Thin-walled Structures 61, 22–26.

Product Technical Information, ICBO ER-4943P, 2001. Steel Stud Manufacturers Association. http://www.ssma.com.

Schafer, B.W., Ádány, S., 2006. Buckling analysis of cold-formed steel members using CUFSM: conventional and constrained finite strip methods. In: Proceedings of the Eighteenth International Specialty Conference on Cold-formed Steel Structures, pp. 39–54.

Schafer, B., Peköz, T., 1999. Laterally braced cold-formed steel flexural members with edge stiffened flanges. Journal of Structural Engineering 125 (2), 118–127.

Schafer, B.W., Li, Z., Moen, C.D., 2010. Computational modeling of cold-formed steel. Thin-walled Structures 48 (10–11), 752–762.

Sharp, M.L., 1966. Longitudinal stiffeners for compression members. Journal of the Structural Division 92 (5), 187–212.

Silvestre, N., Camotim, D., 2002a. First-order generalised beam theory for arbitrary orthotropic materials. Thin-walled Structures 40 (9), 755–789.

Silvestre, N., Camotim, D., 2002b. Second-order generalised beam theory for arbitrary orthotropic materials. Thin-walled Structures 40 (9), 791–820.

Silvestre, N., Camotim, D., 2003. Nonlinear generalized beam theory for cold-formed steel members. International Journal of Structural Stability and Dynamics 3 (4), 461–490.

Silvestre, N., Camotim, D., 2004a. Distortional buckling formulae for cold-formed steel C and Z-section members: part I derivation. Thin-walled Structures 42 (11), 1567–1597.

Silvestre, N., Camotim, D., 2004b. Distortional buckling formulae for cold-formed steel C- and Z-section members: part II validation and application. Thin-walled Structures 42 (11), 1599–1629.

Silvestre, N., 2005. Generalised Beam Theory: New Formulations, Numerical Implementation and Applications. Technical University of Lisbon.

Simulia, 2012. Abaqus/User Subroutines. Dassault Systemes Simulia Corp, RI, USA.

Yu, C., Schafer, B., 2007. Effect of longitudinal stress gradients on elastic buckling of thin plates. Journal of Engineering Mechanics 133 (4), 452–463.

Ziemian, R.D. (Ed.), 2010. Guide to Stability Design Criteria for Metal Structures, sixth ed. Wiley.

Optimization techniques for structural design of cold-formed steel structures

6

J. Leng
Postdoctoral Fellow, Mechanical Engineering Department, McGill University, Montreal, Canada

6.1 Introduction and problem formulation

Structural optimization is the search for a structural design that is optimal for a certain design criterion while satisfying other constraints. It is a specific case of general optimization problems in mathematics. A well-defined optimization problem should include design variables and usually one objective function, often with constraints formulated using the design variables. In terms of structural optimization, objective functions in general can be the minimization of compliance or weight, or the maximization of capacity. It is customary to formulate the objective as a minimization. Multiobjective optimization problems also exist, but one is unlikely to identify a solution that can achieve optimal values for every single objective at the same time. Solutions of practical optimization problems usually have to satisfy equality and inequality constraints developed from design, manufacturability, and end-use requirements, making constrained optimization problems more difficult to solve than unconstrained problems. Another important feature of structural optimization is that it usually requires solving equilibrium problems of the structure every time the objective function needs to be evaluated. Accordingly, structural optimization problems are usually simulation-based: modeling and structural analysis tools are typically embedded in the optimization framework. The objective function evaluation can be computationally costly because of the simulation involved. Arora (2004) is a book specially dedicated to the application of optimization theory in design problems with detailed examples.

Typically, structural optimization incorporates optimization of dimension, shape, and topology (Leng, 2015). Dimension optimization operates on a fixed topology by adjusting characteristic dimensions of the topology only; topology optimization has the most flexibility for developing an optimized configuration within a given domain of material—connectivity can be modified by introducing voids (Leng, 2015). Shape optimization requires a shape generation subroutine to update the geometry of the design for better performance, but it may not be possible to change the connectivity and topology greatly.

In dimension optimization, design variables are usually characteristic dimension(s) of the dictated topology and should be positive real numbers within certain bounds

Recent Trends in Cold-Formed Steel Construction. http://dx.doi.org/10.1016/B978-0-08-100160-8.00006-2

(Leng, 2015). For topology optimization, design variables can ideally be zero (void) or one (solid) for a given discretized domain. However, to allow the use of the gradient-based method, which requires continuity and differentiability, design variables can be treated as a continuous number between zero and one in some topology optimization implementations (Leng, 2015). A classic monograph on topology optimization was presented by Bendsøe and Sigmund (2003). For shape optimization, design variables are parameters that can uniquely define the shape and are dependent on the shape generation algorithm.

For the scale level of structural optimization, research efforts focus on member and layout optimization of structural systems. These two levels can be combined, so the structural design can be significantly automated by generating a highly efficient layout with optimized selection of members. The history of research on optimization of cold-formed steel (CFS) framed structures started with dimension optimization of designated types of members, and has achieved some inspiring progress on shape optimization on the member level. Initial efforts on CFS framing layout optimization have recently been reported.

So far, the great majority of CFS structural optimization research has been performed on the member level—mostly columns under axial compression, since the stress distribution is assumed to be uniform and no further effort of neutral axis determination is needed (Leng, 2015). Given that buckling is one important limit state for CFS members, the method used for determining nominal axial capacity, P_n, from design specifications can have a significant influence on the formulation of optimization problems (Leng, 2015). P_n can be an objective of maximization, or it can be a constraint that must be satisfied by the optimal design. Classical design procedure, as documented in early versions of American Iron and Steel Institute (AISI) specifications, adopted the long-established effective width method (AISI, 2007a,b), which requires reducing a plate under nonlinear longitudinal stress into a plate with effective width under constant stress. However, the process can be tedious and cumbersome for complicated cross-sections. The direct strength method (DSM) (AISI, 2007a,b) is much simpler, requiring only the critical load in local (P_{crl}), distortional (P_{crd}), and global buckling (P_{cre}), and the load at yield (P_y). Design equations in DSM are simple forms of elementary functions and can be coded easily into simulation-based optimization programs. Detailed discussion on CFS member design methods is presented in Chapters 3 and 4.

Previous researchers have applied various combinations of design-code-based capacity evaluation methods and search algorithms in dimension and shape optimization of CFS members (Leng, 2015). The early work on numerical optimization by Seaburg and Salmon (1971) applied the gradient-based steepest descent (SD) method to explore the dimensions of hat sections with the effective width method of the AISI specification (AISI, 1968). Tran and Li (2006) solved the optimization of a lipped channel beam using the trust region method; the design should satisfy various failure modes in British code BS 5950-5 (British Standards Institution, 1998) and Eurocode 3 (ECCS, 1996). Tian and Lu (2004) performed optimization of cross-section dimensions of channel columns utilizing sequential quadratic programming (SQP) according to BS 5950-5 (British Standards Institution, 1998). Algorithms in these works belong to mathematical programming that requires first- or second-order (partial) derivatives of the objective function. In terms of derivative-free heuristic methods, Adeli et al. developed a computational neural network model (Adeli and Karim, 1997) and applied it to the optimization of CFS beams with hat, I, and Z sections (Karim and Adeli, 1999) following AISI Allowable Stress Design (AISI, 1989) and Load Resistance Factor Design (AISI, 1991) specifications, and space trusses with lipped channel sections (Tashakori and Adeli, 2002), according to AISI Allowable Stress Design specification (AISI, 1996). Lu (2003) optimized Z-section dimensions under the effective width design of Eurocode 3 (EuroCode-3, 1996), with CUFSM (see Chapter 5) as the simulation tool. Lee et al. (2005, 2006) modified genetic algorithms (GAs) in their search for optimized channel cross-section dimensions of CFS columns under axial compression and beams under uniformly distributed loads. Magnucki and Paczos (2009) numerically solved the dimension optimization of CFS channel beams with open or closed profile of drop flanges subject to strength and buckling constraints. Analytical solutions of the problems of global and local buckling are presented in their paper. Kripka and colleagues (Chamberlain Pravia and Kripka, 2012; Kripka and Martin, 2013) utilized simulated annealing (SA) to optimize the dimension of a C-section column following the effective width method in the AISI (2007a,b) specification. In all these works, dimensional optimization was performed on a predetermined cross-section by adjusting its web depth, flange width, lip lengths, etc., with constraints on the member's capacity. Most researchers searched for minimum-weight designs, but Lu maximized load efficiency (Leng, 2015).

An earlier traceable work of shape optimization of CFS members is credited to Liu et al. (2004). They maximized the capacity of CFS columns by applying Bayesian classification trees as the optimizer, and used DSM and CUFSM to evaluate axial capacity P_n. Kolcu et al. (2010) combined Mindlin–Reissner finite strips and SQP in maximizing the critical load P_{cr} of CFS members. Leng et al. (2011, 2012, 2013, 2014) systematically studied shape optimization of CFS columns in a search for cross-sections that can maximize the axial capacity with a given steel sheet. Firstly, an unconstrained optimization problem was investigated (Leng et al., 2011). The same perimeter, thickness, and material properties of the steel sheet used by Liu et al. (2004) were adopted for better comparison. DSM and CUFSM were packaged to evaluate the axial capacity of columns. Several novel cross-sections were identified for intermediate-length (4 ft, 1.22 m) and long-length (16 ft, 4.88 m) columns with more than 100% capacity increase. In the process, they also compared the performance of gradient-based (SD) and stochastic search algorithms (SA and GA). Subsequently, end-use and manufacturability constraints were introduced so that optimized cross-sections are more readily fabricated and useful in practice (Leng et al., 2012). The case of a limited number of rollers was also taken into account as a reflection of fabrication cost (Leng et al., 2013, 2014). The modified code uses SA to perform stochastic search on the constrained design space. The resulting optimized cross-sections include singly symmetric "Σ"-like sections for short (2 ft, 0.61 m) and intermediate-length (4 ft, 1.22 m) columns, and point-symmetric squashed "S"-like sections for long (16 ft, 4.88 m) columns. These optimized shapes offer from 50% to over 200% improvement in P_n over reference lipped C-sections, indicating that significant enhancements can be obtained through cross-section optimization without loss of practicality and manufacturability (Leng, 2015). Comparable work by other researchers was also able to identify similar optimized sections. Gilbert et al. developed "self-shape optimization" based on floating-point-type GA and applied the algorithm in CFS columns (Gilbert et al., 2012a,b) with CUFSM for buckling analysis. Wang et al also tried to introduce certain manufacturing rules as constraints into their algorithm (Wang et al., 2014a,b). Moharrami et al. (2014) and Gargari et al. (2013) used GA and a recent version of CUFSM to maximize the capacity of CFS columns with consideration of general boundary conditions other than simple supports (see Chapter 5 for recent development of CUFSM). Franco et al. (2014) designed a MATLAB code with a graphical user interface using shape grammar to generate cross-sections, DSM and CUFSM to evaluate capacity, and GA to optimize design of CFS columns and beams. Sharafi et al. (2014) developed their code using graph theory to generate open or close sections and identified optimized ones for the two-objective problem: minimum mass and maximum strength (represented by elastic section modulus about the major axis in their examples). Their solver was an ant colony optimization (ACO) algorithm; some geometric, behavioral, and serviceability constraints were applied.

In general, the formulation for the optimization of a CFS framing structure, with its objective, design variables, and constraints, can be presented in the form below:

$$\begin{aligned} &\min f(\boldsymbol{x}) \\ &s.t. c_{\mathrm{i}}(\boldsymbol{x}) \leq 0 \quad \text{for } i = 1, \cdots, N \end{aligned} \tag{6.1}$$

where $\boldsymbol{x}$ is the vector of design variables and N is the number of constraints, expressed through constraint functions c_{i}. Features of some typical optimization algorithms that have been applied in CFS structural optimization are addressed in the next section.

6.2 Algorithms for design optimization of CFS framing structures

This section introduces several types of algorithms in CFS framing design optimization. Since the development and study of algorithms are generally in the scope of numerical analysis and computer science, the intention here is to present the basic ideas of common algorithms and their implementation in structural optimization. More mathematically rigorous monographs are included as references for readers with further interest.

6.2.1 Gradient-based algorithms

Gradient-based algorithms have a solid mathematical background, in that Karush–Kuhn–Tucker (KKT) conditions are necessary for local minimal solutions. Under certain circumstances (for example, if the objective function is convex defined on a convex set), they can also be sufficient conditions. However, solving KKT conditions directly is usually very cumbersome, in that the equations are nonlinear, so practical algorithms aim to decrease the objective function value step by step instead (Leng, 2015). For unconstrained optimization, line search and trust region methods are most widely used. For more complicated constrained optimization, SQP, penalty, and projection methods can be utilized (Leng, 2015). Duality theory can also be a powerful tool by reformulating the primal problem in the dual space; the correspondent dual function is always concave, and sometimes the dual problem is much easier to solve. Nevertheless, all these methods require the gradient of a certain function with respect to its variables (Leng, 2015). An analytical form of the gradient is not guaranteed to exist, or could be very complicated. On the other hand, finite difference approximation of the gradient can be computationally costly for simulation-based optimization due to the increased number of objective function evaluations, which is usually characteristic of structural optimizations (Leng, 2015). For theoretical background in functional analysis, see Luenberger (1969). For numerical optimization theory and description of algorithms, see Nocedal and Wright (2006). Luenberger and Ye (2008) is also a recognized reference on mathematical programming theory.

SD is a simple, gradient-based optimizer that uses only first-order derivatives at the current design point to guide search. The well-established theoretical background of SD is quite intuitive: negative gradient is the SD direction in the neighborhood of the current point. SD is widely used on its own, and can also be integrated into other algorithms. The iterative scheme of SD with design variables updated at an iteration k is as follows:

$$\boldsymbol{x}_{k+1} = \boldsymbol{x}_k + \alpha_k \boldsymbol{d}_k \tag{6.2}$$

where $\boldsymbol{x}_k$ is the vector of design variables at iteration k, $\boldsymbol{d}_k$ is the vector of design variable change direction, and scalar α_k is a step length control parameter used to ensure improvement in the objective function. For SD, the step direction is the negative of the gradient of the function to be optimized.

$$\boldsymbol{d}_k = -\nabla f(\boldsymbol{x}_k) \tag{6.3}$$

The gradient ∇f, formed by partial derivatives of f with respect to each component of $\boldsymbol{x}$, can be evaluated analytically or approximated by finite difference. When the vector norm of ∇f is zero, the necessary condition of a local minimum is satisfied and the algorithm has converged.

Given a step direction, a line search algorithm is used to identify the step size α_k that produces a maximum reduction of the objective function. This is equivalent to finding the optimal value of scalar α for a function of only one variable, defined as

$$\min \bar{f}(\alpha) = f(\boldsymbol{x}_{k+1}) = f(\boldsymbol{x}_k + \alpha \boldsymbol{d}_k) \tag{6.4}$$

Various forms of line search have been developed; see Nocedal and Wright (2006) and Luenberger and Ye (2008) for additional details on line search algorithms.

Trust region method is another iterative algorithm effective for unconstrained numerical optimization. The terms "trust region" relates to a quadratic model m_k of the objective function near the design point $\boldsymbol{x}_k$ and the radius of that neighborhood Δ_k, developed using the Taylor-series expansion:

$$\begin{aligned} f(\boldsymbol{x}_k+\boldsymbol{p}) &= f_k + g_k^{\mathrm{T}}\boldsymbol{p} + \frac{1}{2}\boldsymbol{p}^{\mathrm{T}}\nabla^2 f(\boldsymbol{x}_k+t\boldsymbol{p})\boldsymbol{p} \\ m_k(\boldsymbol{p}) &= f_k + \boldsymbol{g}_k^{\mathrm{T}}\boldsymbol{p} + \frac{1}{2}\boldsymbol{p}^{\mathrm{T}}\boldsymbol{B}_k\boldsymbol{p} \end{aligned} \qquad [6.5]$$

where $f_k = f(\boldsymbol{x}_k)$, $\boldsymbol{g}_k = \nabla f(\boldsymbol{x}_k)$, and t is some scalar in the interval (0, 1). In the quadratic model function m_k, an approximation $\boldsymbol{B}_k$ of the Hessian matrix $\nabla^2 f(\boldsymbol{x}_k)$ (second-order partial derivatives of the objective function) is adopted.

The model function m_k is minimized locally within the neighborhood, characterized by its radius Δ_k, ie, solving the subproblem:

$$\begin{aligned} &\min_{p\in R^n} m_k(\boldsymbol{p}) = f_k + \boldsymbol{g}_k^{\mathrm{T}}\boldsymbol{p} + \frac{1}{2}\boldsymbol{p}^{\mathrm{T}}\boldsymbol{B}_k\boldsymbol{p} \\ &s.t. \|\boldsymbol{p}\| \le \Delta_k \end{aligned} \qquad [6.6]$$

Once a step $\boldsymbol{p}_k$ is available, the performance of the quadratic model m_k is evaluated using the ratio ρ_k as the predicted reduction of the objective function over the actual reduction of the model function:

$$\rho_k = \frac{f(\boldsymbol{x}_k) - f(\boldsymbol{x}_k+\boldsymbol{p}_k)}{m_k(\boldsymbol{0}) - m_k(\boldsymbol{p}_k)} \qquad [6.7]$$

If ρ_k is close to 1, indicating a good agreement between the model m_k and the objective f, it is safe to expand the trust region for the next iteration. If ρ_k is positive but significantly smaller than 1, the radius of the trust region will remain the same; if ρ_k is close to zero or negative, the trust region will shrink. Also, the design point will be updated as $\boldsymbol{x}_{k+1} = \boldsymbol{x}_k + \boldsymbol{p}_k$ if ρ_k is large enough.

The subproblem (Eq. [6.6]) is a constrained optimization problem, but it has some unique features that make it easier to solve. A straightforward method that is related to the idea of SD is to find $\boldsymbol{p}_k$ along the direction of $-\boldsymbol{g}_k$, called the Cauchy point. More detailed discussion on the trust region method is beyond the scope of this chapter but can be found in Nocedal and Wright (2006).

The trust region method has shown the power of using the quadratic model function in the optimization of general nonlinear objective functions. For general nonlinear programming problems with nonlinear objective functions and nonlinear equality and inequality constraints, the logic of the SQP method is inductive: at iteration k, an approximated quadratic programming is proposed as:

$$\min_{p} f_k + \nabla f_k^{\mathrm{T}} \boldsymbol{p} + \frac{1}{2} \boldsymbol{p}^{\mathrm{T}} \nabla_{xx}^2 \mathcal{L}_k \boldsymbol{p}$$

$$s.t.\ \ \nabla c_i(\boldsymbol{x}_k)^{\mathrm{T}} \boldsymbol{p} + c_i(\boldsymbol{x}_k) = 0, \quad i \in \mathcal{E} \tag{6.8}$$

$$\nabla c_i(\boldsymbol{x}_k)^{\mathrm{T}} \boldsymbol{p} + c_i(\boldsymbol{x}_k) \geq 0, \quad i \in \mathcal{I}$$

where $\nabla_{xx}^2 \mathcal{L}_k$ is the matrix of second-order partial derivatives of the Lagrangian $\mathcal{L}(\boldsymbol{x}, \lambda) = f(\boldsymbol{x}) - \sum_{i \in \mathcal{I} \cup \mathcal{E}} \lambda_i c_i(\boldsymbol{x})$ with respect to $\boldsymbol{x}$; equality (index set $\mathcal{E}$) or inequality (index set $\mathcal{I}$) constraint $c_i(\boldsymbol{x})$ is of the generic form. Note that in Eq. [6.8] the constraints are linearized. As in the trust region method, the $\nabla_{xx}^2 \mathcal{L}_k$ can be approximated by a symmetric matrix $\boldsymbol{B}_k$.

The solution of Eq. [6.8] can be obtained by the active-set or interior-point methods, which cannot be further explained due to space limitations. The new iterate at $k + 1$ is given by $(\boldsymbol{x}_k + \boldsymbol{p}_k, \lambda_{k+1})$, where $\boldsymbol{p}_k$ and λ_{k+1} are the solution and Lagrange multiplier of Eq. [6.8]. The solution of Eq. [6.8] can further be modified using linear search technique as $(\boldsymbol{x}_k + \alpha_k \boldsymbol{p}_k, \lambda_k + \alpha_k \boldsymbol{p}_\lambda)$, since Eq. [6.8] is an approximated problem for $\boldsymbol{p}_k$. Methods of a trust region type have also been developed to solve Eq. [6.8].

Solving constrained mathematical programming problems is a challenging and fascinating task that is still under development. Other well-documented methods, like penalty and augmented Lagrangian methods and interior-point methods, can be found in Nocedal and Wright (2006) and Luenberger and Ye (2008).

6.2.2 Stochastic search algorithms

Stochastic search algorithms are designed for problems with inherent random noise or deterministic problems solved by injected randomness. In structural optimization, these are problems with uncertainties of design variables or those where adding random perturbation to deterministic design variables is the method to perform the search (Leng, 2015). The search favors designs with better performance. An important feature of stochastic search algorithms is that they can carry out broad search of the design space and thus avoid local optima. Also, stochastic search algorithms do not require gradients to guide the search, making them a good fit for discrete problems. However, there is no necessary condition for an optimum solution and the algorithm must run multiple times to make sure the attained solutions are robust. To handle constraints, penalties can also be applied on designs that violate constraints. For constraints that are difficult to be formulated explicitly, a true/false check is straightforward to implement. Randomly perturbed designs are checked against constraints, and only those passing the check will enter the stage of performance evaluation. Stochastic search can be applied on one design or on a population of them

(Leng, 2015), using for example SA or GA, respectively. Arora (2004) depicts the logic of SA and GA for convenience of application. A monograph devoted to stochastic search and optimization (Spall, 2003) provides further details on a broad scope, including mathematical theory, algorithm design, and applications in simulation and control.

SA is a mimic of the natural process of annealing in metallurgy (Kirkpatrick et al., 1983). The algorithm performs iteratively; the code generates a new candidate design by randomly perturbing the variables of the current elite design that performs best. A unique feature of SA is that its "hill-climbing" property allows inferior designs to be accepted in place of elite ones to expand the search space and prevent the algorithm from becoming trapped in a low-quality local minimum (Leng, 2015). The probability that a suboptimal design is accepted is a function of the magnitude of performance loss and a user-selected parameter. This parameter is tightened as the optimization progresses, reducing the probability of accepting suboptimal designs (Leng, 2015). Two influential parameters of the algorithm are initial "temperature" T_0 and the rate at which this temperature is reduced, referred to as the "cooling rate" r (Leng, 2015). The reduction of T occurs when a certain number ($k_{\max}$) of qualified designs has been evaluated. SA terminates after the temperature has been reduced $m_{\max}$ times. The product of $k_{\max}$ and $m_{\max}$ is the maximum number of objective function evaluations, commonly used as an indicator of algorithm efficiency (Leng, 2015). Convergence is said to occur if the elite design does not change over a large number of iterations ($n_{\max}$). See Spall (2003) and Arora (2004) for more detailed discussions of SA.

As an example of using SA in CFS member optimization, Fig. 6.1 shows a flow chart from Leng et al. (2011). Their objective was to maximize the axial capacity P_n of a CFS column by updating its cross-section shape. The only feasibility constraint is the openness of the cross-section. As shown in the figure, if the new design is feasible and offers a higher $P_{\rm n}$, it becomes the elite design. The constraint of open cross-sections is realized by the true/false check on perturbed designs (ie, perturbed designs with overlapping will be rejected). The flow chart is also informative on general use of SA for other objective functions and constraints.

GAs are popular stochastic search algorithms based on the idea of Darwin's evolution theory (Holland, 1975; Golberg, 1989). Rather than operating on a single design and its perturbation, as in SA, GA operates on a population of designs (Leng, 2015). The designs are then analyzed and ranked according to their objective function performance (fitness). The generation of a new design population includes random selection of two designs (parents) and random exchange of a portion of properties of them (reproduction). Occasionally, a design is also randomly perturbed (mutation). This process is repeated until an entire new population (children) is formed (Leng, 2015). Designs with higher fitness have a higher probability of being selected as parents, and thus the performance of the population as a whole should improve as the optimization progresses (Leng, 2015). Similar to other stochastic search algorithms, GA terminates if either a maximum number of iterations is achieved ($k_{\max}$), or convergence is detected. Convergence is said to occur if the elite design does not change over a large number of iterations ($n_{\max}$) (Leng, 2015). Further development of GA under the

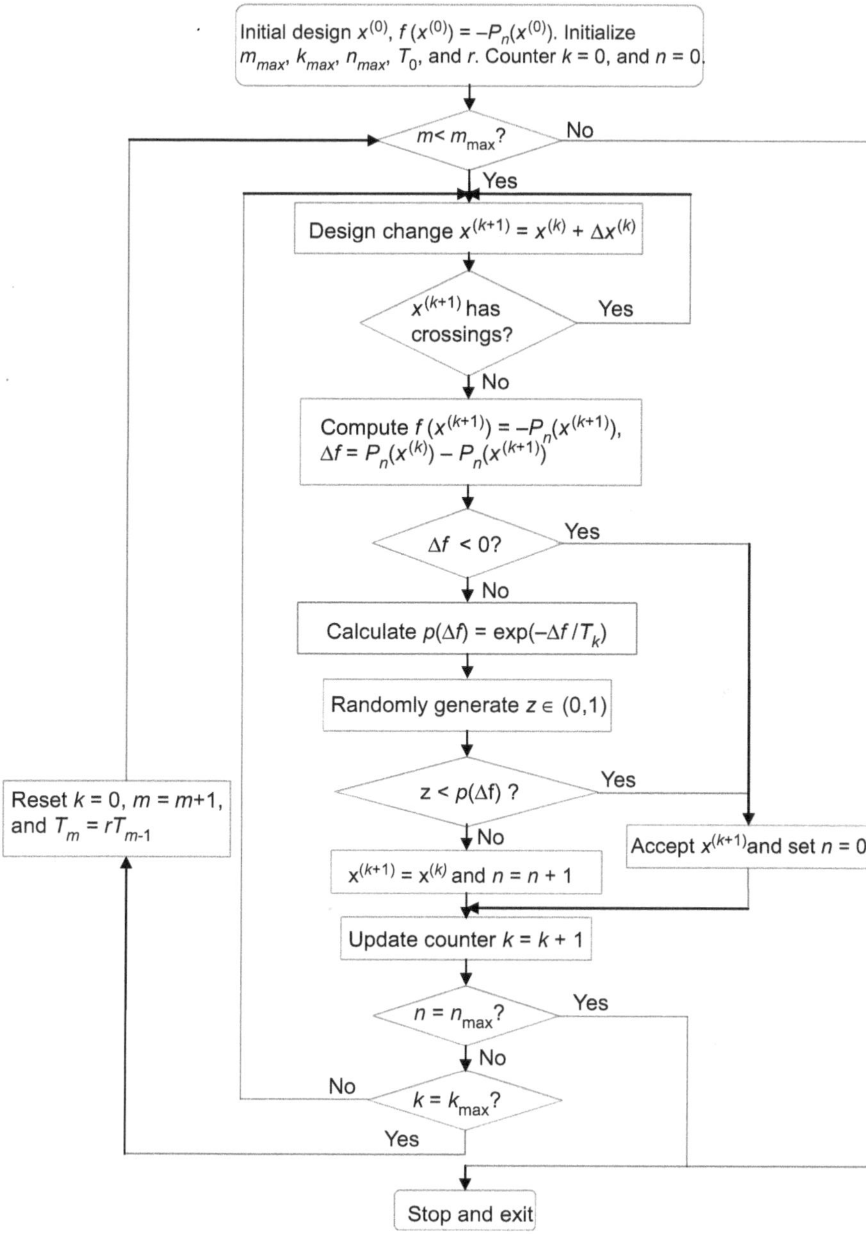

Figure 6.1 Flow chart of SA algorithm for unconstrained shape optimization of CFS columns. Reprinted from Leng J., Guest J.K., Schafer B.W., 2011. Shape optimization of cold-formed steel columns. Thin-walled Structures 49 (12), 1492–1503.

general category of evolution computation and specific details on its application are available in Spall (2003).

A summary of GA in unconstrained optimization of CFS columns by Leng et al. (2011) is shown as a flow chart in Fig. 6.2. Like the flow chart in Fig. 6.1, the objective again is to maximize the column's axial capacity. In the formulation, the vector $\boldsymbol{x}$ of a given design is rounded to a user-specified precision and converted into a binary string. This is a straightforward process and facilitates the exchange of information between designs. Parent selection is based on the roulette wheel algorithm, and single-point cross-over is used to exchange information between two parents. To handle the constraint on overlapping, the penalty method is used. If element crossing is detected in a new design, the computed strength is penalized by subtracting a large number. The procedure in the flow chart can be readily applied to other design objectives.

6.2.3 Other algorithms

Other derivative-free algorithms mentioned in the literature on CFS framing optimization include the artificial neural network (ANN) and ACO. They are briefly described on the basis of reference works.

ANN is a popular model in machine learning and data mining, and mimics biological neural networks. Generally, an ANN consists of three elements: an organized topology of interconnected nodes (network), a method of encoding information (learning), and a method of recalling information (retrieving) (Adeli and Park, 1995). The network usually has several layers of nodes. The role of a node is like a neuron: it can receive and send information to other nodes, and even to itself (recurrent connection). The connection from a node to other nodes can be interlayer (between nodes on different layers, forward or backward) or intralayer (between nodes on the same layer). Also, the connection can be excitatory, like an amplifier, or inhibitory, like a suppressor. The cross-link between nodes is able to achieve the goal of passing useful information and suppressing interference.

The computational neural network model developed by Adeli and Karim (1997) for structural optimization first describes the process of solving for an optimal solution as an integral of a dynamic system:

$$\boldsymbol{x} = \int \dot{\boldsymbol{x}}\, dt \qquad [6.9]$$

where $\boldsymbol{x}$ is the vector of design variables. The integration is evaluated numerically by the Euler method from iteration k to $k+1$ for every one of N_{v} design variables:

$$x_i^{k+1} = x_i^k + \int I_{\mathrm{v}i}\, dt \quad i = 1, \cdots, N_{\mathrm{v}} \qquad [6.10]$$

The direction of variable update, $I_{\mathrm{v}i}$, is determined using the ANN. The ANN of Adeli and Karim (1997) has a variable layer of N_{v} nodes and N_{L} (the number of load cases) constraint layers, where each constraint layer has a number of constraint

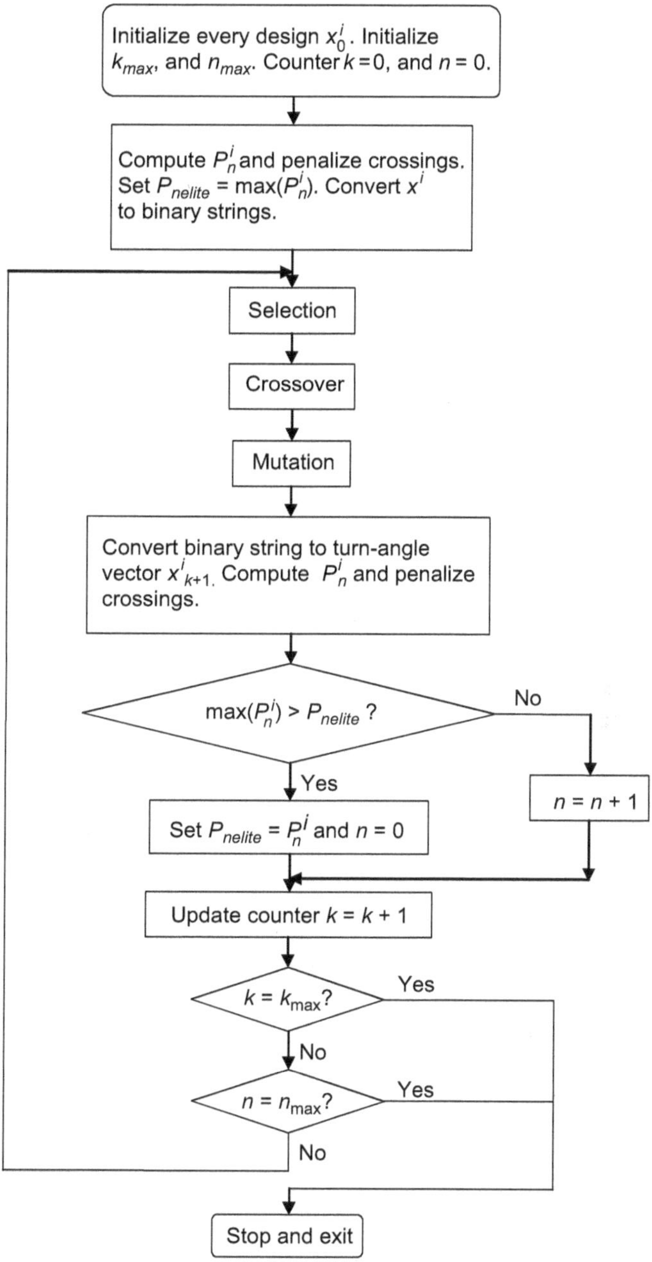

Figure 6.2 Flow chart of GA for unconstrained shape optimization of CFS columns. Reprinted from Leng J., Guest J.K., Schafer B.W., 2011. Shape optimization of cold-formed steel columns. Thin-walled Structures 49 (12), 1492–1503.

nodes corresponding to a specific load case. Weight factors for the connections make use of the gradient information of the objective function and constraints. There is also an information server that provides supportive data like the cross-section shape and formulation of constraints, and runs structural analysis so the iteration can be initialized and proceed. Termination is achieved by checking KKT conditions. Since the ANN is more closely tied to artificial intelligence, most available monographs and research papers, though informative, are beyond the focus of this chapter.

The ACO is another type of metaheuristic algorithm, inspired by the behavior of ants seeking a path between their colony and a source of food. An ant randomly looks for food and returns to the colony once a food source is located; it leave a chemical called pheromone on its trail so that other ants are likely to follow this path and reach the food. However, the pheromone trail evaporates over time. As a result, a shorter pheromone trail will be more frequently visited before it evaporates and thus the pheromone density becomes higher. The idea of ACO mimics this process, so the optimal solution will be found like the identification of the shortest pheromone trail. The logic of ACO is depicted by a pseudocode (Sharafi et al., 2014):

```
procedure ACO
    1 - Set parameters, initialize pheromone trails
    while (termination condition not met) do
      2 - Construct Ants Solutions
      3 - Update Pheromones
    end
end
```

Sharafi et al. (2014) formed the ant colony by placing "simulated ants" at nodes of the grid of feasible paths. The route of an ant depicts a cross-section shape, and the pheromone density and its update are tied to the performance of the member with the generated cross-section.

ACO belongs the general category of swarm intelligence algorithms, among others like particle swarm optimization, artificial bee colony (ABC) algorithm, bat algorithm, etc. For a general discussion on swarm intelligence, see Kennedy et al. (2001).

6.3 Application examples

This section presents optimization results of representative research papers. The results are organized by the scale levels of optimization problems: CFS member optimization results are shown first, followed by examples of CFS framing layout optimization.

6.3.1 *Examples of shape optimization of CFS members*

As discussed, dimension optimization of CFS members requires a designated cross-section shape so that the physical quantities characterizing the shape become design variables. The selection of design variables is dependent on the specific shape, and minimum weight is usually seen as the objective. Optimal designs are identified for several span lengths or load cases, and the sensitivity of the objective with respect to design variables can be predicted. Since the designated cross-section shapes, span lengths, and load cases vary from one publication to another, direct comparison of past results is not feasible.

The results of shape optimization of CFS members are more illustrative, since optimized shapes are identified. Although a rigorous benchmark problem is not yet available, a typical problem investigated by several researchers since Liu et al. (2004) is the search for an optimal cross-section shape with a fixed length of steel sheet so the CFS column's axial capacity is maximized. Leng et al. (2011, 2012, 2013, 2014) tackled this by first using unconstrained shape optimization for more expansive exploration of the design space, and then implementing a series of manufacturability and end-use constraints to identify constrained optimal designs with much improved practicality. This research influenced other scholars whose results are presented in later publications, so it is introduced here as an archetype.

The initial design is a lipped channel section (Fig. 6.3(a)) for unconstrained and constrained shape optimization. However, the discretization of cross-sections and formulation of design variables differ: a cross-section is meshed by equal-width finite strip elements in CUFSM so that the turn angles between them form the design vector of unconstrained optimization (Fig. 6.3(b)), but both strip widths and turn angles are design variables that account for the limit on the number of rollers in constrained optimization (Fig. 6.4).

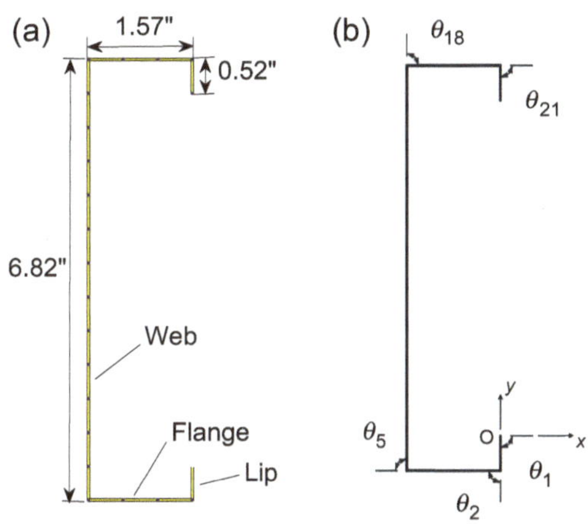

Figure 6.3 Lipped channel section and design variables of unconstrained optimization. (a) Dimensions and mesh of lipped channel section; (b) nonzero turn angles of lipped channel section.

Reprinted from Leng J., Guest J.K., Schafer B.W., 2011. Shape optimization of cold-formed steel columns. Thin-walled Structures 49 (12), 1492–1503.

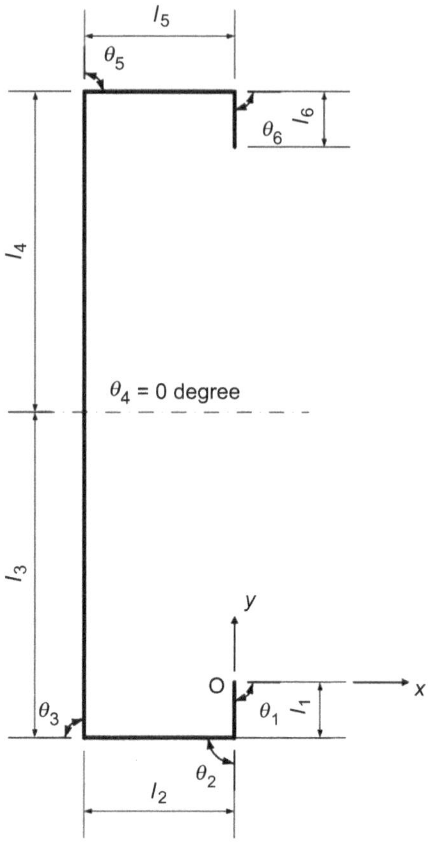

Figure 6.4 Design variables of constrained shape optimization.
Reprinted from Leng J., Li Z., Guest J.K., Schafer B.W., 2014. Shape optimization of cold-formed steel columns with fabrication and geometric end-use constraints. Thin-walled Structures 85, 271–290.

The algorithms for unconstrained shape optimization include gradient-based (SD) and stochastic search algorithms (SA and GA). The intention was to make a thorough search for optimized designs close to global optima, and to compare the performance of algorithms. For the purpose of comparison, SD was also applied at initial cross-sections other than the lipped channel, like hat and sigma sections. SA and GA were performed with 10 replications to check the robustness of solutions. The manufacturability and end-use constraints include symmetry and antisymmetry, parallel flanges and dimension constraints, utility pass-through allowance, open section, and limit on the number of folds (rollers). Most of these constraints cannot be formulated analytically, so the true/false check approach was adopted and coded under the framework of SA for solving constrained optimization. (See Leng (2015) and Leng et al. (2014) for more explanation of constraints and their implementation.) To study the influence of controlling buckling mode on optimization results, unconstrained

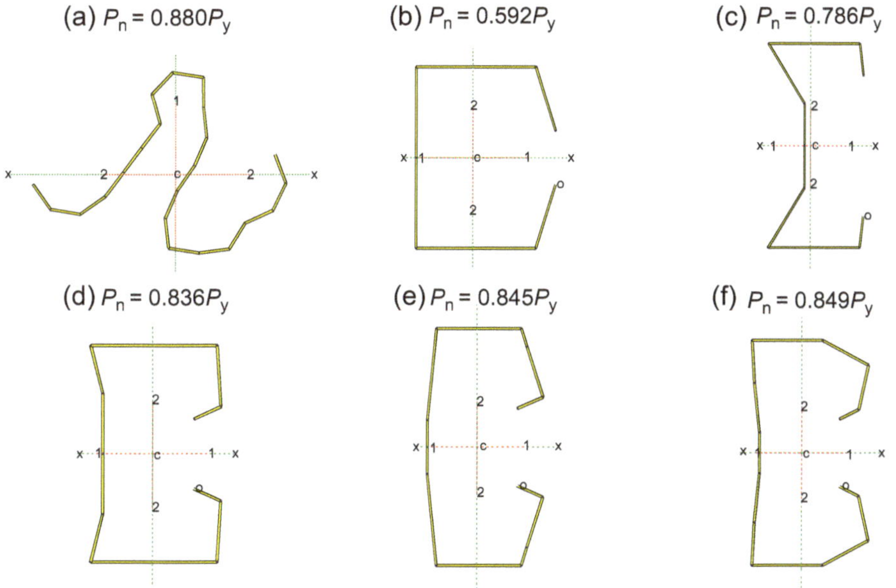

Figure 6.5 Optimized cross-sections for 4 ft (1.22 m) long columns. (a) Unconstrained; (b) constrained, four rollers; (c) constrained, six rollers; (d) constrained, eight rollers; (e) constrained, 10 rollers; (f) constrained, 12 rollers.
Reprinted from Leng J., Guest J.K., Schafer B.W., 2011. Shape optimization of cold-formed steel columns. Thin-walled Structures 49 (12), 1492–1503; Leng J., Li Z., Guest J.K., Schafer B.W., 2014. Shape optimization of cold-formed steel columns with fabrication and geometric end-use constraints. Thin-walled Structures 85, 271–290.

optimization considered 4 ft (1.22 m) and 16 ft (4.88 m) long columns, and a 2 ft (0.61 m) column length is further included in constrained optimization.

The results and conclusions from this work are comprehensive. Firstly, a number of novel cross-sections were identified by unconstrained optimization. The "bobby pin" and squashed "S" sections in Figs. 6.5(a) and 6.6(a) are examples. Secondly, the implementation of manufacturability and end-use constraints successfully drives the search toward much more practical designs (see Figs. 6.5–6.7) that are still able to retain a significant increase in axial capacity (compare the first row with other rows for each part in Table 6.1). Thirdly, constrained optimization results suggest that practical optimal designs can be produced with only marginal increase of manufacturing cost. As shown in Table 6.1, constrained optimized sections can still achieve 50% capacity growth from the initial lipped channel (C) section even with only four rollers. Last but not least, comparison of algorithm performance indicates SD is a local optimizer that is dependent on the initial design; while SA and GA are both capable of searching the design space globally, but SA is preferred for this problem since it runs only on one design instead of a population and saves computational cost (Leng, 2015; Leng et al., 2014).

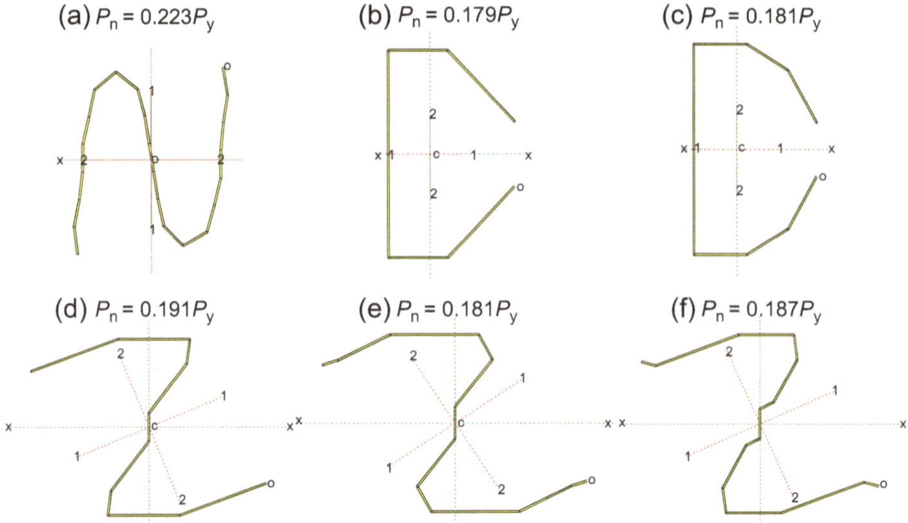

Figure 6.6 Optimized cross-sections for 16 ft (4.88 m) long columns. (a) Unconstrained; (b) constrained, four rollers; (c) constrained, six rollers; (d) constrained, eight rollers; (e) constrained, 10 rollers; (f) constrained, 12 rollers.
Reprinted from Leng J., Guest J.K., Schafer B.W., 2011. Shape optimization of cold-formed steel columns. Thin-walled Structures 49 (12), 1492–1503; Leng J., Li Z., Guest J.K., Schafer B.W., 2014. Shape optimization of cold-formed steel columns with fabrication and geometric end-use constraints. Thin-walled Structures 85, 271–290.

6.3.2 Examples of system-level optimization of CFS framing structures

There have not been many examples so far in the field of layout optimization of CFS framed structures. However, the work by Saka et al. (2015) is noticeable. Their design objective was to optimize the selection of CFS framing members from the list of C-sections in the AISI Design Manual (AISI, 2008) such that the design satisfies a series of constraints and the weight (mass) of the frame is minimized. Constraints from a frame design perspective are taken into account, including serviceability constraints related to story deflection/drift, strengths constraints of beam-columns (tension/compression coupling with bending), and allowable slenderness ratio constraints. Constraints also include requirements on the geometry of C-sections so that connections between consecutive columns or beams and columns are feasible.

In terms of optimization algorithms, since the problem runs on discrete variables they explored the performance of five more recently developed metaheuristic algorithms: the firefly algorithm (FFA in Table 6.2), cuckoo search algorithm (CSA), ABC algorithm with Levy flight, biogeography-based optimization algorithm (BBO), and teaching-learning-based optimization algorithm (TLBO) (Saka et al.,

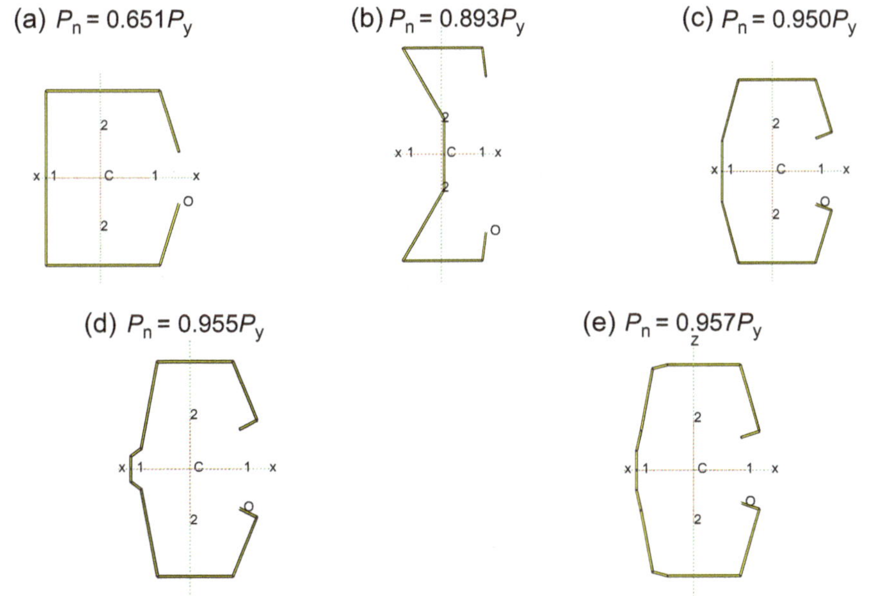

Figure 6.7 Optimized cross-sections for 2 ft (0.61 m) long columns. (a) Constrained, four rollers; (b) constrained, six rollers; (c) constrained, eight rollers; (d) constrained, 10 rollers; (e) constrained, 12 rollers.
Reprinted from Leng J., Li Z., Guest J.K., Schafer B.W., 2014. Shape optimization of cold-formed steel columns with fabrication and geometric end-use constraints. Thin-walled Structures 85, 271–290.

2015). The algorithms share some commonalities with SA, GA, and ACO, introduced above, have some evolutionary features, and do not rely on gradient information to guide the search.

The design example of Saka et al. (2015), a two-story, 1211-member lightweight CFS space frame, is shown in Fig. 6.8. The members are categorized into 14 independent groups, and the C-section types of these groups are taken as the design variables. Three load combinations of dead, live, wind, and snow load are investigated.

The optimum designs determined by these algorithms are listed in Table 6.2 (optimized member types are not shown). The optimum designs are very close to each other. The difference of mass between the lightest (BBO) and the heaviest (CSA) optimum designs is only 1.2% (Saka et al., 2015). Strength constraints seem dominant, as shown in the row for the maximum strength ratio. The table also suggests that displacement and interstory drift constraints are much less than their upper bounds (Saka et al., 2015). The convergence histories are compared in Fig. 6.9. The convergence curves of BBO, TLBO, and CSA drop faster than the others, getting close to their optimum designs earlier. Generally speaking, all five algorithms are capable of solving the example problem, and the results demonstrate the robustness of optimum designs.

Table 6.1 **Design performance comparisons of optimized cross-sections**

Member length	Section	Type of optimization and constraint	No. of rollers	P_n/P_y	Max. no. of P_n evaluation
0.61 m (2 ft)	C	None	20	0.424	1
0.61 m (2 ft)	C	End use and manufacturability	4	0.651	10,000
0.61 m (2 ft)	"Σ"	End use and manufacturability	6	0.893	10,000
0.61 m (2 ft)	"Σ"	End use and manufacturability	8	0.950	10,000
0.61 m (2 ft)	"Σ"	End use and manufacturability	10	0.955	10,000
0.61 m (2 ft)	"Σ"	End use and manufacturability	12	0.957	10,000
1.22 m (4 ft)	C	None	20	0.365	1
1.22 m (4 ft)	"Bobby pin"	Unconstrained	20	0.880	2600
1.22 m (4 ft)	C	End use and manufacturability	4	0.592	10,000
1.22 m (4 ft)	"Σ"	End use and manufacturability	6	0.786	10,000
1.22 m (4 ft)	"Σ"	End use and manufacturability	8	0.836	10,000
1.22 m (4 ft)	"Σ"	End use and manufacturability	10	0.845	10,000
1.22 m (4 ft)	"Σ"	End use and manufacturability	12	0.849	10,000
4.88 m (16 ft)	C	None	20	0.072	1
4.88 m (16 ft)	Squashed "S"	Unconstrained	20	0.223	10,000
4.88 m (16 ft)	C	End use and manufacturability	4	0.179	10,000
4.88 m (16 ft)	C	End use and manufacturability	6	0.181	10,000
4.88 m (16 ft)	Squashed "S"	End use and manufacturability	8	0.191	10,000
4.88 m (16 ft)	Squashed "S"	End use and manufacturability	10	0.181	10,000
4.88 m (16 ft)	Squashed "S"	End use and manufacturability	12	0.187	10,000

Modified from Leng J., Li Z., Guest J.K., Schafer B.W., 2014. Shape optimization of cold-formed steel columns with fabrication and geometric end-use constraints. Thin-walled Structures 85, 271−290.

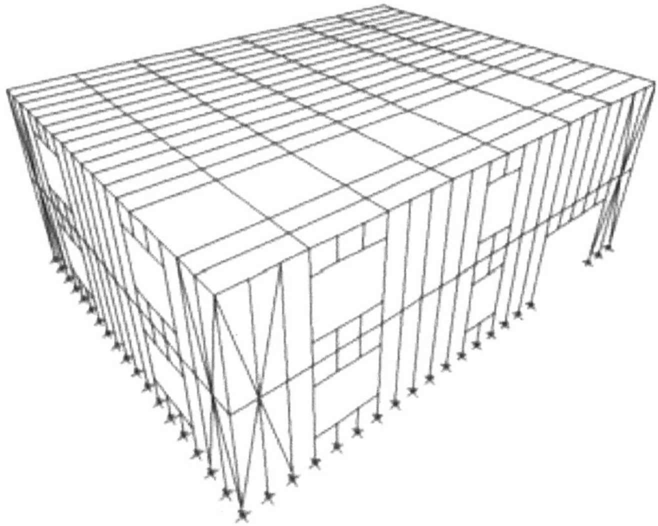

Figure 6.8 Three-dimensional view of the 1211-member lightweight CFS frame from front and left shots.
Reprinted from Saka M.P., Carbas S., Aydogdu I., Akin A., Geem Z.W., 2015. Comparative study on recent metaheuristic algorithms in design optimization of cold-formed steel structures. In: Lagaros N.D., Papadrakakis M. (Eds.), Engineering and Applied Sciences Optimization. Cham, Switzerland: Springer International Publishing, 145–173.

Table 6.2 **Optimum design results of 1211-member lightweight steel frame**

Algorithm	BBO	TLBO	FFA	ABC	CSA
Min. mass (kg)	5464.05	5473.53	5502.61	5522.91	5529.74
Max. top-story drift (mm)	9.158	7.885	8.774	7.284	7.568
Max. interstory drift (mm)	1.759	2.433	2.022	2.761	1.882
Max. deflection (mm)	0.199	0.247	0.198	0.336	0.239
Max. strength ratio	0.938	0.998	0.937	0.999	0.938
Max. number of iterations	20,000	20,000	20,000	20,000	20,000

Excerpted from Saka M.P., Carbas S., Aydogdu I., Akin A., Geem Z.W., 2015. Comparative study on recent metaheuristic algorithms in design optimization of cold-formed steel structures. In: Lagaros N.D., Papadrakakis M. (Eds.), Engineering and Applied Sciences Optimization. Cham, Switzerland: Springer International Publishing, 145–173.

6.4 Future research directions

Despite the achieved progress with the dimension and shape optimization of CFS members and the layout optimization of CFS systems, more work can still be accomplished in this area to improve the efficiency of the design of CFS framed structures

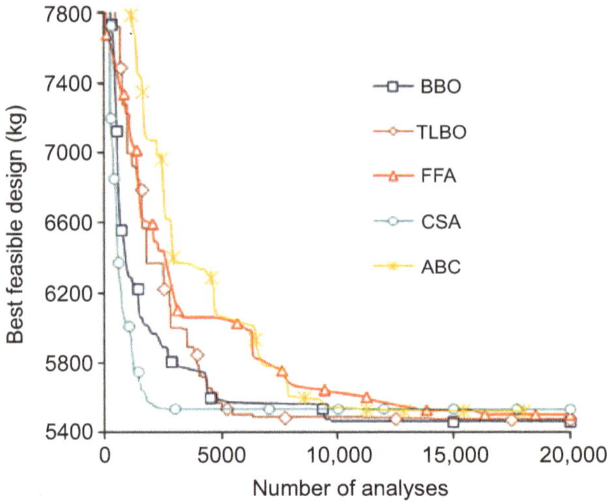

Figure 6.9 Search histories of 1211-member lightweight CFS frame.
Reprinted from Saka M.P., Carbas S., Aydogdu I., Akin A., Geem Z.W., 2015. Comparative study on recent metaheuristic algorithms in design optimization of cold-formed steel structures. In: Lagaros N.D., Papadrakakis M. (Eds.), Engineering and Applied Sciences Optimization. Cham, Switzerland: Springer International Publishing, 145–173.

further. Three promising research directions are proposed, to the best of the author's knowledge.

Firstly, the shape optimization of CFS members *per se* can be further improved. Past research has compared the performance of various optimization algorithms, but the shape generator itself also plays an important role since it dictates the form of design variables and determines the scope of the design space. It will be interesting to explore other tools and algorithms for cross-section shape generation and compare their capability with past findings. Also, the construction industry will benefit greatly if the more practical optimized cross-section shapes are developed into standard families of commercial products. The task may involve biobjective optimization for bending and axial loads, and sensitivity of the discovered shapes should be studied to identify and standardize important geometric features.

Secondly, the layout and configuration optimization of CFS systems should proceed, and member optimization needs to be integrated into the package. The layout optimization can be applied on midrise and highrise CFS framings. Furthermore, other load combinations, especially the earthquake load, should be included; the structure's natural period/frequency can also be explicitly considered as a constraint or a design objective. If families of other cross-section shapes, like Z sections, hat sections, and some novel sections, are included in the pool for members in layout optimization, the structure's performance will have further improvement. The long-term goal will be the integration of advanced optimization modules in commercial design software.

Thirdly, topology optimization of CFS framed structures is a virgin territory. Topology optimization is a powerful tool since it can greatly update the design domain, and it has been successfully applied in other engineering fields. The search for even better cross-section shapes will be an interesting example in the two-dimensional domain. The layout optimization of three-dimensional structures will enable members to be removed or introduced with the help of topology optimization. Technical difficulties may still exist regarding the discretization, algorithms, and some *ad hoc* design problems, but the results will be very fruitful for researchers and engineers once we overcome the barriers.

References

Adeli, H., Karim, A., 1997. Neural network model for optimization of cold-formed steel beams. Journal of Structural Engineering 123 (11), 1535–1543.

Adeli, H., Park, H.S., 1995. A neural dynamics model for structural optimization—theory. Computers and Structures 57 (3), 383–390.

American Iron and Steel Institute, 1968. AISI Specification for the Design of Cold-formed Steel Structural Members. American Iron and Steel Institute, Washington, DC.

American Iron and Steel Institute, 1989. AISI Specification for the Design of Cold-formed Steel Structural Members, 1986 Ed. with 1989 Addendum ed. American Iron and Steel Institute, Washington, DC.

American Iron and Steel Institute, 1991. AISI Load and Resistance Factor Design Specification for Cold-formed Steel Structural Members. American Iron and Steel Institute, Washington, DC.

American Iron and Steel Institute, 1996. AISI Specification for the Design of Cold-formed Steel Structural Members. American Iron and Steel Institute, Washington, DC.

American Iron and Steel Institute, 2007a. North American Specification for the Design of Cold-formed Steel Structural Members, 2007 edition. AISI S100–07 ed. American Iron and Steel Institute, Washington, DC.

American Iron and Steel Institute, 2007b. North American Specification for the Design of Cold-formed Steel Structural Members, 2007 edition with supplement 2. AISI S100-07 w/S2-10 ed. American Iron and Steel Institute, Washington, DC.

American Iron and Steel Institute, 2008. Cold-formed Steel Design Manual. D100–08 ed. American Iron and Steel Institute, Washington, DC.

Arora, J.S., 2004. Introduction to Optimum Design, second ed. Academic Press, Waltham, MA.

Bendsøe, M.P., Sigmund, O., 2003. Topology Optimization: Theory, Methods and Applications. Springer Verlag, Berlin, Germany.

British Standards Institution, 1998. Structural Use of Steel in Building: Part 5: Code of Practice for Design of Cold Formed Sections. BS 5950 ed. British Standards Institution, London, UK.

Chamberlain Pravia, Z.M., Kripka, M., 2012. Cross section optimization using simulated annealing of cold-formed steel channel columns. In: Proceedings of the Annual Stability Conference Structural Stability Research Council; Grapevine, TX.

European Convention for Constructional Steelwork, 1996. Eurocode 3: Design of steel structures, Part1-3: General rules for cold formed thin gauge members and sheeting. European Convention for Constructional Steelwork, Brussels, Belgium.

Franco, J.M.S., Duarte, J.P., Batista, E.M., Landesmann, A., 2014. Shape grammar of steel cold-formed sections based on manufacturing rules. Thin-walled Structures 79, 218–232.

Gargari, M., Louhghalam, A., Tootkaboni, M., 2013. Optimal cross sections for cold formed steel members under compression. In: Proceedings of 10th World Congress on Structural and Multidisciplinary Optimization; Orlando, FL.

Gilbert, B.P., Teh, L.H., Guan, H., 2012a. Self-shape optimisation principles: optimisation of section capacity for thin-walled profiles. Thin-walled Structures 60, 194–204.

Gilbert, B.P., Savoyat, T.J.M., Teh, L.H., 2012b. Self-shape optimisation application: optimisation of cold-formed steel columns. Thin-walled Structures 60, 173–184.

Golberg, D.E., 1989. Genetic Algorithms in Search, Optimization, and Machine Learning. Addison-Wesley Publishing Company, Boston, MA.

Holland, J.H., 1975. Adaptation in Natural and Artificial Systems: An Introductory Analysis With Applications to Biology, Control, and Artificial Intelligence. University of Michigan Press, Ann Arbor, MI.

Karim, A., Adeli, H., 1999. Global optimum design of cold-formed steel hat-shape beams. Thin-walled Structures 35 (4), 275–288.

Kennedy, J., Eberhart, R.C., Shi, Y., 2001. Swarm Intelligence. Academic Press, London, UK.

Kirkpatrick, S., Gelatt, C.D., Vecchi, M.P., 1983. Optimization by simulated annealing. Science 220 (4598), 671–680.

Kolcu, F., Ekmekyapar, T., Özakça, M., 2010. Linear buckling optimization and post-buckling behavior of optimized cold formed steel members. Scientific Research and Essays 5 (14), 1916–1924.

Kripka, M., Martin, Z., 2013. Cold-formed steel channel columns optimization with simulated annealing method. Structural Engineering and Mechanics 48 (3), 383–394.

Lee, J., Kim, S., Park, H., Woo, B., 2005. Optimum design of cold-formed steel channel beams using micro genetic algorithm. Engineering Structures 27 (1), 17–24.

Lee, J., Kim, S., Seon Park, H., 2006. Optimum design of cold-formed steel columns by using micro genetic algorithms. Thin-walled Structures 44 (9), 952–960.

Leng, J., Guest, J.K., Schafer, B.W., 2011. Shape optimization of cold-formed steel columns. Thin-walled Structures 49 (12), 1492–1503.

Leng, J., Li, Z., Guest, J.K., Schafer, B.W., 2012. Constrained shape optimization of cold-formed steel columns. In: Proceedings of 21st International Specialty Conference on Cold-formed Steel Structures; St. Louis, MO.

Leng, J., Li, Z., Guest, J.K., Schafer, B.W., 2013. Shape optimization of cold-formed steel columns with manufacturing constraints and limited number of rollers. In: Proceedings of the Annual Stability Conference-structural Stability Research Council; St. Louis, MO.

Leng, J., Li, Z., Guest, J.K., Schafer, B.W., 2014. Shape optimization of cold-formed steel columns with fabrication and geometric end-use constraints. Thin-walled Structures 85, 271–290.

Leng, J., 2015. Simulation of Cold-formed Steel Structures. Johns Hopkins University, Baltimore, MD.

Liu, H., Igusa, T., Schafer, B., 2004. Knowledge-based global optimization of cold-formed steel columns. Thin-walled Structures 42 (6), 785–801.

Lu, W., 2003. Optimum Design of Cold-formed Steel Purlins Using Genetic Algorithms. Helsinki University of Technology, Helsinki, Finland.

Luenberger, D.G., Ye, Y., 2008. Linear and Nonlinear Programming, third ed. Springer Science & Business Media, New York, NY.

Luenberger, D.G., 1969. Optimization by Vector Space Methods. John Wiley & Sons, New York, NY.

Magnucki, K., Paczos, P., 2009. Theoretical shape optimization of cold-formed thin-walled channel beams with drop flanges in pure bending. Journal of Constructional Steel Research 65 (8), 1731–1737.

Moharrami, M., Louhghalam, A., Tootkaboni, M., 2014. Optimal folding of cold formed steel cross sections under compression. Thin-walled Structures 76, 145–156.

Nocedal, J., Wright, S., 2006. Numerical Optimization, second ed. Springer, New York.
Saka, M.P., Carbas, S., Aydogdu, I., Akin, A., Geem, Z.W., 2015. Comparative study on recent metaheuristic algorithms in design optimization of cold-formed steel structures. In: Lagaros, N.D., Papadrakakis, M. (Eds.), Engineering and Applied Sciences Optimization. Springer International Publishing, Cham, Switzerland, pp. 145–173.
Seaburg, P.A., Salmon, C.G., 1971. Minimum weight design of light gage steel members. Journal of the Structural Division 97 (1), 203–222.
Sharafi, P., Teh, L.H., Hadi, M.N.S., 2014. Shape optimization of thin-walled steel sections using graph theory and ACO algorithm. Journal of Constructional Steel Research 101, 331–341.
Spall, J.C., 2003. Introduction to Stochastic Search and Optimization: Estimation, Simulation, and Control. John Wiley and Sons, Hoboken, NJ.
Tashakori, A., Adeli, H., 2002. Optimum design of cold-formed steel space structures using neural dynamics model. Journal of Constructional Steel Research 58 (12), 1545–1566.
Tian, Y., Lu, T., 2004. Minimum weight of cold-formed steel sections under compression. Thin-walled Structures 42 (4), 515–532.
Tran, T., Li, L., 2006. Global optimization of cold-formed steel channel sections. Thin-walled Structures 44 (4), 399–406.
Wang, B., Gilbert, B.P., Molinier, A.M., Guan, H., Teh, L.H., 2014a. Shape optimisation of cold-formed steel profiles with manufacturing constraints - Part I: Algorithm. In: Proceedings of 22nd International Specialty Conference on Cold-formed Steel Structures; St. Louis, MO.
Wang, B., Gilbert, B.P., Molinier, A.M., Guan, H., Teh, L.H., 2014b. Shape optimisation of cold-formed steel profiles with manufacturing constraints - Part II: Applications. In: Proceedings of 22nd International Specialty Conference on Cold-formed Steel Structures; St. Louis, MO.

Part Three

The structural performance of cold-formed steel buildings

Thermal performance of cold-formed steel structures in fire

7

J.C. Batista Abreu
Bucknell University, Lewisburg, Pennsylvania, United States

7.1 Introduction

In general, current fire protection for cold-formed steel (CFS) buildings is provided through prescriptive design because, among other reasons presented in this chapter, the thermal performance of CFS members, subsystems, and systems is not well understood. Standardized tests are used to assess the behavior of wall and floor assemblies under predetermined conditions, using a prescribed time—temperature curve that is unlikely to represent the duration and severity of a real fire these days. Results from standard fire tests are evaluated in terms of failure criteria, and then categorized according to the amount of time the assembly was able to withstand fire action. Design solutions are thus limited, and fully tied to test results.

Alternative engineering-based analysis and design are evolving, with the objective of providing more freedom and efficiency in design solutions. Recent research explores the necessity for new and the feasibility of current methodologies extrapolated to structural fire engineering. Rationalizing fire resistance beyond prescriptive testing is a challenging endeavor. The general vision seeks to enable structural fire engineering of CFS structures by codified standards, completed by computational analysis, allowing engineers to design complete structures rationally and optimize occupant protection and structural reliability, including under unconventional multihazard scenarios such as fire following earthquake or blast.

This chapter offers an overview on new research findings, including the development of numerical models for heat transfer and thermomechanical analyses, and the use of analysis tools and design methods for fire design. A major portion of the chapter is dedicated to studies on the thermal and mechanical properties of CFS, as a fundamental building block for understanding fire response of CFS systems. The chapter ends with two design examples related to the strength of CFS members under uniform and nonuniform elevated temperatures.

7.2 Numerical models of heat transfer

After the fire demand on a building is approximated, heat transfer analysis is used to estimate the temperature field on structural and nonstructural components. A typical scenario of study includes a compartment containing a localized fire. The temperatures

Recent Trends in Cold-Formed Steel Construction. http://dx.doi.org/10.1016/B978-0-08-100160-8.00007-4

inside the compartment starts increasing after ignition, potentially affecting the performance of the entire structure. Ideally, the compartment should be able to contain the fire, avoiding the spread of smoke and flames to other compartments. Also, structural instability and significant increase of temperature in adjacent rooms need to be avoided. These integrity, stability, and insulation criteria are intended to be satisfied through the fire protection system.

In CFS buildings, walls and floors are essential components designed to guarantee fire protection for (at least) a certain amount of time. A common assembly is constructed with CFS framing members (both studs and tracks) and sheathing material (usually fire-rated gypsum boards).

The heat transfer through a CFS assembly is complex. For instance, when a wall is exposed to fire on one side, conduction occurs through the exposed gypsum boards, screws, and then CFS members. Inside the cavity, radiation and convection cause the temperature on the unexposed board to increase. Conduction then occurs through the unexposed board, and radiation and convection will govern the temperatures on the unexposed surface. In addition, gypsum boards experience dehydration when heated, and change their thermomechanical properties. Moisture then migrates inside and outside the wall cavity, creating thermal gradients. The paper sheets enclosing the gypsum core burn, facilitating cracking and ablation processes, consequently leading to fall-off of the boards. Also, CFS studs experience thermal bowing toward the heat source, potentially creating joint openings between gypsum boards. Cracks and joint openings may allow the passage of smoke and flames to other compartments, compromising the fire safety of the structure and its occupants.

Several thermal response models have been proposed to analyze the case of CFS walls exposed to fire on one side (Sultan, 1996; Alfawakhiri, 2001; Shahbazian and Wang, 2013; Chen et al., 2014). In general, these models adopt multiple assumptions to scale the complex heat-transfer problem to one dimension. Hence the presence of the CFS frame and its thermomechanical response are ignored; and moisture migration, gypsum shrinkage, cracking, and thermal ablation processes are indirectly considered by calibrating the material properties and model geometry (eg, specific heat, thermal conductivity, and thickness of gypsum boards). The advantage of these one-dimensional methods is their efficiency compared to costly computational fluid dynamics simulations.

Heat-transfer analysis requires accurate modeling of material properties; however, limited data are available in the literature. For instance, large discrepancies are observed in the temperature-dependent thermal expansion/contraction coefficient of gypsum, generally attributed to differences in chemical composition of samples, differences in test methods, and variability associated with the orientation of the sample (eg, short and long directions of the board).

Validation of thermal response models relies on comparisons against temperature measurements during standard fire tests. Models can be evaluated when comparing time–temperature curves from heat-transfer analysis and limited numbers of thermocouple measurements obtained in a few locations on wall specimens during testing.

Recent trends include the use of thermographic cameras to collect information over the entire unexposed surfaces of assemblies during fire tests (Lázaro et al., 2015). This technology allows broader data acquisition in real time, useful for extensive validation of numerical simulations.

Generally, the heat-transfer analysis is followed by structural analysis based on the temperature distribution from the former. Normally, the thermal and mechanical analyses are sequentially coupled, instead of fully coupled. It implies that the structural response does not have an impact on the heat transfer. For instance, the thermal bowing of CFS studs leading to joint opening between gypsum boards and subsequent passage of hot gases does not affect the heat transfer. For this reason, these sequentially coupled models may exhibit adequate accuracy as long as the mechanical behavior has not taken a leading role in the overall response of the system. Finite element packages previously validated for thermal analysis of wall systems include ANSYS (2015), ABAQUS (2013), SAFIR (2003), and TASEF (Sterner and Wickstrom, 1900).

In summary, two main challenges constrain the improvement of current engineering-based analysis: understanding temperature-dependent material behavior, and coupling thermal and mechanical responses.

7.3 Experimental studies on thermomechanical performance of materials

Thermomechanical properties of CFS and gypsum boards are necessary for the understanding of the performance of light-gauge systems in a fire. However, significant variability in limited material testing results at elevated temperatures is observed.

7.3.1 Thermal properties

Studies suggest that thermal properties of gypsum boards vary according to the chemical formulation (Mehaffey et al., 1994); however, differences might not be significant for similar materials (Park et al., 2009). Thermal properties of gypsum are available in tabular form or through sets of linear equations (Chen et al., 2014; Thomas, 2002; Sultan, 1996). Commonly, these data are calibrated and manipulated for use in one-dimensional thermal response models. Material models assume the volume of gypsum remains constant with increasing temperature; thus the mass loss is directly proportional to the density reduction. Material densities slightly decrease with increasing temperature. At 1000°C gypsum and steel densities decay to about 80% and 96% of their values at ambient conditions, respectively.

In thermal response models, moisture migration and crack development processes in gypsum boards are indirectly considered by exaggerating the first peak in the specific heat, and increasing the thermal conductivity at high temperatures (Fig. 7.1). In general, thermal properties of steel are standardized for analysis purposes (Fig. 7.2).

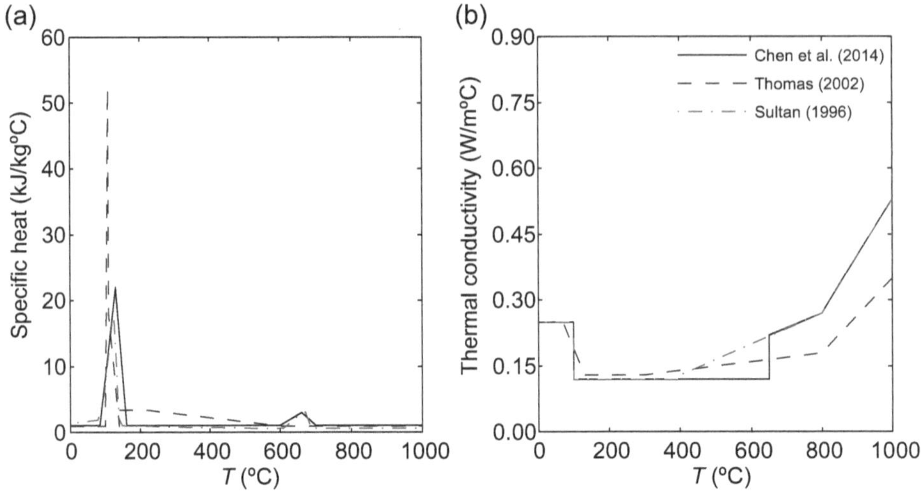

Figure 7.1 Thermal properties of gypsum at elevated temperatures.

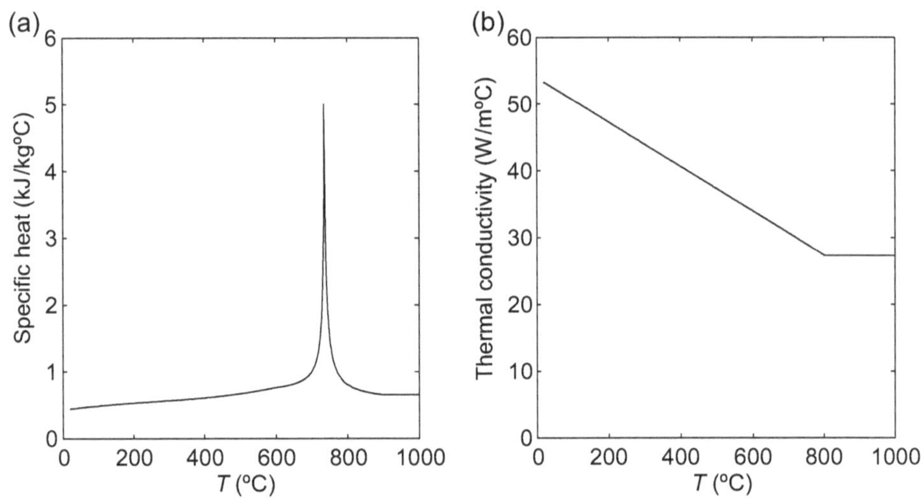

Figure 7.2 Thermal properties of steel at elevated temperatures.
From EN 1993-1-2 (2005).

7.3.2 Mechanical properties

Increasing temperatures alter the mechanical response of CFS systems by (generally) degrading the strength and stiffness of structural and nonstructural components. Therefore, understanding the mechanical properties of building material is essential.

Mechanical properties are typically obtained through steady-state or transient-state tension tests. During steady-state tests, flat CFS specimens are heated up from ambient temperature (Ta) to a target temperature (Tt) at a predetermined heating rate (usually around 10°C/min), allowing free thermal expansion (Fig. 7.3(a)). Then the specimen is kept at a constant temperature for about 20 min to guarantee uniform temperature distribution. Next, tensile load is applied (usually adopting a displacement-controlled test with displacement rates around 0.25 mm/min) until fracture occurs (Fig. 7.3(b)). Experimental results include the measured force and extension within the gauge length of the specimen. Strain—stress curves can then be constructed at different temperatures. This method is simple and quite independent of the heating rate. However, the procedure adopted (ie, loading after heating) is opposite to the process observed during a fire, when the structure is loaded under normal conditions and then heated.

In a transient-state test the specimen is statically subjected to a target stress by axially loading it at about 3 kN/min (Fig. 7.4(a)). Next, temperature is gradually increased at a constant rate (eg, 10°C/min), while keeping constant the target stress until a failure criterion is met (Fig. 7.4(b)). Extensions within the gauge length and material temperatures are monitored during the test. Strain—temperature curves at a constant stress can be constructed, and then used to build stress—strain curves at a

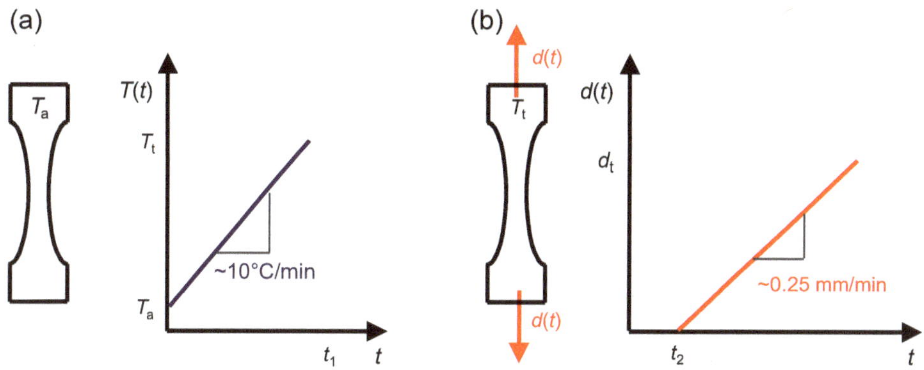

Figure 7.3 Steady-state test: (a) heating; (b) loading.

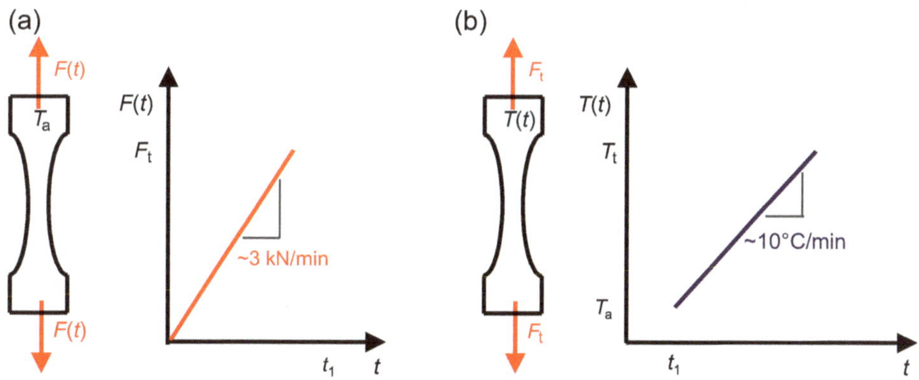

Figure 7.4 Transient-state test: (a) loading; (b) heating.

certain temperature. This method reflects the condition of a structure that is loaded before being heated during a fire. However, a larger number of experiments is needed to build accurate stress–strain curves when using a transient-state test. Many experiments at small stress levels are necessary to estimate precisely the elastic modulus of the material. Also, keeping a constant stress during experimentation can be challenging. Furthermore, the heating rate should be slow enough to allow material damage and capture accurate strain measures. Additional challenges are to measure real-time temperatures precisely, and ensure uniform temperature distribution throughout the specimen.

Both steady-state and transient-state tests have advantages and limitations. So which method is better to characterize material properties to be included in design codes and standards? This is a question currently debated among the research community and industry.

Recent trends include the use of nondestructive tests to measure material properties at elevated temperatures. For instance, Ahmadi et al. (2013) adopted a sonic resonance method to estimate the elastic and shear moduli of CFS up to 351°C. Because of the nature of the test, full stress–strain curves are not developed. Potentially, nondestructive tests could be use for postfire evaluation of structures to estimate remaining load-carrying capacity.

Stress–strain curves change with increasing temperature. CFS shows a clear yield plateau at ambient conditions (Fig. 7.5). However, this characteristic is lost at elevated temperatures, thus the stress–strain curves do not show a clear yield stress (eg, at 400°C), and yield and ultimate stresses exhibit more similar magnitudes (eg, at 600°C). Several constitutive models have been proposed, generally based on modified Ramberg–Osgood equations (Ramberg and Osgood, 1943).

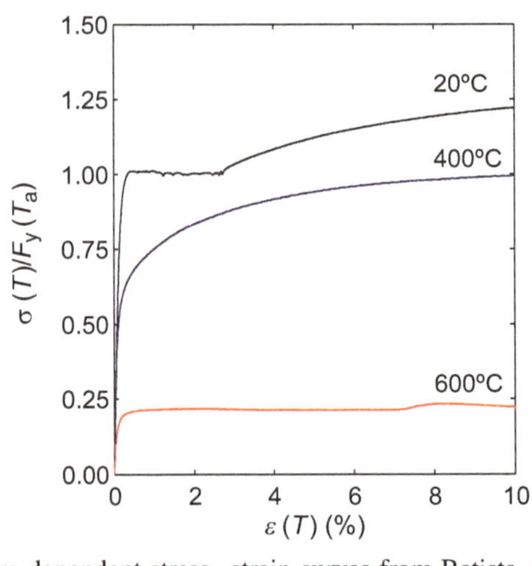

Figure 7.5 Temperature-dependent stress–strain curves from Batista-Abreu (2015) based on steady-state tension tests on CFS with nominal yield stress of 345 MPa at ambient temperature.

Experimental results show the thermal expansion coefficient of CFS increases with rising temperature (Fig. 7.6(a)), and the ultimate strain is temperature-dependent (Fig. 7.6(b)). Mechanical properties (eg, elastic modulus, yield, and ultimate stresses) are obtained from stress–strain curves. The elastic modulus is calculated by fitting the initial slope of the stress–strain curve, while the 0.2% offset method is commonly used to estimate the yield stress. Retention factors are calculated using mechanical properties at elevated temperatures normalized with respect to mechanical properties at ambient conditions (usually 20°C). Retention factors may vary according to the material tested, test method and parameters (eg, heating and loading rates), and data post-processing criteria.

Fig. 7.7(a) shows retention factors for the elastic modulus of CFS from multiple research groups. The overall trend of the data is consistent: the elastic modulus degrades with increasing temperature. However, there is significant scatter in the data. The shear modulus of CFS shown in Fig. 7.7(b) was estimated through nondestructive testing, assuming isotropic material and a constant Poisson's ratio equal to 0.3. Significant scatter is also observed in retention factors for the yield and ultimate stresses of CFS at elevated temperatures (Fig. 7.8).

Several research groups have proposed prediction equations to estimate mechanical properties of CFS at elevated temperatures, based on deterministic models. A recent trend in the structural steel literature, aligned with the growth of performance-based design, is the development of probabilistic models to determine mechanical properties at elevated temperatures. These probabilistic models account for uncertainty, and can be subsequently used to perform probabilistic analysis of structures subjected to fire (Khorasani et al., 2015). Similar efforts have not yet been made in the CFS research field.

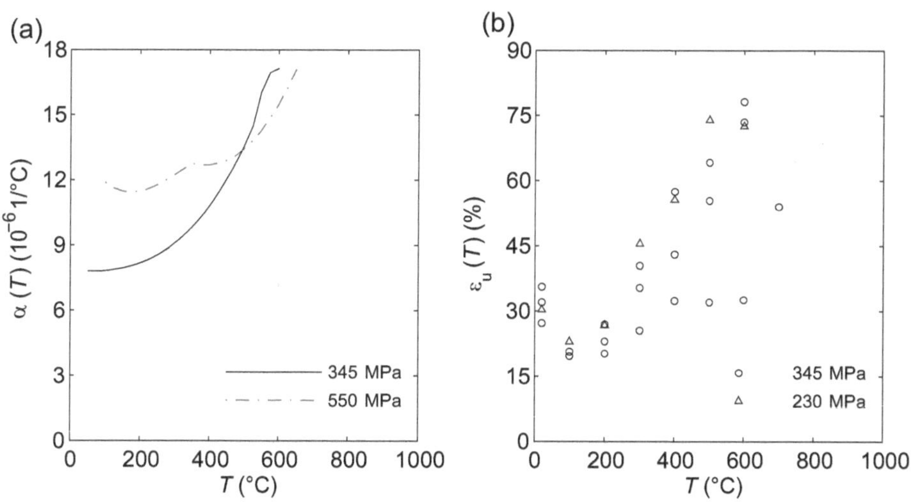

Figure 7.6 CFS temperature-dependent (a) thermal expansion coefficient from Chen and Ye (2012), and Ye and Chen (2013); (b) ultimate strain based on proprietary data.

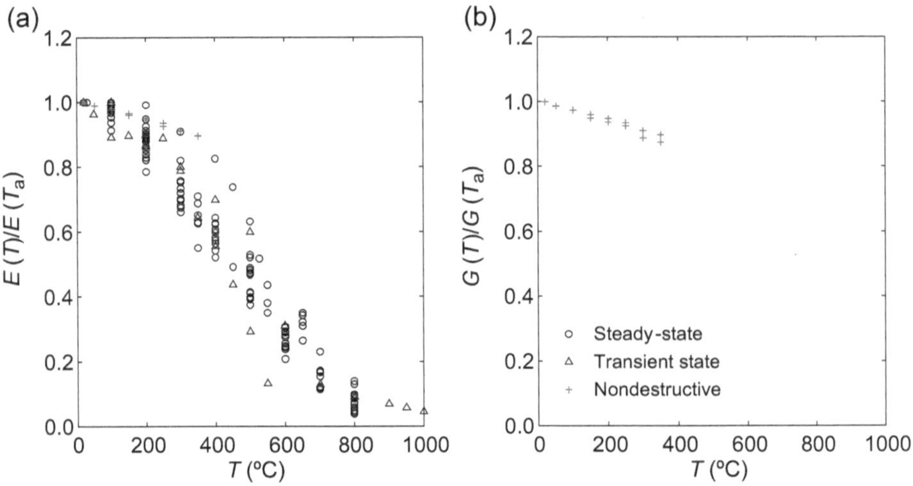

Figure 7.7 Retention factors for (a) elastic modulus and (b) shear modulus of CFS at elevated temperatures.
From Lee et al. (2003), Outinen (2006), Ranawaka and Mahendran (2009), Kankanamge and Mahendran (2011), Chen and Ye (2012), and Ahmadi et al. (2013).

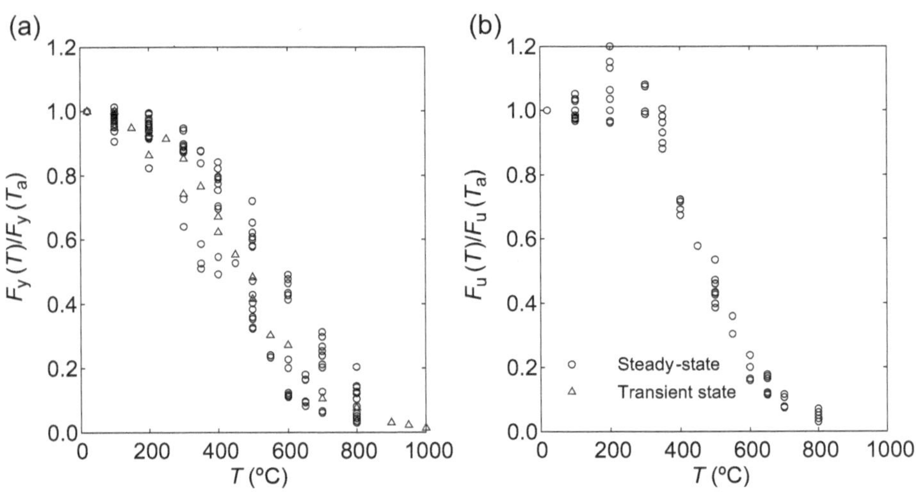

Figure 7.8 Retention factors for (a) yield stress and (b) ultimate stress of CFS at elevated temperatures.
From Lee et al. (2003), Outinen (2006), Ranawaka and Mahendran (2009), Kankanamge and Mahendran (2011), and Chen and Ye (2012).

7.4 Inclusion of thermal effects

In addition to altering mechanical properties, elevated temperatures may induce strain and stresses in structural components, depending on the temperature distribution and boundary conditions. For instance, a CFS member free to deform experiences uniform expansion under uniform temperature increase. However, thermal bowing occurs when a thermal gradient is developed through the cross-section of a member, since the elements with higher temperatures will expand more than the rest.

A simple case is illustrated with a structural member (eg, a pin-ended lipped channel column) subjected to uniform temperature increase. Due to end restraints, the CFS member is not allowed to expand axially during the heating process, thus axial forces are gradually developed. Simultaneously, CFS material is degraded as the elastic modulus and the yield and ultimate stresses decrease. In this scenario, the demand on the member increases as its capacity decreases. At a certain time (or temperature), the demand will reach the capacity of the thin-walled member, leading to failure.

A more complex case involves the same structural member being subjected to thermal gradient through its cross-section. With pinned ends, the member will progressively bow toward the hotter side as the temperature difference between elements increases. Axial forces cause bending moments in the bowed member. In addition, due to differences in temperatures in the cross-section and therefore differences in material properties, the center of resistance shifts away from the hotter side. In summary, CFS columns experiencing thermal gradients through the cross-section develop combined axial force and bending moment due to thermal bowing and the shift of the center of resistance.

Thermal effects in CFS members and systems have been studied experimentally and analytically to formulate appropriate engineering design methods for structural fire design. The next section provides a summary of experimental programs and proposed analysis methods for CFS subjected to elevated temperatures.

7.5 Tests, analysis, and results

Most studies focus on the behavior of CFS structural members or assemblies, and/or the suitability of current design methods for fire engineering applications. The study of CFS structures includes understanding important factors relevant to the nature of thin-walled members, such as the impact of geometric imperfections, residual stresses and strains, and buckling modes and their interactions, and the effects of (end and lateral) restraints. Although most available experimental data come from column (Feng et al., 2003a,b,c; Lee, 2004; Ranawaka, 2006; Shahbazian and Wang, 2011a,b, 2012) and sheathed stud and wall testing (Manzello et al., 2008; Gunalan and Mahendran, 2010; Kolarkar and Mahendran, 2012; Chen et al., 2012; Batista-Abreu et al., 2015), stronger efforts are being dedicated to the behavior of beams

(Lu et al., 2010; Chen and Jin, 2008), floor assemblies (Bénichou et al., 2009; Baleshan and Mahendran, 2010), and screwed and bolted connections (Yan and Young, 2011, 2012a,b; Yu et al., 2008; Lim and Young, 2007).

At ambient conditions, the impact of geometric imperfections on the load-carrying capacity of CFS thin-walled members is significant, and therefore these imperfections have been characterized (Zeinoddini and Schafer, 2012). Consistently, geometric imperfections are considered to affect the stiffness and load-carrying capacity of CFS members at elevated temperatures (Kaitila, 2002). For instance, short columns are more sensitive to local buckling imperfections, whereas global imperfections have a pronounced impact on long columns, as expected (Feng et al., 2004). Recent trends indicate a preference for the use of geometric imperfections based on spectral representations and measured data, in contrast to classical approaches based on superposition of eigenmode imperfections.

Mechanics-based prediction methods to estimate residual stresses and plastic strains based on experimental data are proposed and used in numerical studies of CFS members when corner information is of concern (Moen et al., 2008). At high temperatures the effects of residual stresses and plastic strains are commonly ignored (Lee, 2004), especially due to minor impact of the behavior of slender members (Batista-Abreu and Schafer, 2014).

Limit states of CFS members at ambient conditions may include local, distortional, and global buckling, as well as their interactions. Experimental data from CFS specimens suggest that stronger modal interactions may exist at elevated temperatures, and such interactions vary over time according to the temperature distribution. Modal interactions at elevated temperature have been described qualitatively (Feng et al., 2003a), but visual identification of buckling modes can be challenging. Li et al. (2014) further developed the constrained finite strip method as a tool for modal identification to assess coupled instabilities. This method has been extended to the analysis of members under thermal gradients. Potentially, the constrained finite strip method paired with geometry measurements could be used to evaluate the response of CFS members subjected to fire.

The effective width method (EWM) has served as a traditional approach to designing CFS thin-walled members. This design method has been explored, and found to be adequate for fire design in some cases. Lee (2004) concluded that the EWM is adequate for estimating the load-carrying capacity of CFS members with low strength, but inaccurate for analyzing high-strength CFS members susceptible to local buckling. Ranawaka (2006) identified multiple limitations related to the EWM and its ability to predict distortional buckling.

Alternatively, the direct strength method (DSM) estimates the load-carrying capacity of CFS thin-walled members based on elastic buckling loads and the squash load (AISI, 2012). Current DSM equations along with temperature-dependent mechanical properties were found accurate for analyzing CFS members. A general agreement is found regarding the accuracy of current DSM equations to compute the local and global buckling strength of CFS members at uniform elevated temperatures. Ranawaka (2006) proposed a modified DSM equation to account for distortional buckling of members under uniform elevated temperatures; however,

Shahbazian and Wang (2012) found current DSM equations are adequate when distortional buckling controls. For cases including nonuniform temperature distributions, Shahbazian and Wang (2011b) proposed modified DSM equations and the computation of an "effective squash load" to predict the load-carrying capacity of CFS members accurately. Batista-Abreu and Schafer (2014) proposed the use of current DSM equations to estimate the strength of CFS members under nonuniform elevated temperatures by defining the squash load as the load that will cause first yield in the cross-section (similar to beam design). In general, the literature indicates that DSM is practical and suitable to analyze CFS members at elevated temperatures.

The development of accurate temperature-dependent constitutive relations and the feasibility of current engineering-based analysis method for fire design applications shape the basis for the envisioned performance-based framework for fire design. Traditional prescriptive design, although consistent, impedes innovation and progress, while engineering-based analysis methods allow exploration of a broader spectrum of solutions without extensive and costly experimentation. Current prescriptive design methods only account for member and (at most) subsystem-level mechanisms, without considering the interaction of single elements or assemblies with other components of the system. In contrast, performance-based design allows the assessment of fire effects on structures considering fire as the primary hazard, but also allows using multihazard scenarios when fire occurs as a secondary effect after earthquake or blast (for example). This integration of hazards permits a reasonable and consistent evaluation of structural performance, providing a means to reenvision structural optimization aiming toward sustainability.

7.6 Design examples of thermomechanical performance

In these examples, temperature-dependent mechanical properties are used to determine the elastic buckling behavior of CFS members at elevated temperatures, and then current DSM column equations are used to estimate their axial strength.

7.6.1 *Example #1: strength of CFS column at uniform elevated temperature*

The objective of this example is to estimate the load-carrying capacity of a lipped-channel column at 400°C. It is assumed that the temperature is uniformly distributed and the column has pinned ends. DSM equations found in AISI S100-2012 are used to solve the problem. Elastic buckling loads based on the geometry of the thin-walled member and temperature-dependent mechanical properties are needed. CUFSM is utilized to obtain these buckling loads (Li and Schafer, 2010).

From Figs. 7.7(a) and 7.8(a), the minimum and maximum retention factors at 400°C are 0.521 and 0.826 for the elastic modulus and 0.492 and 0.842 for the yield stress, respectively. Mechanical properties at elevated temperature are estimated assuming nominal values of 203.4 GPa (29,500 ksi) and 345.0 MPa (50.0 ksi) for the elastic modulus and yield stress at ambient conditions, respectively. Therefore the elastic modulus of the heated column ranges between 106.0 GPa (15,370 ksi) and 168.0 GPa (24,367 ksi); and the yield stress ranges between 169.6 MPa (24.6 ksi) and 290.3 MPa (42.1 ksi) at 400°C. Poisson's ratio of 0.3 is assumed.

Two extreme cases are considered. First, the column strength is computed assuming the minima elastic modulus and yield stress. In the second case, maximum values of mechanical properties are used. Centerline dimensions of the web, flanges, lips, and corner radius are 152.4 mm (6 in.), 50.8 mm (2 in.), 12.7 mm (0.5 in.), and 6.35 mm (0.25 in.), respectively. The thickness of the lipped-channel section is 1.52 mm (0.06 in.). These dimensions are the default values in CUSFM version 4.05. The physical length of the column is 2.44 m (8 ft).

Fig. 7.9 shows the signature curves obtained through finite strip analysis, considering the maximum and minimum elastic moduli. Results are provided in Table 7.1. Since homogeneous material properties are assumed in both cases, the length corresponding to local and distortional minima does not change for different elastic moduli.

The squash load is computed as the product of the cross-sectional area times the temperature-dependent yield stress. Following current DSM column equations

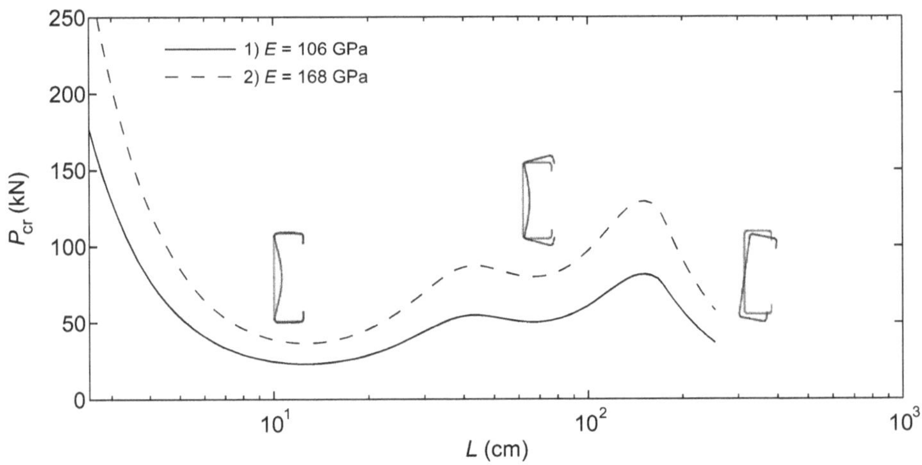

Figure 7.9 Signature curve showing elastic buckling load (P_{cr}) versus length (L).

Table 7.1 Elastic buckling loads of CFS lipped channel at 400°C

	E = 106.0 GPa (15,370 ksi)		**E = 168.0 GPa (24,367 ksi)**	
Buckling mode	**L, cm (in.)**	**P_{cr}, kN (Kips)**	**L, cm (in.)**	**P_{cr}, kN (Kips)**
Local	12.34 (4.86)	23.00 (5.17)	12.34 (4.86)	36.52 (8.21)
Distortional	65.91 (25.95)	50.31 (11.31)	65.91 (25.95)	79.80 (17.94)
Global	243.84 (96.00)	39.46 (8.87)	243.84 (96.00)	62.54 (14.06)

(AISI S100, Appendix 1), the nominal axial strength of the column is 25.53 kN (5.74 Kips) if assuming minimum retention factors, or 40.66 kN (9.14 Kips) if assuming maximum retention factors. Finally, local–global interaction controls the column behavior.

Results vary significantly according to the retention factors used to estimate the mechanical properties of CFS. Thus it is essential to base calculations on reliable material properties obtained from testing on representative material samples using proper testing methods and adequate data postprocessing.

7.6.2 Example #2: strength of CFS column subjected to thermal gradient

In this example it is assumed that the CFS member analyzed in Section 7.6.1 is subjected to temperature increase that creates a thermal gradient throughout the cross-section. The temperature of the flange and lip directly exposed to fire is 400°C, and the temperature of the unexposed flange and lip is 300°C. Linear temperature variation is assumed in the web (decaying from the exposed to the unexposed flange). For instance, the temperature in the middle of the web is 350°C. The objective of this example is to estimate the axial strength of the CFS member under the specified temperature distribution.

Current DSM column equations are adopted. CUFSM is used to determine elastic buckling loads, considering temperature-dependent mechanical properties. Using average values from Figs. 7.7(a) and 7.8(a), it is assumed that the elastic modulus decays linearly with increasing temperature from 151.4 GPa (21,960 ksi) at 300°C to 124.6 GPa (18,065 ksi) at 400°C, and the yield stress decays linearly from 291.6 MPa (42.3 ksi) at 300°C to 244.8 MPa (35.5 ksi) at 400°C.

The signature curve is provided in Fig. 7.10. The minimum local buckling load is significantly lower than the distortional and global buckling minima, so local–global

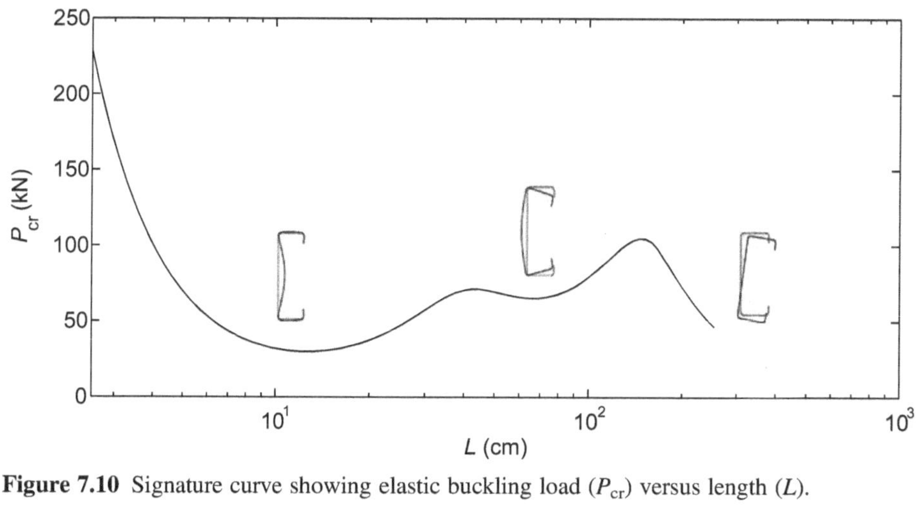

Figure 7.10 Signature curve showing elastic buckling load (P_{cr}) versus length (L).

Table 7.2 **Elastic buckling loads of CFS lipped channel under thermal gradient**

Buckling mode	L, cm (in.)	P_{cr}, kN (Kips)
Local	12.93 (5.09)	29.89 (6.72)
Distortional	65.91 (25.95)	65.12 (14.64)
Global	243.84 (96.00)	50.00 (11.24)

interaction is expected to dominate (Table 7.2). The mode shapes show a slight asymmetry in cross-section deformations due to nonuniform temperature distribution and mechanical properties (Batista-Abreu and Schafer, 2013).

The squash load was computed as the product of the cross-sectional area and the minimum yield stress of the cross-section (ie, 244.8 MPa on the exposed flange and lip), following a simple approach by Batista-Abreu and Schafer (2014). Using current DSM column equations, the axial strength of the CFS member under thermal gradient is 32.78 kN (7.37 Kips), controlled by local–global interaction.

Other calculation methods based on the EWM and modified DSM are detailed in Feng and Wang (2005) and Shahbazian and Wang (2014).

Sources of further information

Most relevant papers related to fire behavior and design of CFS are cited in this chapter. Additional comprehensive sources of information are presented by Gunalan and Mahendran (2013) and Batista Abreu et al. (2014).

References

American Iron and Steel Institute, AISI Standard, AISI S100–2012. North American Specification for the Design of Cold-formed Steel Structural Members. Appendix 1.

ABAQUS, 2013. ABAQUS 6.13 Documentation (Providence, RI, USA).

Ahmadi, M., Zhang, H., Yu, C., Wahrmund, J., 2013. Determining elastic and shear moduli of cold-formed steel at elevated temperatures using a new sonic resonance method. Nondestructive Testing and Evaluation 29, 1–13.

Alfawakhiri, F., 2001. Behaviour of Cold-formed-steel-framed Walls and Floors in Standard Fire Resistance Tests. Doctor of Philosophy in Civil Engineering, Carleton University.

ANSYS, 2015. Academic Research, Release 16.0, Help System, Coupled Field Analysis Guide. ANSYS, Inc.

Baleshan, B., Mahendran, M., 2010. Full scale fire tests of a new light gauge steel floor-ceiling system. In: 4th International Conference on Steel & Composite Structures. Sydney.

Batista Abreu, J., Vieira, L., Abu-Hamd, M., Schafer, B., 2014. Review: development of performance-based fire design for cold-formed steel. Fire Science Reviews 3, 1.

Batista-Abreu, J.C., 2015. Fire performance of cold-formed steel walls. Doctor of Philosophy. The Johns Hopkins University.

Batista-Abreu, J.C., Schafer, B.W., 2013. Stability of cold-formed steel compression members under thermal gradients. St. Louis, Missouri. In: Annual Stability Conference Structural Stability Research Council, pp. 136–154.

Batista-Abreu, J.C., Schafer, B.W., 2014. Stability and load-carrying capacity of cold-formed steel compression members at elevated temperatures. In: Annual Stability Conference Structural Stability Research Council (Toronto, Canada).

Batista-Abreu, J.C., Vieira, J.R., Abu-Hamd, L.C.M., Schafer, B.W., 2015. Experiments on cold-formed steel sheathed studs at elevated temperatures. In: Annual Stability Conference Structural Stability Research Council (Nashville, TN).

Bénichou, N., Su, J.Z., Bwalya, A.C., Lougheed, G.D., Taber, B.C., Leroux, P., Kashef, A.H., Thomas, J.R., 2009. Fire Performance of Houses. Phase I: Study of Unprotected Floor Assemblies in Basement Fire Scenarios. Part 3 – Results of Test UF-04 (Light-gauge Steel C-joists). IRC-RR-248. NRC-CNRC. Institute for Research in Construction - Fire Research Program.

Chen, J., Jin, W.L., 2008. Behaviour of cold-formed stainless steel beams at elevated temperatures. Journal of Zhejiang University: Science A 9, 1507–1513.

Chen, W., Ye, J., 2012. Mechanical properties of G550 cold-formed steel under transient and steady state conditions. Journal of Constructional Steel Research 73, 1–11.

Chen, W., Ye, J., Bai, Y., Zhao, X., 2012. Thermal and mechanical modeling of load-bearing cold-formed steel wall systems in fire. Journal of Structural Engineering 140.

Chen, W., Ye, J., Bai, Y., Zhao, X., 2014. Thermal and mechanical modeling of load-bearing cold-formed steel wall systems in fire. Journal of Structural Engineering 140.

EN 1993-1-2, 2005. Eurocode 3: Design of Steel Structures – Part 1–2: General Rules. Structural fire design, Brussels.

Feng, M., Wang, Y., 2005. An analysis of the structural behaviour of axially loaded full-scale cold-formed thin-walled steel structural panels tested under fire conditions. Thin-Walled Structures 43, 291–332.

Feng, M., Wang, Y., Davies, J., 2003a. Structural behaviour of cold-formed thin-walled short steel channel columns at elevated temperatures. Part 1: experiments. Thin-Walled Structures 41, 543–570.

Feng, M., Wang, Y., Davies, J., 2004. A numerical imperfection sensitivity study of cold-formed thin-walled tubular steel columns at uniform elevated temperatures. Thin-Walled Structures 42, 533–555.

Feng, M., Wang, Y.C., Davies, J.M., 2003b. Axial strength of cold-formed thin-walled steel channels under non-uniform temperatures in fire. Fire Safety Journal 38, 679–707.

Feng, M., Wang, Y.C., Davies, J.M., 2003c. Structural behaviour of cold-formed thin-walled short steel channel columns at elevated temperatures. Part 2: design calculations and numerical analysis. Thin-Walled Structures 41, 571–594.

Gunalan, S., Mahendran, M., 2010. Structural and fire behaviour of a new light gauge steel wall system. In: Kodur, V., Franssen, J.M. (Eds.), Structures in Fire: Sixth International Conference. Pennsylvania, USA.

Gunalan, S., Mahendran, M., 2013. Review of current fire design rules for cold-formed steel wall systems. Journal of Fire Sciences 32.

Kaitila, O., 2002. Finite Element Modelling of Cold-formed Steel Members at High Temperatures. Licentiate of Science in Technology, Helsinki University of Technology.

Kankanamge, N.D., Mahendran, M., 2011. Mechanical properties of cold-formed steels at elevated temperatures. Thin-Walled Structures 49, 26–44.

Khorasani, N., Gardoni, P., Garlock, M., 2015. Probabilistic fire analysis: material models and evaluation of steel structural members. Journal of Structural Engineering 0, 04015050.

Kolarkar, P., Mahendran, M., 2012. Experimental studies of non-load bearing steel wall systems under fire conditions. Fire Safety Journal 53, 85–104.

Lázaro, D., Puente, E., Lázaro, M., Lázaro, P.G., Peña, J., 2015. Thermal modelling of gypsum plasterboard assemblies exposed to standard fire tests. Fire and Materials. http://dx.doi.org/10.1002/fam.2311.

Lee, J.H., 2004. Local Buckling Behaviour and Design of Cold-formed Steel Compression Members at Elevated Temperatues. Doctor of Philosophy, Queensland University of Technology.

Lee, J.H., Mahendran, M., Makelainen, P., 2003. Prediction of mechanical properties of light gauge steels at elevated temperatures. Journal of Constructional Steel Research 59, 1517–1532.

Li, Z., Schafer, B.W., 2010. Buckling analysis of cold-formed steel members with general boundary conditions using CUFSM: conventional and constrained finite strip methods. In: Twentieth International Specialty Conference on Cold-formed Steel Structures Saint Louis, Missouri, USA.

Li, Z., Batista Abreu, J.C., Leng, J., Ádány, S., Schafer, B.W., 2014. Review: constrained finite strip method developments and applications in cold-formed steel design. Thin-Walled Structures 81, 2–18.

Lim, J.B.P., Young, B., 2007. Effects of elevated temperatures on bolted moment-connections between cold-formed steel members. Engineering Structures 29, 2419–2427.

Lu, W., Makelainen, P., Outinen, J., 2010. Numerical investigation of cold-formed steel purlin in fire. Journal of Structural Mechanics 43, 12–24.

Manzello, S.L., Grosshandler, W.L., Mizukami, T., 2008. Furnace Testing of Full-scale GypsumSteel Stud Non-load Bearing Wall Assemblies: Results of Multi-laboratory Testing in Canada, Japan and USA. Structures in Fire (SiF'08) (Singapore).

Mehaffey, J.R., Cuerrier, P., Carisse, G., 1994. A model for predicting heat transfer through gypsum-board/wood-stud walls exposed to fire. Fire and Materials 18, 297–305.

Moen, C.D., Igusa, T., Schafer, B.W., 2008. Prediction of residual stresses and strains in cold-formed steel members. Thin-Walled Structures 46, 1274–1289.

Outinen, J., 2006. Mechanical Properties of Structural Steel at High Temperatures and after Cooling Down. Doctor of Science in Technology Doctoral Dissertation, Helsinki University of Technology.

Park, S.-H., Manzello, S.L., Bentz, D.P., Mizukami, T., 2009. Determining thermal properties of gypsum board at elevated temperatures. Fire and Materials 34, 237–250.

Ramberg, W., Osgood, W.R., 1943. Description of Stress-strain Curves by Three Parameters. Technical Notes. National Advisory Committee for Aeronautics, Washignton.

Ranawaka, T., 2006. Distortional Buckling Behaviour of Cold-formed Steel Compression Members at Elevated Temperatures. Doctor of Philosophy, Queensland University of Technology.

Ranawaka, T., Mahendran, M., 2009. Experimental study of the mechanical properties of light gauge cold-formed steels at elevated temperatures. Fire Safety Journal 44, 219–229.

SAFIR, 2003. A thermal/structural program modelling structures under fire. In: Franssen, J.-M. (Ed.), Proceedings NASCC 2003, AISC Inc., Baltimore, April 2–4.

Shahbazian, A., Wang, Y.C., 2011a. Application of the direct strength method to local buckling resistance of thin-walled steel members with non-uniform elevated temperatures under axial compression. Thin-Walled Structures 49, 1573–1583.

Shahbazian, A., Wang, Y.C., 2011b. Calculating the global buckling resistance of thin-walled steel members with uniform and non-uniform elevated temperatures under axial compression. Thin-Walled Structures 49, 1415–1428.

Shahbazian, A., Wang, Y.C., 2012. Direct strength method for calculating distortional buckling capacity of cold-formed thin-walled steel columns with uniform and non-uniform elevated temperatures. Thin-Walled Structures 53, 188–199.

Shahbazian, A., Wang, Y.C., 2013. A simplified approach for calculating temperatures in axially loaded cold-formed thin-walled steel studs in wall panel assemblies exposed to fire from one side. Thin-Walled Structures 64, 60–72.

Shahbazian, A., Wang, Y.C., 2014. Performance-based fire resistance design method for wall panel assemblies using thin-walled steel sections. The Structural Engineer 92 (1).

Sterner, E., Wickstrom, U., 1990. TASEF – Temperature Analysis of Structures Exposed to Fire - User's Manual. SP Report 1990:05. Swedish National Testing and Research Institute (SP), Boras.

Sultan, M., 1996. A model for predicting heat transfer through noninsulated unloaded steel-stud gypsum board wall assemblies exposed to fire. Fire Technology 32, 239–259.

Thomas, G., 2002. Thermal properties of gypsum plasterboard at high temperatures. Fire Materials 26, 37–45.

Yan, S., Young, B., 2011. Tests of single shear bolted connections of thin sheet steels at elevated temperatures—Part I: steady state tests. Thin-Walled Structures 49, 1320–1333.

Yan, S., Young, B., 2012a. Screwed connections of thin sheet steels at elevated temperatures – Part I: steady state tests. Engineering Structures 35, 234–243.

Yan, S., Young, B., 2012b. Screwed connections of thin sheet steels at elevated temperatures – Part II: transient state tests. Engineering Structures 35, 228–233.

Ye, J., Chen, W., 2013. Elevated temperature material degradation of cold-formed steels under steady- and transient-state conditions. Journal of Materials in Civil Engineering 25, 947–957.

Yu, H., Burgess, I.W., Davison, J.B., Plank, R.J., 2008. Numerical simulation of bolted steel connections in fire using explicit dynamic analysis. Journal of Constructional Steel Research 64, 515–525.

Zeinoddini, V., Schafer, B., 2012. Simulation of geometric imperfections in cold-formed steel members using spectral representation approach. Thin-Walled Structures 60, 105–117.

Acoustic performance of cold-formed steel buildings

8

C.J. Wang, D.J. Mynors
University of Sussex, Brighton, United Kingdom

8.1 Introduction

Sound is a vibration that can be transmitted through media such as air, water, and solids. In buildings, the acoustic performance of floors and walls is a very important consideration: party walls and floors must achieve certain sound reduction values required by building regulations. Cold-formed steel (CFS) profiles, gypsum plasterboard, and mineral wool are widely used in building systems. Light steel framings are used to construct load-bearing walls, floors, separating walls, internal partitions, and external infill walls. For example, plasterboard is supported by CFS studs on both sides to form a double-leaf partition wall. Although the studs act as sound bridges transferring sound energy, high levels of acoustic performance of CFS buildings have been achieved using multilayer constructions, acoustic CFS studs, and resilient bars. Many studies have been conducted to understand the sound transmission behavior in double-leaf walls. Hongisto et al. (2002) reported the stiffness of the stud and the spacing of the screws as the most important parameters in sound insulation, with flexible studs performing better than rigid studs at low and middle frequencies, but the difference reduced at high frequencies.

Sound pressure is the difference between the pressure in a sound wave and the ambient pressure at the same point. Sound pressure level is measured as the logarithm of the ratio of the root-mean-square sound pressure of a sound to the reference pressure. According to ISO 10140 (2010) and shown in Eq. [8.1], average sound pressure level, L_p, in a room is defined as 10 times the common logarithm of the ratio of the space and time average of the squared sound pressure, p, to the square of the reference sound pressure, p_{ref}, (20 μPa for air).

$$L_p = 10 \log \frac{p^2}{p_{ref}^2} = 20 \log \frac{p}{p_{ref}} \quad [8.1]$$

The space average over the entire room does not include parts such as the direct radiation of a sound source or the near field of the boundaries, due to their significant influence.

Sound insulation values in buildings represent the reduction of the sound pressure level from a source room to a receiving room caused by the separating construction. In buildings, sound insulation can be divided into two types: impact and airborne.

Recent Trends in Cold-Formed Steel Construction. http://dx.doi.org/10.1016/B978-0-08-100160-8.00008-6

Impact sound is mainly relevant to floors, and may be caused by people walking on the floor. The sound pressure level is measured in the room below the floor to determine the impact sound insulation performance of the floor. A tapping machine is used to strike the floor repeatedly at a standard rate in different positions; the resulting sound level in the receiving room is measured and named as the impact sound transmission level. A tapping machine has five hammers, each weighing 0.5 kg, which strike from a height of 40 mm with an impact frequency of 10 Hz. Measurements in the receiving room are taken in one-third octave bands between 100 and 3150 Hz. ISO 717-2 (2013) specifies the rules on rating impact sound insulation and the weighted standardized impact sound pressure level in the receiving room.

Airborne sound insulation is the difference between the sound pressure levels of the source room and the receiving room; the sound source should be in the larger room.

The sound reduction index, R, or sound transmission loss of a partition is defined in ISO 10140 as

$$R = L_1 - L_2 + 10\log_{10}\frac{S}{A} \qquad [8.2]$$

where L_1 and L_2 are the sound pressure levels in the source room and receiving room respectively, S is the area of the partition, and A is the absorption area of the receiving room.

Sound insulation is normally measured at a number of different frequencies; generally the sound reduction index increases with frequency, and the higher the frequency the greater the sound reduction index. A single-number rating is normally required for sound insulation in buildings and of building elements such as walls, floors, doors, and windows. The single numbers, in accordance with ISO 717-1 (2013), are intended to rate the airborne sound insulation and simplify the formulation of acoustic requirements in building codes. The weighted sound reduction index, R_W, is a single-number rating system to represent the R values at one-third octaves between 100 and 3150 Hz. R_w values can be calculated with reference to ISO 717-1. To take into consideration the different sound level spectra of various noise sources, A-weighted pink noise, C, and A-weighted urban traffic noise, C_{tr}, are also calculated and specified alongside the weighted sound reduction index, R_W. ISO 717-1 contains rules for determining these spectrum adaptation quantities from measurements carried out using one-third octave or octave bands in accordance with ISO 10140. There are five parts, under the general title of *Acoustics—Laboratory measurement of sound insulation of building elements*, in ISO 10140.

Hongisto et al. (2015) compared the subjective acoustic satisfaction in residential buildings of a similar weighted sound reduction index but with different wall constructions: monolithic concrete walls and lightweight staggered steel stud double walls. Although lightweight constructions had lower sound reduction at low frequencies, they found the subjective sound satisfaction was the same, and concluded that the single-number quantities of sound insulation like R_W, can explain well the subjective sensation of sound insulation.

Field measurements of sound insulation in buildings and of building elements are specified in ISO 16283 (2014), which has three parts covering airborne sound insulation, impact sound insulation, and façade sound insulation. The American Society for Testing and Materials (ASTM) also provides standard test methods relating to sound transmission in buildings.

In CFS buildings, to reduce sound transmission by building elements such as walls or floors, one method is to increase the mass of the partition by using several layers of board materials. A second method is to increase the depth of cavity within the partition; a third method is to install sound-absorptive material in the partition cavity; and the fourth method is to decouple the structures, such as by using two separate sets of studs and resilient bars.

8.2 Acoustic insulation performance requirements

One of the main types of sound-absorptive materials in buildings is porous materials such as glass fiber and mineral fiber, which perform well at higher frequencies and provide high levels of sound absorption. Another type is panel absorbers, which perform well at low frequencies. As reported by Cambridge et al. (2013), the airflow resistivity of porous sound-absorptive materials can improve the effectiveness of a material's sound reduction properties, but the more rigid materials in the cavity may increase the structural sound transmission. Airflow resistivity of 5000 Ns/m^4 and density of 40 kg/m^3 are recommended. However, the optimum amount of sound-absorptive materials used in the cavity is dependent on the mass of wall panels and the structural connections.

Thick and heavy materials can be used for sound reduction; high-density materials restrict sound vibrations inside the materials. The stiffness of a partition can affect the sound reduction at certain frequencies due to resonances. Heavyweight structures normally transmit less sound than lightweight structures.

Rasmussen (2010) summarized airborne and impact sound insulation requirements for dwellings in 24 European countries, and revealed that there were significant differences in descriptors and required levels. As building technology has gradually changed from heavyweight toward more lightweight constructions, it is important to have an evaluation system that can work for both. Rasmussen and Rindel (2010) discussed the source of the different descriptors, and suggested a single set of harmonized descriptors should be applied for airborne and impact sound insulation criteria, for example using $D_{nT;w} + C_{50\text{-}3150}$ for airborne sound insulation.

In England and Wales the acoustic performance requirement of residential and school buildings is given in *Building Regulations, Approved Document E.* Airborne sound reduction must be greater than 45 dB for separating walls and floors between dwellings, and 40 dB for internal walls. Impact sound transmission must be less than 62 dB for separating floors.

According to ISO 717-1, a statement of performance of building elements shall be presented as a single number with the two spectrum adaptation terms, $R_w(C;\ C_{tr})$.

Requirements for acoustic performance of buildings shall be given a single-number quantity. For examples, $R'_w + C_{tr} \geq 45$ dB, this means that the weighted apparent sound reduction with A-weighted urban traffic noise must be greater than or equal to 45 dB. $D_{nt,W} + C \geq 45$ dB, which means that the standardized weighted level difference for airborne sound with A-weighted pink noise must be greater than or equal to 45 dB.

In North America the Sound Transmission Class (STC) is normally used in building acoustics; it is a single-number rating that indicates the effectiveness of a wall or floor to retard the transmission of airborne sound. The STC is similar to the weighted sound reduction index R_W, and rates a partition's resistance to airborne sound transmission at 16 different sound frequencies between 125 and 4000 Hz, which is consistent with the frequency range of speech. The higher the STC number, the better the sound insulation. The Impact Insulation Class (IIC) is another single-number rating that indicates the amount of impact sound insulation provided by a wall or floor assembly. The International Building Code (2012) requires STC and IIC ratings of 50 dB for floor and ceiling assemblies if tested in a laboratory or 45 dB if tested in the field.

8.3 Building design for acoustic performance

To ensure a suitable acoustic standard is achieved for buildings, building design in relation to sound transmission between rooms should be dealt with during the design stage. Buildings should be designed and constructed to resist sound transmission from adjoining buildings and other parts of the same building. Small gaps in building partition elements are detrimental to sound reduction, thus it is important to ensure adequate sealing around floors and partitions.

Although flexible materials with high mass will provide better sound insulation, flexibility in a wall or a floor is structurally not desirable. CFS sections used for improving acoustic performance include acoustic studs and resilient bars made from light-gauge steel. Single-leaf partition walls can further reduce sound transmission by 5−6 dB by doubling their mass. However, the sound reduction index of double-leaf partition walls shown in Fig. 8.1 is not twice the value of single-leaf partition walls due to the air cavities and studs dynamically coupling the two leaves. Three sound transmission paths through a double-leaf wall are shown in Fig. 8.1.

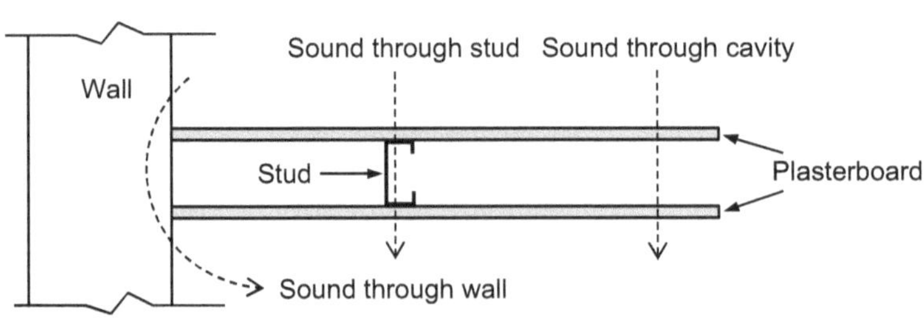

Figure 8.1 Sound transmission paths between two rooms.

CFS studs used in lightweight double-leaf walls may increase the sound transmission when compared to those without studs. The studs provide a direct sound transmission path, and change the vibration behavior of the leaves. Staggered-stud and double-stud walls can greatly reduce sound transmission. To study the effect of studs on the sound reduction index of lightweight double-leaf walls, several approaches have been proposed. Craik and Smith (2000) studied structure-borne sound transmission through lightweight parallel plates: statistical energy analysis models were created to analyze the coupling of two plates by either point connection or line connection. Fahy (1989) treated studs as rigid connections, while Wang et al. (2005) analyzed the studs as springs having linear and rotational stiffness. Uris et al. (2002) studied the influence of screw spacing on the sound reduction index of lightweight partitions constructed using steel studs. They found that screw spacing only affects double-leaf walls. Partitions with gypsum plasterboard layers screwed at 600 mm centers have a 2 dB higher sound reduction index than partitions with screw spacing of 300 mm.

To reduce sound transmission through metal studs, acoustic studs have been designed by manufacturers. Acoustic studs may increase the sound reduction index by 3 dB when compared to standard studs. With regard to stud design, Poblet-Puig et al. (2009) explained that acoustic studs have a larger perimeter than standard studs; at higher frequencies this increases the effective damping in the vibration transmission path and the vibration wavelength is smaller than the characteristic length of the stud.

Resilient bars have been designed to reduce sound transmission; these can increase the sound insulation by absorbing vibrations and isolation of the plasterboard. They are manufactured from light-gauge steel with a stepped profile, and are fixed between the light steel wall studs and the wall linings. However, models used to predict sound reductions with resilient bars have not been studied extensively.

To estimate acoustic performance of assembled walls, Warnock (2008) presented several regression equations derived from experimental test data to predict STC in various types of walls. These equations use parameters related to the unit mass of board, cavity depth, insulation material thickness, density of the sound-absorbing material, and stud spacing. The acoustic performance of a floor with light steel joists and profiled steel decking is dependent on the whole floor construction and floor treatment, but no equations are available to predict the sound performance of floors due to lack of floor test results. In the UK, Robust Details (2015) were developed to guide the design and construction of walls and floors. Following this guide correctly, buildings should satisfy the required acoustic performance.

8.4 Tests, performance tables, and insulation values

Building elements can be tested in acoustic laboratories. Two adjacent rooms are connected by an aperture containing a test panel of the building element, and the rooms are isolated against flanking transmission. The sound reduction index R of the test panel is calculated using Eq. [8.1]. The samples should be mounted, positioned, and tested following the ISO 10140-2/3 guidelines. These codes specify rules for laboratory measurements of sound insulation in buildings and of building elements. Broadband noise is produced in the source room, and the resulting sound pressure levels in the source and receiving rooms are sampled, filtered into one-third octave band widths, integrated, and averaged. The equivalent sound pressure levels for the rooms are obtained, and the change in the levels is used to evaluate the sound reduction index. In laboratory-based experimental tests of building components, the main sound transmission path is the test sample. However, in field tests sound may transmit through different paths, resulting in a smaller sound reduction value than that measured in the laboratory.

Various models were developed to predict the influence of periodically spaced structural links on sound transmission through lightweight double-panel structures. Decoupled approaches assume that the sound energy transfer from the source panel to the receiving panel is via two paths: the structure-borne path through the studs and the airborne path through the cavity are independent and additive. Theoretical modeling of sound transmission loss through double-leaf lightweight partitions stiffened with periodically placed studs was studied by Wang et al. (2005). Legault and Atalla (2009) examined various models for predicting the influence of periodically spaced structural links on sound transmission through light double-panel structures, and found that the model proposed by Davy is most acceptable. The model by Davy (2009) can be found in a paper predicting the sound insulation of walls, and more papers by Davy on refining the models have recently been published.

Numerical simulations can be conducted to study the acoustic behavior of building elements: these methods could be employed in the development of new products, thus reducing the number of experimental tests. Maluski and Gibbs (2000) used the finite element (FE) method to investigate sound transmission between dwellings—good agreement was obtained between measured and predicted sound pressure levels.

Poblet-Puig et al. (2009) created a two-dimensional FE model of two rooms separated by a double-leaf wall, and showed that the stud shapes are not important at low frequencies but had a large effect on sound transmission at higher frequencies. To find an efficient numerical method to predict sound transmission loss through a multilayer wall, Coz Diaz et al. (2010) used two-dimensional FE models to obtain sound pressures at one-third octave bandwidths between 100 and 5000 Hz; fluid and structure interactions between air and solid were included in the models.

Two- and three-dimensional FE models were developed by Arjunan et al. (2013, 2014) to predict the sound reduction index of double-leaf partition walls with CFS studs. Acoustic performance of the double-leaf wall was simulated using a harmonic acoustic FE model. The simulation was executed using real-time coupling between the structural and fluid elements accounting for fluid–structure interaction. The frequency-dependent pressure distribution and sound pressure level in the source and receiving rooms were obtained, and acoustic performances of roll-formed steel studs were predicted. A schematic representation of a two-dimensional FE model of a double-leaf wall is shown in Fig. 8.2. The CFS stud and plasterboard were modeled using solid elements, the air in the two rooms and the cavity were modeled using fluid elements, and the fluid and structure interactions were specified in the fluid elements which are in contact with the solid elements. Although the maximum element side

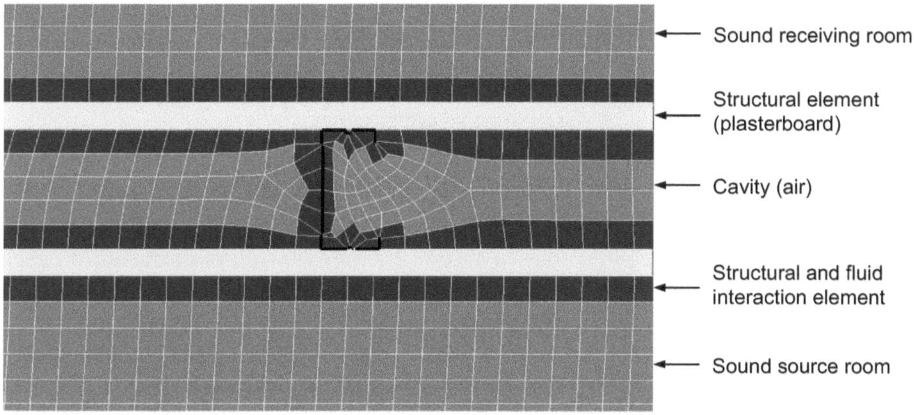

Figure 8.2 Two-dimensional FE model of a metal stud in a partition wall.

length is inversely proportional to the highest frequency in the simulation, the current increasing computing power means that very large FE models can be created to predict the effect of stud shape on sound transmission. Sound reduction values predicted using the FE models developed by Arjunan et al. (2013, 2014) showed good agreement with the experimental data. A double-leaf wall consists of a CFS stud of 0.5 mm thickness and 70 mm depth placed between two 15 mm thick gypsum plasterboards without insulation material in the cavity: a weighted sound reduction index of 39 dB was predicted using the FE model. To predict sound reduction using FE models properly, material damping parameters related to the dissipation of acoustic energy and frequency-dependent boundary admittance values need to be included in the model.

Recent publication of acoustic data for CFS floors, walls, and roof assemblies by the Steel Framing Alliance (2013) provides a good reference for the acoustic performance of different constructions, and summarizes the STC and IIS data relevant to residential and light commercial constructions.

Another book on acoustic detailing for steel construction by Way and Couchman (2008) provides construction details of floors and walls, and expected airborne and impact sound performance values.

8.5 Design examples of acoustic performance in buildings

Staggered or double CFS stud walls can achieve a higher sound reduction index value than single stud walls. Fig. 8.3 shows a construction of staggered stud double-leaf wall: C studs with resilient bars are placed to both sides, and one layer of plasterboard is installed at each side with absorptive material used in the cavity.

In a guide published by the Steel Framing Alliance (2013), a tested STC of 45 dB is reported in a nonload-bearing wall which consists of 64 × 35 × 0.5 mm thick C steel

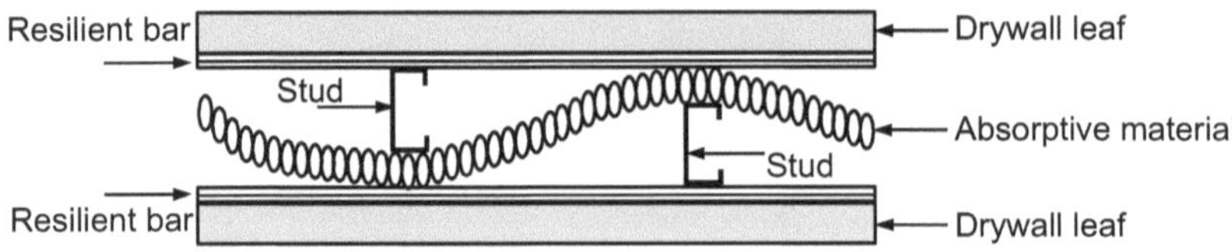

Figure 8.3 Staggered studs, absorptive materials, and resilient metal bars.

studs spaced at 600 mm centers, one layer of 12.7 mm gypsum plasterboard on each side, and 38 mm of mineral wool insulation.

In the acoustic detailing for steel constructions by Way and Couchman (2008), a typical separation wall made of a single acoustic stud 45–125 mm deep with one layer of gypsum plasterboard on each side has an expected airborne sound reduction index of 39–43 dB. To increase the sound reduction index, the separation wall can include resilient bars, two layers of gypsum board on each side of the stud, and a mineral wool quilt placed between the studs. For a wall of typical thickness of 140–220 mm, the expected acoustic performance is 59–62 dB.

Floors can be constructed with CFS joists, wood-based floor decking, a plasterboard ceiling, a resilient bar, and absorptive material, as shown in Fig. 8.4. Impact sound transmission of floors mainly depends on the floor finish and construction: a floating floor with a soft surface finish can greatly reduce the impact sound.

A tested floor shows an IIC of 51 dB in the guide published by the Steel Framing Alliance (2013); this floor consists of 19 mm plywood as floor decking, and 203 mm deep steel joists with 1.22 mm gauge thickness spaced at 610 mm centers. It has 100 mm thick cellulose fiber insulation material on both sides of the joist and 94 mm on the underside of the floor deck, and resilient metal channels are spaced at 610 mm centers, and two layers of 12.7 mm thick gypsum board are used on the ceiling side.

A floor construction using light steel joists described by Way and Couchman (2008) can achieve a weighted standardized impact sound pressure level of 44–58 dB. The floor depth ranges from 250 to 380 mm with floor treatment such as mineral wool between battens, and the ceiling has two layers of gypsum board attached to the resilient bars and fixed to the underside of the light steel joints.

It is difficult to predict flanking sound, as it is transmitted through junctions between building construction elements such as windows and external walls. To reduce flanking sound transmission it is important to avoid direct contact between building construction elements, and ensure adequate sealing is provided around the structural

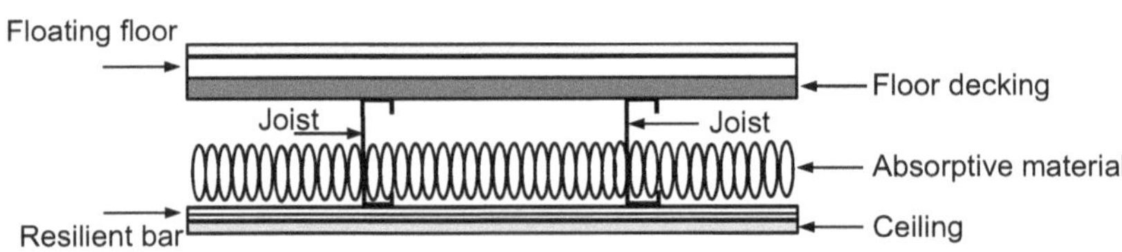

Figure 8.4 Floating floor with metal joists and resilient metal bars.

elements—for example floors and walls. Small gaps and holes can transmit airborne sound and reduce sound insulation, so these should be sealed with tape or filled with acoustic sealant. The Steel Construction Institute in the UK has published a number of case studies on the acoustic performance of residential buildings erected using different forms of steel construction, including design details of the floors and walls. The measured airborne and impact sound performance satisfies the building regulation requirements. Acoustic performance of partitions and floors can be acquired from British Gypsum, which has published literature on the design and construction details of building systems using CFS profiles.

Sources of further information

ASTM Standard E90-09, 2009. Standard Test Method for Laboratory Measurement of Airborne Sound Transmission Loss of Building Partitions and Elements. ASTM International, West Conshohocken, PA, USA. www.astm.org.

ASTM Standard E492-09, 2009. Standard Test Method for Laboratory Measurement of Impact Sound Transmission Through Floor-ceiling Assemblies Using the Tapping Machine. ASTM International, West Conshohocken, PA, USA. www.astm.org.

Building Regulations, 2010. Approved Document E, Resistance to the Passage of Sound. HM Government, UK.

Vigran, T.E., 2008. Building Acoustics. Taylor & Francis, Abingdon UK.

References

Arjunan, A., Wang, C.J., Yahiaoui, K., Mynors, D., Morgan, T., Nguyen, V.B., English, M., 2014. Development of a 3D finite element acoustic model to predict the sound reduction index of stud based double-leaf walls. Journal of Sound and Vibration 333 (23), 6140–6155.

Arjunan, A., Wang, C.J., Yahiaoui, K., Mynors, D.J., Morgan, T., English, M., 2013. Finite element acoustic analysis of a steel stud based double-leaf wall. Building and Environment 67, 202–210.

British Gypsum, Gotham Road, East Leake, Loughborough, LE12 6HX. www.british-gypsum.com.

Cambridge, J.E., Davy, J.L., Pearse, J., 2013. The influence of the wall cavity on the transmission loss of wall systems – experimental trends. Journal of Building Acoustics 20 (2), 87–106.

Coz Diaz, J.J., Alvarez Rabanal, F.P., Garcia Nieto, P.J., Serrano Lopez, M.A., 2010. Sound transmission loss analysis through a multilayer lightweight concrete hollow brick wall by FEM and experimental validation. Building and Environment 45, 2373–2386.

Craik, R.J.M., Smith, R.S., 2000. Sound transmission through lightweight parallel plates. Part II: structure-borne sound. Journal of Applied Acoustics 61 (2), 247–269.

Davy, J.L., 2009. Predicting the sound insulation of walls. Journal of Building Acoustics 16, 1–20.

Fahy, F., 1989. Sound and Structural Vibration. Academic Press, London.

Hongisto, V., Lindgren, M., Helenius, R., 2002. Sound insulation of double walls – an experimental parametric study. Acta Acustica 88 (6), 904–923.

Hongisto, V., Mäkilä, M., Suokas, M., 2015. Satisfaction with sound insulation in residential dwellings – the effect of wall construction. Building and Environment 85, 309–320.

IBC, 2012. International Building Code, Chapter 12 Interior Environment, Section 1207 Sound Transmission. International Code Council, 500 New Jersey Avenue, NW, 6th Floor, Washington, DC 20001, USA.

ISO 10140, 2010. Acoustics – Laboratory Measurement of Sound Insulation of Building Elements. BSI, Chiswick, UK.

ISO 16283, 2014. Acoustics – Field Measurement of Sound Insulation in Buildings and of Building Elements. BSI, Chiswick, UK.

ISO 717-1, 2013. Acoustics – Rating of Sound Insulation in Buildings and of Building Elements, Part 1: Airborne Sound Insulation. BSI, Chiswick, UK.

ISO 717-2, 2013. Acoustics – Rating of Sound Insulation in Buildings and of Building Elements, Part 2: Impact Sound Insulation. BSI, Chiswick, UK (Chiswick, UK).

Legault, J., Atalla, N., 2009. Numerical and experimental investigation of the effect of structural links on the sound transmission of a lightweight double panel structure. Journal of Sound and Vibration 324, 712–732.

Maluski, S.P.S., Gibbs, B.M., 2000. Application of a finite-element model to low-frequency sound insulation in dwellings. Journal of the Acoustical Society of America 108 (4), 1741–1751.

Poblet-Puig, J., Rodriguez-Ferran, A., Guigou-Carter, C., Villot, M., 2009. The role of studs in the sound transmission of double walls. Acta Acustica United With Acustica 95 (3), 555–567.

Rasmussen, B., 2010. Sound insulation between dwellings – requirements in building regulations in Europe. Journal of Applied Acoustics 71, 373–385.

Rasmussen, B., Rindel, J.H., 2010. Sound insulation between dwellings – descriptors applied in building regulations in Europe. Journal of Applied Acoustics 71, 171–180.

Robust Details, 2015. Robust Details Handbook, Robust Details Limited, Block E, Bletchley Park Science and Innovation Centre. Milton Keynes, Buckinghamshire, MK3 6EB, UK.

Steel Construction Institute. Silwood Park, Ascot, SL5 7QN UK. www.steel-sci.com.

Steel Framing Alliance, 2013. A Guide to Fire & Acoustic Data for Cold-formed Steel Floor, Wall & Roof Assembles. Steel Framing Alliance, 25 Massachusetts Avenue, NW, Suite 800, Washington, DC, USA.

Uris, A., Sinisterra, J., Bravo, J.M., Llinares, J., Estelles, H., 2002. Influence of screw spacings on sound reduction index in lightweight partitions. Journal of Applied Acoustics 63, 813–818.

Wang, J., Lu, T.J., Woodhouse, J., Langley, R.S., Evans, J., 2005. Sound transmission through lightweight double-leaf partitions: theoretical modelling. Journal of Sound and Vibration 286, 817–847.

Warnock, A.C.C., 2008. Estimation of Sound Transmission Class and Impact Insulation Class Rating for Steel Framed Assemblies. Research report RP-08–7. American Iron and Steel Institute/Steel Framing Alliance.

Way, A.G.J., Couchman, G.H., 2008. Acoustic Detailing for Steel Construction. The steel Construction Institute, Ascot, UK.

Floor vibration in cold-formed steel buildings

9

L. Xu
University of Waterloo, Waterloo, ON, Canada

9.1 Introduction

Cold-formed steel (CFS) has become an increasingly popular building material for residential and commercial construction in recent years. This increased use can be attributed to the numerous advantages that CFS has over traditional residential building materials. As CFS has the highest strength-to-weight ratio of any building material, its characteristics of high strength and stiffness give the advantage of achieving longer spans economically in floor construction. However, similar to lightweight floors supported by engineered wood joists, CFS floors with longer span and lighter weight are likely to be susceptible to annoying vibrations induced by normal human activity such as walking.

To meet the ever-increasing demand from industry for design standards and guidelines for CFS construction, the America Iron and Steel Institute (AISI) has developed a series of standards and design guides for CFS framing. The current edition of the design standard series was published in 2012; among them is a standard for design of floor and roof systems (AISI, 2012). The issue of floor vibration associated with human walking is not addressed in the standard, however, primarily due to lack of research and appropriate design guidelines.

Floor vibration as a serviceability concern has not been well addressed in design and construction practice for lightweight floors. Historically, most North American homebuilders in constructing lightweight floors follow the recommendation of the US National Association of Home Builders, which limits the span deflection to $L/480$ under specified uniform live loads, where L is the span length. This recommendation was established based on long-term practice in residential floors with solid lumber joists, which provided floor systems with limited span lengths. However, timber floor systems based on such an oversimplified design criterion may still be susceptible to annoying floor vibrations induced by human activities, especially for midspan to long-span floors. The issue of vibration serviceability of lightweight floors supported by timber or engineered wood joists was addressed by the wood construction industry in North America (Onysko, 1985; Chui, 1988; CWC, 1996; Hu and Chui, 2004).

Designing a CFS floor system to control annoying vibrations can be challenging, and correcting inadequacies after construction is usually very costly. Applying design methods that were developed for traditional lightweight wood or structural steel frame systems to the CFS floor system can also be problematic, as the design methods have not been adequately adapted for the specific characteristics of CFS members.

Recent Trends in Cold-Formed Steel Construction. http://dx.doi.org/10.1016/B978-0-08-100160-8.00009-8

9.2 Evaluation of vibration performance of CFS floors

The serviceability issues related to vibrations in floor systems are long-standing. The vibration performance evaluation of CFS floors consists of two aspects: selecting an acceptability criterion or criteria related to human perception of vibrations, and developing a procedure to evaluate the floor responses based on the selected criterion or criteria. Currently, the analytical background of vibrations is well developed. Sophisticated and rigorous computational tools such as the finite element method are available to analyze static and dynamic responses of both simple and complex structures. However, design engineers still struggle with the poor correlation between the outcome of computations at the design stage and the response of the floor constructed accordingly. To the author's knowledge, an accurate, practical, and comprehensive design procedure is still currently unavailable for effectively controlling vibration of CFS floors induced by human walking.

9.2.1 Classification of human perceptibility to floor vibrations

Human perception of floor vibration is a combination of movement of the floor system, physical perception, and the psychological reaction to vibration. Occupants in different settings will be more sensitive to exactly the same vibration than others, depending on frequency of occurrence, duration, and time of day. Humans are most sensitive to vibration frequencies in the range of 4–8 Hz, due to resonance within the body cavity itself (Grether, 1971). In addition, human perception depends on the activity of a person at the time of the perceived event. Individuals who are sleeping or sedentary will be more sensitive than those who are walking, running, dancing, or doing aerobics.

The classification of uncomfortable vibration is subjective. Determining acceptability can be challenge due to the large number of factors upon which vibration perception depends. Reiher and Meister (1931) conducted a study in human vibration perception by applying a steady-state vibration to several individuals situated on floor systems in various orientations. It was found that sensitivity to vibration decreases as the excitation frequency increases. The so-called Reiher–Meister scale was developed to classify vibration acceptability based on floor frequency and vibration amplitude, categorizing an occupant's perception as "not perceptible," "slightly perceptible," "distinctly perceptible," and "strongly perceptible." Recognizing that floor vibrations from typical use are transient in nature, Lenzen (1966) investigated human perception of floor vibration in a similar fashion to the testing conducted by Reiher and Meister, but inducing a transient vibration instead of the steady-state vibration. Lenzen investigated the dynamic and subjective response of steel joist and concrete slab floors, and examined the influence of damping and fundamental frequency on occupant comfort. The original Reiher–Meister scale was made less stringent by a factor of 10 to account for the nature of human perception of transient excitations. This updated criterion is known as the modified Reiher–Meister scale, and is shown in Fig. 9.1. The modified

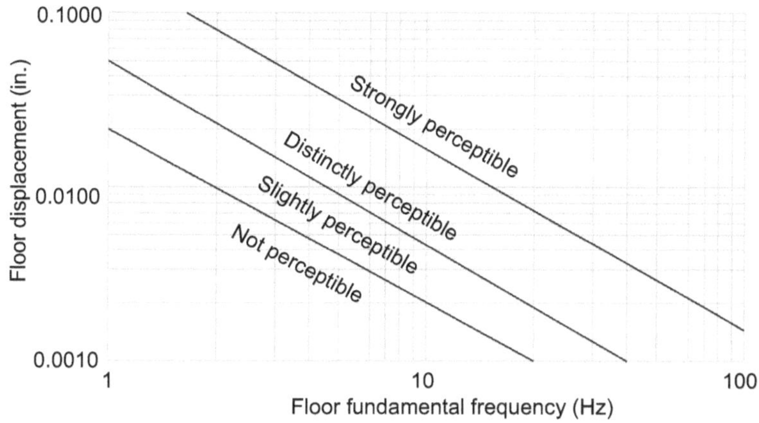

Figure 9.1 Modified Reiher–Meister scale.

Reiher–Meister scale has long served as an acceptance criterion for controlling vibration in practice.

The International Organization for Standardization (ISO) developed a limiting criterion for floor vibration based on a maximum acceptable root mean squared (RMS) acceleration for a given fundamental frequency of a structure, as part of the ISO 2631 "Standard for Mechanical Vibration and Shock—Evaluation of Human Exposure to Whole-Body Vibration" (ISO, 1989). The specific criterion applicable to floor vibration serviceability provides a limiting RMS acceleration for all fundamental frequencies as a baseline curve. Multipliers are introduced to account for the variation of occupant sensitivity and event frequency in different types of occupancies, as shown in Fig. 9.2. The shape of the baseline curve indicates the lowest tolerable acceleration levels are 4–8 Hz. This is for two reasons: human physiology makes occupants more sensitive in the 4–8 Hz range (Grether, 1971), and normal walking excitations contain harmonics of 4, 6, and 8 Hz, which will lead to more frequent resonant events in this range (Allen et al., 1999). This extra sensitivity at vibrations in the range of 4–8 Hz is important to note. Floor systems with a fundamental frequency within this range should be avoided, or measures taken to ensure that RMS acceleration is below the applicable limiting values.

Human response to vibration in buildings is complex. In many circumstances the degree of annoyance and complaint cannot be explained directly by the magnitude of monitored vibration alone. Under some conditions of amplitude and frequency, adverse comments may arise while measured amplitude is lower than the perception level (ISO 2631-2, 2003). To evaluate human perception, many additional effects need to be considered, such as the environment surrounding the person and the person's activity and psychological reaction. Thus the second edition of ISO 2631-2 (2003) does not state the acceptable magnitudes of vibration; instead, it provides guidelines for collecting data concerning complaints about building vibration to develop acceptable magnitudes.

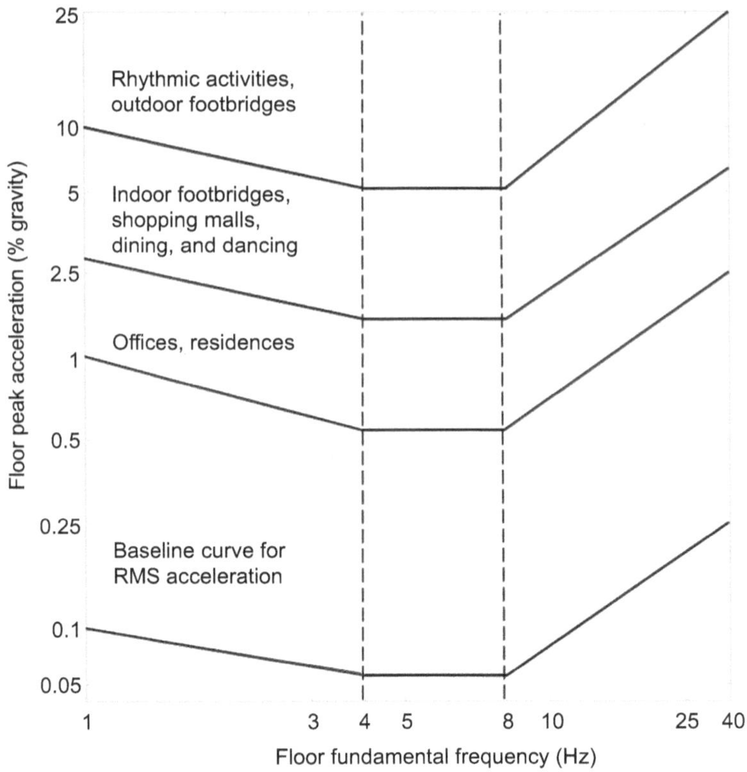

Figure 9.2 ISO acceleration limits (ISO 2631-2, 1989).

9.2.2 Classification of acceptability criteria for floor vibration

Acceptability criteria of floor vibration due to human walking are primarily developed based on experimental tests which account for different structural properties such as material, self-weight, span length, etc. Floor response to human walking is largely influenced by the mass and fundamental frequency of the floor. For floors with large mass and low fundamental frequency, the vibration performance is primarily dominated by the natural vibration; whereas for floors associated with small mass and high fundamental frequency, the vibration performance is likely to be influenced by the deflection induced by steps of human walking. Generally, acceptability criteria concerning floor vibration can be categorized as follows.

1. Criteria for heavy floors (low-frequency floors).
2. Criteria for lightweight floors (high-frequency floors).

Low-frequency floors are susceptible to resonant responses due to walking excitation, while high-frequency floors are likely to dissipate the individual impulses generated by footfalls without undergoing resonance. However, the categorization of low- and high-frequency floors is not consistent among researchers. For example, Burstrand and Talja (2001) defined 8 Hz as the limit of fundamental frequency for low- and high-frequency floors. Classifying floors with fundamental frequencies of less than 10 Hz and greater than 12 Hz as low and high frequency, respectively, Brownjohn and Middleton (2008) also stated that floors with fundamental frequency values between 10 and 12 Hz have been found to fall into a gray area where both high- and low-frequency behavior occurs. Although it is not explicitly stated, the Applied Technologies Council (ATC) design guide (Allen et al., 1999) employed 8 Hz as the limit of fundamental frequency for low- and high-frequency floors, with a caution if the fundamental frequency is less than approximately 10 Hz there can be resonance amplification of the impulse vibrations.

9.2.2.1 Acceptability criteria and design method for CFS floors

A limited number of recommended design criteria for lightweight residential floors are available (Onysko, 1985; Ohlsson, 1988; Chui, 1988; Smith and Chui, 1988; Hu et al., 2001), most of which are primarily focused on timber floor applications.

Onysko (1985) conducted an extensive field survey to evaluate the vibration performance of residential floor systems, involving the assessment of 646 wood floor systems. The assessment was based on subjective evaluations made by homeowners, not on testing data. Results from the study showed that the dynamic response due to an impact load (eg, heel drop or sandbag drop) and deflection due to a concentrated static load were the two parameters that correlated best with perceived vibration acceptability, and these were used to develop a satisfactory design criterion. The criterion was used by the National Building Code of Canada (NBCC, 1990) to develop the allowable span table for lightweight floors built with solid lumber joists, and subsequently modified by the Canadian Construction Materials Center to take account of the use of engineered wood joists (CWC, 1996). The criterion was also adopted by the ATC (Allen et al., 1999) in its guide on floor vibration design. The criterion, shown in Fig. 9.3, is applicable for lightweight floors with a fundamental frequency greater than 8 Hz by limiting the maximum deflection under a 1 kN concentrated load placed at the center of the floor to ensure adequate stiffness in the floor system.

$$\Delta_p \leq 0.61 + 2.54e^{-0.59(L-1.95)} \leq 2.0\ \text{mm} \qquad [9.1]$$

where Δ_p = maximum floor deflection (mm) and L = span of floor (m).

The maximum floor deflection, Δ_p, from Eq. [9.1] is calculated as

$$\Delta = \frac{C_{pd}}{N_{eff}} \frac{PL^3}{48\ \mathrm{EI_{eff}}} \qquad [9.2]$$

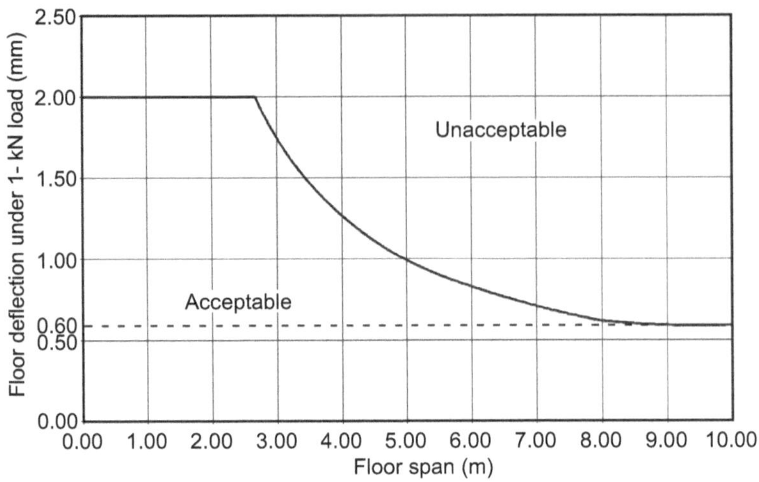

Figure 9.3 Floor deflection limit under 1 kN concentrated load.

where

C_{pd} = continuity factor for concentrated load
= 1.0 for simply supported joists
= 0.7 for joist continuous over two spans

EI_{eff} = effective flexural stiffness (N mm^2)

L = span of floor (mm)

N_{eff} = effective number of joists

P = 1000 N.

The effective composite bending stiffness, EI_{eff}, is computed according to Eq. [9.3]. It allows accounting for different floor details, such as presence or absence of ceiling, glued subfloor, and top finishes (eg, concrete topping, OSB).

$$EI_{eff} = \frac{EI}{1 + \frac{\gamma EI}{C_{pd} EI_m}} \qquad [9.3]$$

where

γ = ratio of shear deflection to flexural deflection

C_{pd} = continuity factor (1.0 for simple supports, 0.7 if joists are continuous at one or two supports)

EI = flexural stiffness of the floor panel (N mm^2)

EI_m = flexural stiffness of the structural member (N mm^2).

According to the ATC (Allen et al., 1999), the ratio of shear deflection to flexural deflection is zero for CFS C-joists. Therefore Eq. [9.3] may be rewritten as

$$EI_{eff} = EI \qquad [9.4]$$

and EI is computed according to Eq. [9.5], which accounts for partial composite action between joists and panel decking.

$$EI = EI_m + EI_{top} + EA_m y^2 + EA_{top}(h_{top} - y)^2 \qquad [9.5]$$

where

EA_c = axial stiffness of concrete (N/mm)
EA_{flr} = axial stiffness of the floor deck (N/mm) of width $s = s \times EA_{wpar}$
EA_{wpar} = unit axial stiffness of wood subflooring parallel to the member (N/mm)
EA_m = axial stiffness of the structural member (N)
EA_{top} = effective axial stiffness of the floor deck parallel to the member (N) $= \frac{EA_{flr}}{1+\frac{10EA_{flr}}{S_{flr}L_{flr}^2}}$
E_c = modulus of elasticity of concrete (N/mm²)
EI_c = unit flexural stiffness of concrete (N mm²/mm) $= 1.2E_cI_c$ (note: multiplying by 1.2 accounts for dynamic loading)
EI_m = flexural stiffness of structural member (N mm²)
EI_{top} = flexural stiffness of floor deck parallel to the member (N mm²)
$= EI_{wpar} + EI_c + \frac{EA_{wpar}EA_c h_{cw}^2}{EA_{wpar}+EA_c}$ (for concrete topping on wood deck)
$= EI_{wpar}$ (for no concrete topping)
EI_{wpar} = unit flexural stiffness of subfloor parallel to the member (N mm²/mm)
h_{cw} = half the total deck thickness (mm)
h_{top} = distance to the centroid of the floor deck from centroid of member (mm)
I_c = moment of inertia of concrete topping per unit width (mm³)
L = span of joist (mm)
L_{flr} = width of floor panel along the span of the structural member
= 1220 mm or span L for concrete-topped floors
= L (for concrete-topped floors)
s = spacing of members (or effective width of deck) (mm)
S_{flr} = slip modulus for floor deck to member connection (N/mm/mm)
Y = distance to the centroid of the system (Fig. 9.4) from the centroid of the member (up is positive) (mm) $= \frac{EA_{top}h_{top}}{EA_m+EA_{top}}$.

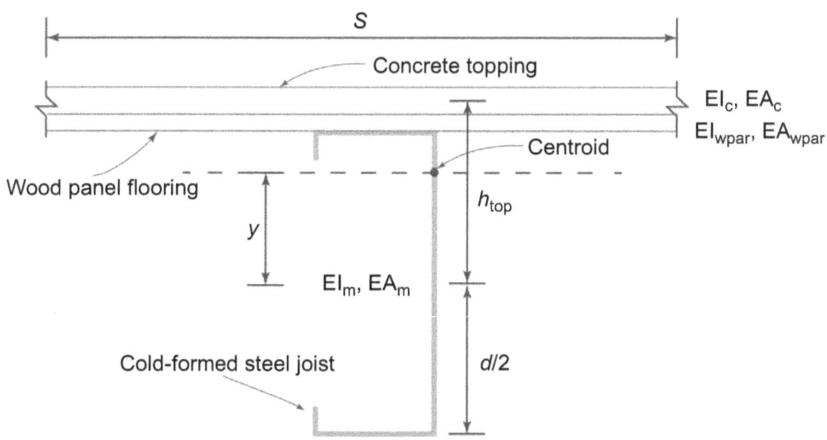

Figure 9.4 Floor cross-section.

The effective number of joists, N_{eff}, shown in Eq. [9.2] is computed as

$$1/N_{eff} = DF_b - DF_v \qquad [9.6]$$

in which DF_b and DF_v are distribution factors for bending elements (ie, subfloor with or without concrete topping, strapping, gypsum ceiling) and shear elements (ie, solid blocking, cross-bridging), respectively.

$$DF_b = 0.0294 + 0.536K_1^{1/4} + 0.516K_1 - 0.31K_1^{3/4} > 0 \qquad [9.7]$$

$$DF_v = -0.00253 - 0.0854K_1^{1/4} + 0.0797K_2^{1/2} - 0.00327K_2 > 0 \qquad [9.8]$$

where

K_1 = stiffness parameter for transverse flexure $= \frac{K_j}{K_j + \sum K_{bi}}$

K_2 = stiffness parameter for transverse shear $= \frac{\sum K_{vi}}{\sum K_{bi}}$

K_{bi} = stiffness of individual bending members $= 0.585\frac{EI_{bi}L}{s^3}$, for transverse flexural components (ie, panel and deck) (N/mm) $= \left(\frac{2a}{L}\right)^{1.71}\frac{EI_x}{s^3}$, for strongbacks and straps (N/mm)

K_j = joist panel stiffness (N/mm) $= \frac{EI_{eff}}{L^3}$

K_{vi} = stiffness of cross bridging and blocking (N/mm) $= \left(\frac{2a}{L}\right)^{1.71}\frac{GA}{s}$

EI_{bi} = unit flexural stiffness of individual bending members (ie, subfloor without concrete topping or gypsum ceiling) (N mm^2/mm)

$= EI_{wperp} + EI_c + \frac{EA_{wperp} \times EA_s h_{cw}^2}{EA_{wperp} + EA_c}$ (subfloor with concrete topping)

EI_{eff} = effective flexural stiffness, defined above (N mm^2)

EI_x = effective flexural stiffness of strongbacks and straps (Table 4.5) (N mm^2)

a = distance of the element to the closest end of joist (mm)

A = effective shear area of blocking (mm^2)

G = effective shear modulus of blocking (N/mm^2)

h_{cw} = distance from centre of subfloor to centre of concrete topping (mm) $= \frac{t_s}{2} + \frac{t_c}{2}$

L = joist span (mm)

s = joist spacing (mm)

t_c = thickness of concrete topping (mm) ($t_c = 0$ if no concrete is used)

t_s = thickness of subfloor (mm).

In addition to the static criterion, the ATC (1999) provides methods to estimate the fundamental frequency of one-way and two-way floor systems. Eq. [9.8] represents the fundamental frequency of a one-way floor structure.

$$f = 0.18\sqrt{\frac{g}{\Delta_j}} \qquad [9.9]$$

where

Δ_j = deflection of the joist floor panel due to the uniformly distributed weight it supports = $C_{flr}\frac{5wL^4}{384EI_{eff}}$
C_{flr} = continuity factor for determining natural frequency = 1.0 (for simply supported beams and continuous joists of equal span; for continuous joists with unequal span, see ATC Fig. 4.3 (Allen et al., 1999))
EI_{eff} = effective composite bending stiffness, calculated based on Eq. [9.4]
g = acceleration due to Earth's gravity = 9810 mm/s^2
L = span of the joist (mm)
s = joist spacing (mm)
w = effective floor weight (N/mm) = $W \cdot s$
W = weight per unit area of floor supported by the joist (N/mm).

It is worth point outing that the foregoing criterion and design method are based on lightweight wood floors, thus the applicability and accuracy of the design method for CFS floors need to be investigated. Kraus (1997) tested 25 residential floor systems supported by CFS joists—one of the earliest published vibration studies that investigated floors supported by CFS joists. The objective was to assess the predictive ability of the methods proposed by Ohlsson (1988), AS 3623 (1993), Johnson (1994), and CWC (1996). The recommended method for use with CFS joist-supported floors was the CWC method based on agreement with test results and designer usability. The author and his research team at the University of Waterloo have carried out 77 tests since 1999 on lightweight floors supported by CFS C-joists with plywood or oriented strand board (OSB) subfloors. The test results were compared to five design methods: Ohlsson (1988), AS 3623 (1993), Johnson (1994), CWC (1996), and the ATC (Allen et al., 1999). Although considerable differences were found between the test results and predictions of the five analytical methods, the tests indicate that the ATC (1999) method provides better results than the other four methods for predicting natural frequencies and deflections (Xu and Tangorra, 2003). Research on vibration of lightweight steel floor systems was also conducted by Zhao et al. (2003) in Australia.

However, there are a few shortcomings associated with the ATC method. For example, some parameters used in evaluating the effective number of joists to resist 1 kN concentrated load (N_{eff}), such as the distribution factors DF_b and DF_v in Eq. [9.6] and Eq. [9.7], respectively, are obtained from calibration with a large number of tests of wood floors carried out by the Canadian wood construction industry (CWC, 1996), which may not be applicable for CFS floors. In addition, the slip modulus, which accounts for the interaction between subflooring material and floor joists, is not provided for floors supported by CFS joists. The values of effective shear area and shear moduli provided for bridging and blocking conditions are obtained based on the practice of wood construction, and there is considerable difference in construction details for bridging used in wood and CFS floors. When using the equivalent values associated with CFS construction details to evaluate the effective number of joists, unrealistic values can result (Rizwan, 2000).

It is noted that the applicability of the foregoing design criterion and method requires the fundamental frequency of the lightweight floors to be greater than or equal to 8 Hz. Based on experimental investigations carried out the author and his research team at the University of Waterloo of more than 100 floors supported by CFS C-joists, it is found the fundamental frequencies of the floors range approximately between 10 and 35 Hz.

If the fundamental frequency of the floor is less than 8 Hz, which may occur for long-span CFS floors with thick concrete topping, the way to achieve acceptability for floor vibration is to limit the acceleration response a_p due to walking excitation, as shown in Eq. [9.9].

$$\frac{a_p}{g} = \frac{P_o e^{-0.35 f_n}}{\beta W} \leq 0.05 \quad [9.10]$$

where g is acceleration due to the Earth's gravity, P_o is a constant excitation force of 0.29 kN for floors, f_n is the fundamental frequency of the floor, and βW is the product of the modal damping ratio and effective weight of the floor. The acceleration response can be evaluated according to American Institute of Steel Construction procedure (Murray et al., 1997; Allen et al., 1999). The limit value of 0.05 is based on the ISO recommended permissible peak vibration acceleration level for residential and office occupancy, as shown in Fig. 9.2. The in situ tests carried out by the author showed there was considerable discrepancy between evaluated and measured results (Xu, 2011).

9.3 Effects of construction details on floor vibration

The general discussion presented here is primarily based on a multiphase study of the vibration performance of CFS floors undertaken at the University of Waterloo since 1999. Both laboratory and in situ investigations were carried out, and more than 100 floor configurations were tested. The laboratory investigations were conducted in the University of Waterloo Structures Laboratory, while field investigations were carried out at different locations in Canada and the United States. The experimental investigations conducted in Phases I and II were focused on lightweight floors supported by conventional CFS C-shape joists with plywood or OSB subfloors (Xu and Tangorrra, 2007). The influence of ceiling materials, support conditions, subfloor gluing, and blocking, bridging, and screw patterns was investigated. The floors investigated in Phase III were constructed with CFS C-shape joists with large lip-reinforced web openings (Parnell et al., 2010). The three types of subfloors investigated in Phase III were OSB panels, fiber-reinforced cement panels, and CFS form decks with gypsum-based floor underlay topping. The floor framing in Phases I and II was conventional platform framing, while both platform and ledger framing were investigated in Phase III. The floor framing details altered in Phase III tests were floor framing, subfloor sheathing and topping, ceiling, and strongback.

It is the author's opinion that unfortunately the effects of the different construction details discussed here on floor vibration performance have not been appropriately quantified in the foregoing design method or any other methods that may be applicable to CFS floors. However, understanding how the different construction details influence the vibration characteristics of CFS would certainly help engineers to minimize annoying floor vibration in their design practice.

9.3.1 General discussion and comparison of results between laboratory and in situ tests

For more than 100 floor configurations tested in the laboratory and 25 floors tested in situ, the fundamental frequencies of the floors ranged approximately between 10 and 35 Hz, with minimum and maximum values of 9.8 and 35.7 Hz, respectively. Specifically, the fundamental frequencies of the floors built and tested in the laboratory were 9.8−26.3 Hz, while those associated with in situ floors were 9.9−35.7 Hz. It should be noted that the construction of a few in situ floors was only partially finished; the floor joist bridging and drywall ceiling were not installed. The measured deflections under 1 kN load of both laboratory and in situ floors were between 0.17 and 1.89 mm, in which those associated with the in situ floors ranged between 0.2 and 1.20 mm.

The test results validated that lightweight CFS floors behave essentially as a one-way slab system. When the floor is subjected to a single concentrated load, there is a lateral distribution of the load between the loaded joist and the adjacent joists (load sharing). Load sharing is the ability of a CFS joist floor system to transfer load away from a loaded joist to adjacent joists through the flexural stiffness of the floor in the transverse direction. Load-sharing efficiency is related to the subfloor stiffness, the properties of bridging, the properties of joists, and floor construction details. The effectiveness of floor transverse stiffness can be observed from the profile of the measured joist span deflections in the transverse direction of the floor, as shown in Fig. 9.5, while the effectiveness of load sharing in a floor system can be determined by measuring the joist-end reactions, illustrated in Fig. 9.6.

In almost all cases, the in situ floors had greater fundamental frequency and a considerably greater damping ratio; thus the vibration performance of the floors tested in the laboratory is worse than that of the in situ floors. This conclusion is supported by comparison of the results of measured peak acceleration response based on a heel drop test and RMS acceleration response obtained by a walking test conducted in Phase III. It was found that the values for both peak acceleration responses and RMS acceleration responses associated with the floors in situ were considerably smaller than those of the floors tested in the laboratory.

Supported by these experimental investigations, it is reasonable to conclude that the vibration performance of lightweight CFS floors tested in the laboratory is worse than that of in situ floors; thus the results obtained based on "equivalent" laboratory experimental evaluation should be quite conservative. In other words, the floors in a finished dwelling will display considerably better vibration performance compared to their

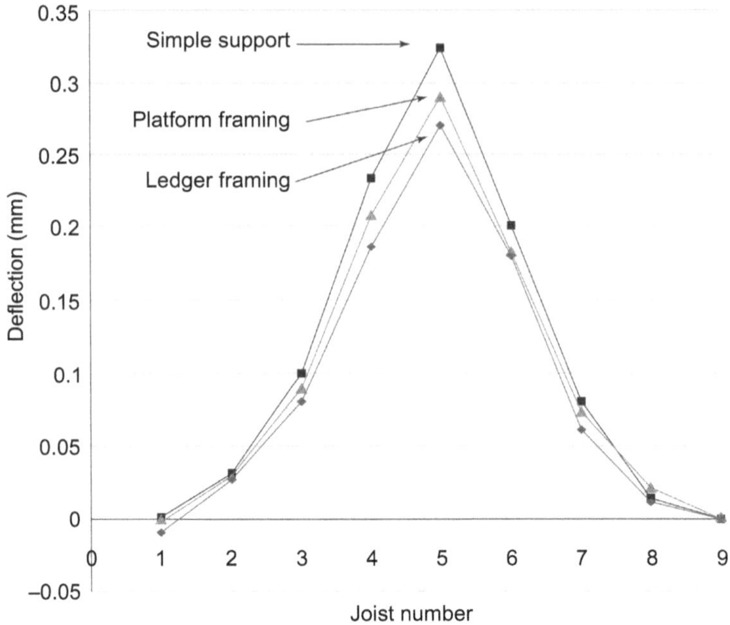

Figure 9.5 Distribution of floor transverse deflection.

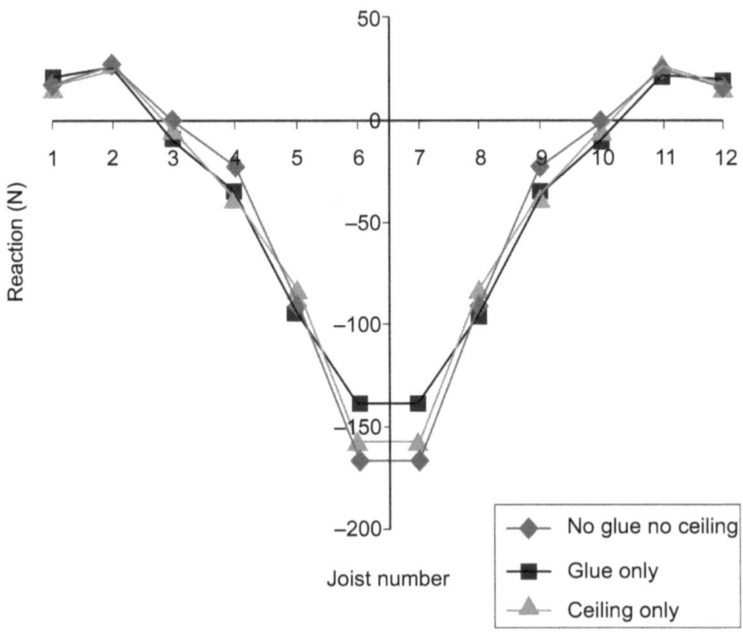

Figure 9.6 Distribution of joist-end reaction.

"laboratory" equivalent, because the in situ floors are supported on all four sides while the floors in the laboratory are only supported on two sides. The increased stiffness from the extra support increases the fundamental frequency and decreases the center deflection of the in situ floors. The increased damping in the in situ floors can be attributed to the presence of support walls. As the in situ floors were constructed with walls above and below the floor edges, the mass of the wall framing and the gypsum board, as well as the friction of wall framing connections, improved the vibration performance of the floors.

The measured static deflections of all the in situ floors subjected to 1 kN concentrated load satisfy Eq. [9.1]. In addition, the RMS acceleration responses of the in situ floors were found to be within the acceptable region of ISO limiting values for residential occupancy shown in Fig. 9.2. Therefore, it appears that all the in situ floors investigated have satisfactory vibration performance.

9.3.2 Effect of floor framing conditions

The effects of three types of floor framing conditions, namely ledger framing, platform framing, and simple support, on vibration characteristics of lightweight CFS floors were investigated in Phase III, while platform framing and simple support cases were studied in Phases I and II. The framing conditions were found to affect fundamental frequency, damping ratio, and center deflection. Ledger framing provided the greatest increase in fundamental frequency and reduction in center deflection. Platform framing also exhibited increased fundamental frequency and reduced center deflection when compared to the simple support condition. Because rotation at the support was partially restrained in some degree for both ledger and platform framing, the bending stiffness of the floor system increased, increasing fundamental frequency and reducing span deflection. Fig. 9.5 shows the typical profiles of midspan transverse deflection distribution of a floor associated with three types of floor framing conditions. The damping ratio appeared to be larger in the simple support condition; the increased flexibility caused greater energy dissipation through friction and displacement-based interactions.

9.3.3 Effect of subfloor sheathing material and topping

For floors without subfloor topping, in comparing OSB and fiber-reinforced cementitious subfloor panels it was observed that floors with cementitious panels had a lower fundamental frequency due to the heaver weight. Increased stiffness reduced deflections in floors with cementitious panels. For lightweight floor systems, annoyance may be dominated by local deflections from walking, making stiffness the primary factor influencing performance.

In comparing floors with fiber-reinforced cementitious panel to those with CFS deck and gypsum-based floor underlay, it was observed that floor systems with a steel deck had a lower fundamental frequency. The thicker gypsum-based floor underlay gives CFS deck floors a greater overall weight, reducing the fundamental frequency. However, there is a significant decrease in center deflection for floors with a CFS deck, because the axis of the CFS deck was perpendicular to the joists, which significantly increased the transverse stiffness of the floor system and the effective width of the joist panel.

With regard to the influence of subfloor topping with gypsum-based floor underlay, the test results showed that floors with subfloor topping yielded lower fundamental frequencies than those without topping. The decrease in frequency is attributed to the added mass when a subfloor topping was added. The floors with subfloor topping were found to be stiffer than the floors without a topping, and a reduction in center deflection was seen in all cases. Though the subfloor topping increased stiffness, the added mass outweighed the stiffness effects and the fundamental frequency was still decreased. The addition of a subfloor topping has both positive and negative impacts on vibration performance of the floors. The decrease in fundamental frequency is seen as a negative impact, as it may bring the fundamental frequency closer to the increased human sensitivity range of 4–8 Hz. The decrease in center deflection is seen as a positive impact on vibration performance because stiffness increases.

9.3.4 Effect of ceiling and gluing subfloor

The effect of the ceiling was investigated in all three phases. When a gypsum board ceiling is fastened directly to the bottom flange of the floor joists or to resilient channels which are fastened to the bottom flange of the floor joists, the ceiling adds mass to the assembly and the resilient channel and gypsum board appear to increase continuity between the joists. It is cautioned that similar stiffening effects are not observed in floors with drop ceilings. The influence of adding a ceiling on the damping ratio was not conclusive based on the results obtained from the tests.

As shown in Fig. 9.6, fastening the OSB subfloor to the joists with glue in addition to screws yields increases the composite action between the OSB subfloor and steel joists and the transverse flexural stiffness of the floor, which consequently increase the floor frequency and decrease deflection. Greater reduced floor frequency and deflection were observed when a gypsum ceiling was attached to a glued floor. Gluing the OSB subfloor also results in a decrease of the floor damping ratio.

9.3.5 Effect of bridging and blocking

In CFS floor framing, a bridging is a flat steel strap or U-shape CFS member fastened to the bottom flange of the joists, while a blocking is a C-shape or U-shape CFS member attached to the web of floor joists and subfloor panels to transfer shear force. Bridging and blocking provide lateral stability to the supporting joists and prevent joist rollover. The effects of floor joist bridging and blocking on floor vibration characteristics were investigated in Phase I and II tests. The bridging tested in Phases I and II was composed of CFS strapping (58×1.44 mm) fastened to the bottom flange of each floor joist. Selected floors were tested with one and three rows of bridging. Two blocking sizes, $41 \times 152 \times 1.22$ mm and $41 \times 203 \times 1.22$ mm CFS track sections, and two blocking patterns, one block installed at every five joist spaces and one block installed at every joist space, were investigated individually.

The tests results showed that increasing the number of steel strapping bridging rows had little impact on the frequency, maximum deflection, and damping ratio of a floor. With regard to blocking, it was found that increasing the depth of blocking from 152 to 203 mm reduced floor deflection and damping but did not greatly influence the frequency. Installing one block at every joist space instead of one block at every five joist spaces resulted in a minor increase in the fundamental frequency but a substantial increase in the third-mode frequency of the floors, a considerable decrease in the maximum deflection, and a minor decrease in damping ratio. These results are expected, as installing one block at every joist space enhances a floor's transverse stiffness and load-sharing capability.

9.3.6 Effect of screw spacing

It is current practice to use a screw spacing pattern of 152 and 305 mm at the edges and in the interior area of a subfloor sheathing, respectively. It is believed that increased floor stiffness may also be obtained through closer screw spacing. This study confirms that closer screw spacing provides additional, though slight, stiffness to a floor. The tests conducted in Phases I and II showed that doubling the screw spacing results in lower floor frequencies and larger deflections, which translate into poorer vibration performance of floor systems even though an increase in damping was observed.

9.3.7 Effect of strongback

A C-shape stiffening member, oriented perpendicular to the joist direction, was added in some floor configurations. The common name of this detail is a strongback. The strongback passes through the web openings in the joists at midspan. Clip angle brackets are screwed in place to connect the web of the strongback and web of the joists at each joist intersection. Floors with strongback installed in Phase III were tested in two configurations: restrained end and free end. The restrained-end configuration replicated a long, narrow floor (high aspect ratio) where all joists are effective right to the outer wall. In this configuration, the strongback was fastened to the web of every joist at midspan and the ends were fastened to a rigid pedestal using a clip angle and five screws, restricting any vertical deflection and partially restraining rotation of the strongback at its ends. The strongback with restrained ends added constraints to the system, which caused the fundamental frequency to increase and the center deflection to decrease, and changed the modes of vibration by imposing a restraint on midspan deflection at the outer joists. Addition of a restrained strongback also increased the damping ratio, possibly due to friction from the added screw connections.

The free-end configuration simulated a short, wide floor (low aspect ratio) where the width of the region defined by the number of effective joists is less than the floor width. In this configuration, the strongback was fastened to the joists in the same fashion as the restrained-end configuration, but its ends were not restrained in any way. In the free-end configuration the center deflection of the floor was decreased, which can be attributed to the added transverse stiffness from the strongback.

However, this configuration had no conclusive effect on the fundamental frequency of the floors or on the damping ratio.

9.3.8 Effect of nonload-bearing partition wall

Last but certainly not least, the presence of nonload-bearing partition walls on the floor, particularly in the transverse direction of floor joists, has the most significant role in enhancing floor vibration performance. Although not tested in laboratory, some in situ tests on partially finished floors (without nonload-bearing partition walls installed) and finished floors (with the partition walls installed) show that the walls increase the floor stiffness, which results in a substantial reduction of floor deflection. To ensure the effectiveness of the partition walls, the bottom track of the walls needs to be firmly connected to the subfloor.

Sources of further information and advice

Considering the fact that designing CFS floors to control vibration induced by human activity has not been addressed well in current design practice, the following guides, although some may not be applicable to CFS structures, provide very useful information on floor vibration characteristics, design criteria and procedures, and remedy measures on controlling vibration induced by human activity, and would certainly be valuable sources of further information for design practice.

1. Murray, T.M., Allen, D.E., Ungar, E.E., 1997. Design Guide 11: Floor Vibrations Due to Human Activity, American Institute of Steel Construction and Canadian Institute of Steel Construction.
2. Allen, D.E., Onysko, D.M., Murray, T.M., 1999. Design Guide 1: Minimizing Floor Vibrations, Applied Technology Council.
3. Bustrand, H., Talja, A., 2001. Design Guide for Light Weight Steel Floors Due to Human Induced Vibrations, Swedish Institute of Steel Construction and VTT Building Technology, Steel and Composite Structures.

Generally speaking, noticeable vibrations occur when a floor has inadequate stiffness, low damping, and low mass. When designing lightweight CFS floors where vibration performance is a concern, it is important to consider the construction materials and framing details that will be implemented. In practice, ensuring adequate floor stiffness is essential to avoid disturbing floor vibration induced by human walking. Specifically, the term "floor stiffness" consists of three aspects: floor joist stiffness, floor transverse stiffness, and subfloor stiffness. Based on the experimental investigation performed on laboratory and in situ floors in this study, the construction details which will effectively enhance vibration performance of CFS floors, in addition to increased floor joist stiffness, are as follows.

1. A nonload-bearing partition wall when it is firmly connected to the floor, particularly placed in the transverse direction of floor joists, noticeably enhances vibration performance of CFS floors.

2. A subfloor topping, such as a lightweight gypsum-based underlay, increases the floor stiffness in both longitudinal and transverse directions, as well as the floor mass. However, the excessively added mass could result in a decrease in natural frequencies of the floors to the human-sensitive region of 4–8 Hz. It is recommended that a lightweight subfloor topping should be applied to floor systems to improve the vibration performance as long as the fundamental frequency remains above 10 Hz.
3. When possible, installing a strongback with the ends fastened to wall studs effectively increases floor transverse stiffness, which increases fundamental frequency and reduces the deflection. However, the effectiveness of a strongback with fastened ends is expected to decrease as the floor width increases. In such cases, installation of multiple rows of strongback is recommended.
4. When possible, placing floor joist spans continuously instead of using a simply supported single span will maximize floor stiffness. Floor joist-end framing conditions were found to affect fundamental frequency, damping ratio, and span deflection through the partial restraint on joist-end rotations. Ledger framing appears to have better vibration performance over platform framing, by increasing fundamental frequency and reducing span deflection of the floor.
5. Installation of a gypsum board ceiling supported by resilient channels which are attached directly to the bottom flange of the floor joist can also effectively reduce the fundamental frequency and span deflection of a floor system. The effect of increasing floor stiffness was not observed in a drop ceiling. If no lightweight subfloor topping is applied, it is more effective to install a gypsum board ceiling and glue the OSB subfloor to the floor joists to increase the composite action between the subfloor and joists.

Floor vibration due to walking as a serviceability concern has not been well addressed in design and construction of lightweight CFS floors. The high strength and stiffness of steel have the advantage of achieving longer floor spans. However, floors with longer span and lighter weight are likely to be susceptible to annoying vibrations induced by normal human activity, such as walking. Designing a lightweight CFS floor to control these vibrations can be difficult due to lack of appropriate design guidelines. In addition, determining acceptability can be a challenge, as human perception of vibration is subjective and variable. Floors designed and constructed to meet specific vibration criteria that satisfy the majority of occupants may not be accepted by a few individuals. Nevertheless, it is important to address the issue of floor vibration in the design stage, as correcting the issue after construction is finished can be quite costly. Fortunately, the in situ investigation of 25 CFS floors conducted by the author has shown a certain degree of confidence in current design and construction practice in North America. The results of the tests show that CFS floors with appropriate design and construction details can perform well to counter floor vibration caused by human walking.

References

AISI, 2012. North American Standard for Cold-formed Steel Framing—Floor and Roof System Design. American Iron and Steel Institute, Washington, DC.

Allen, D.E., Onysko, D.M., Murray, T.M., 1999. Design Guide 1: Minimizing Floor Vibration. Applied Technology Council.

AS 3623, 1993. Domestic Metal Framing Code. Standards Association of Australia, Homebush.
Brownjohn, J., Middleton, C., 2008. Procedures for vibration serviceability assessment of high-frequency floors. Engineering Structures 30 (6), 1548–1559.
Burstrand, H., Talja, A., 2001. Design Guide for Light Weight Steel Floors Due to Human Induced Vibrations. Swedish Institute of Steel Construction and VTT Building Technology, Steel and Composite Structures, Stockholm/Espoo/Lulea.
CWC, 1996. Development of Design Procedures for Vibration Controlled Spans Using Engineered Wood Members (Final Report Prepared for Canadian Construction Materials Center and the Industry Partnership Consortium), Canadian Wood Council.
Chui, Y., 1988. Evaluation of vibrational performance of light-weight wooden floors. In: Proceedings of the 1988 International Conference on Timber Engineering 1988, pp. 102–109.
Grether, W.F., 1971. Vibration and human performance. Human Factors: The Journal of the Human Factors and Ergonomics Society 13 (3), 203–216.
Hu, L.J., Chui, Y., 2004. Development of a design method to control vibrations induced by normal walking action in wood-based floors. In: Proceedings of the 8th World Conference on Timber Engineering 2004, pp. 217–222.
Hu, L.J., Chui, Y.H., Onysko, D.M., 2001. Vibration serviceability of timber floors in residential construction. Progress in Structural Engineering and Materials 3 (3), 228–237.
ISO 2631-2, 1989. Evaluation of Human Exposure to Whole-body Vibration-part 2: Human Exposure to Continuous and Shock-induced Vibrations in Buildings (1 to 80 Hz). International Standards Organization, Geneva.
ISO 2631-2, 2003. Mechanical Vibration and Shock: Evaluation of Human Exposure to Whole-body Vibration-part 2: Vibrations in Buildings (1 Hz to 80 Hz). International Standards Organization, Geneva.
Johnson, J.R., 1994. Vibration Acceptability in Wood Floor Systems. Virginia Polytechnic Institute and State University.
Kraus, C.A., 1997. Floor Vibration Design Criterion for Cold-formed C-shaped Supported Residential Floor Systems. Virginia Polytechnic Institute and State University.
Lenzen, K.H., 1966. Vibration of steel joist-concrete slab floors. Engineering Journal–American Institute of Steel Construction 3 (3), 133–136.
Murray, T.M., Allen, D.E., Ungar, E.E., 1997. Floor Vibrations Due to Human Activity. American Institute of Steel Construction, Chicago.
National Building Code of Canada (NBCC), 1990. Appendix A: Explanatory Material for the NBCC. National Building Code of Canada, Ottawa.
Ohlsson, S.V., 1988. Springiness and Human-induced Floor Vibrations: A Design Guide. Swedish Council for Building Research.
Onysko, D., 1985. Serviceability Criteria for Residential Floors Based on a Field Study of Consumer Response. Forintek Canada Corporation.
Parnell, R., Davis, B.W., Xu, L., 2010. Vibration performance of lightweight cold-formed steel floors. Journal of Structural Engineering, ASCE 136 (6), 645–653.
Reiher, H., Meister, F., 1931. The Effect of Vibration on People. Translated from Forschung auf dem Gebiete des Ingenieurwesens, 2(11) 381–6, by U. S. Air Material Command. Translation F-TS-616-RE. Wright Field. AMC, Ohio, 1946.
Rizwan, U., 2000. Evaluation of Different Floor Vibration Criteria for Cold-formed Steel Residential Construction. University of Waterloo.
Smith, I., Chui, Y.H., 1988. Design of lightweight wooden floors to avoid human discomfort. Canadian Journal of Civil Engineering 15 (2), 254–262.
Xu, L., Tangorra, F.M., 2003. Dynamic Behaviour of Residential Floor Systems Using Cold-formed Steel Joists (Phase II). University of Waterloo, Waterloo, Ontario, Canada.

Xu, L., Tangorra, F., 2007. Experimental investigation of lightweight residential floors supported by cold-formed steel C-shape joists. Journal of Constructional Steel Research 63 (3), 422–435.

Xu, L., 2011. Floor vibration in lightweight cold-formed steel framing. Advances in Structural Engineering 14 (4), 659–672.

Zhao, X., Taplin, G., Alikhail, M., 2003. Connection stiffness and natural frequency of DuraGal lightweight floor systems. Structural Engineering and Mechanics 15 (3), 269–284.

Blast resistance of cold-formed steel buildings

10

N.A. Rahman
The Steel Network, Inc., Durham, North Carolina, United States

10.1 Introduction

This chapter discusses the main design concepts to mitigate effects of exterior blast pressure on cold-formed steel (CFS) buildings. Exterior blast can be the result of an accidental explosion or a terrorist attack, and can release a wide range of energy. The exterior walls and roof of CFS buildings constitute the components or structural systems that first experience the blast pressure and therefore need to be protected from its effects. The main objective of building design against blast effects is to prevent the occurrence of disproportionate collapse, which can cause severe building damage and loss of life among building occupants. A second objective is to reduce the possibility of a breach through the exterior walls and the roof that would allow blast pressure from outside the building to get into the building interior and cause injury to the occupants.

10.1.1 Blast wave effects on buildings

The shock wave created by an explosion travels radially in all directions away from the source of explosion. The resulting pressure at the front of the wave decreases rapidly as the wave expands. When the shock wave arrives at a surface of a building, it creates a peak incident pressure that decays with time to a negative pressure: Fig. 10.1 illustrates a typical pressure–time history curve. The peak incident pressure forms a positive phase, while the negative pressure forms a negative phase of the pressure–time history. The time from peak pressure to zero pressure is the positive phase duration, or simply duration, and the area under the pressure–time curve in the positive phase is the impulse of the blast wave (FEMA, 2003).

10.1.2 Blast wave design parameters

When an incident pressure wave hits a surface of a building that is not parallel to its travel direction, the wave reflects and creates a stronger pressure, called peak reflected pressure. In a CFS building, such a surface can be an exterior wall component or a sloped roof component. The main design parameters of a blast wave can be listed as:

- peak reflected pressure, P_r
- positive phase duration, t_p
- positive phase impulse, I_p.

Recent Trends in Cold-Formed Steel Construction. http://dx.doi.org/10.1016/B978-0-08-100160-8.00010-4

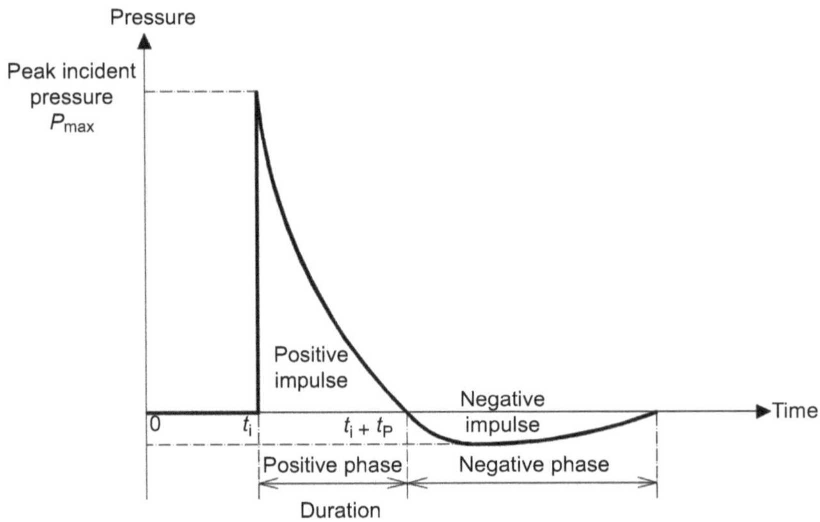

Figure 10.1 Typical pressure—time history curve from a blast event.

Although the negative phase is typically weaker than the positive phase, the negative phase parameters may become important in some design applications. Examples are the design check of anchorages against uplift loads acting on roof trusses or roof rafters, and the design check of fastener edge distance in concrete for anchorage of base connections of exterior walls.

Other design parameters of a blast wave that may be considered based on the application are dynamic pressure, shock wave velocity, and blast wave length (ASCE, 2010).

10.1.3 Codes and standards

Blast research programs started in the United States in the early 1960s and resulted in the development of multiple procedures to evaluate structures exposed to explosive threats. One of the early blast-resistant design manuals developed by the US Department of Defense is TM 5—1300, which included step-by-step analysis and design procedures. TM 5—1300 has been replaced by Unified Facilities Criteria (UFC) 340-02 (DoD, 2008a) and UFC 4-010-01 (DoD, 2013). The Steel Construction Institute in the United Kingdom published several technical notes and design guides for structural components exposed to explosive threats (SCI, 1999). Risk analysis considers multiple variables, including factors such as the value of assets inside the building and the severity of the explosive threat. The American Society of Civil Engineers (ASCE) Standard 59 (ASCE, 2011) and Federal Emergency Management Agency (FEMA) 452 (FEMA, 2005) provide guidance on performing risk analysis for buildings. The Canadian Standards Association (CSA) Standard S850 (CSA, 2012) does not address risk assessment but provides performance criteria to limit the extent of blast damage in buildings.

10.2 Dynamic modeling and performance criteria

Dynamic modeling is the primary engineering method to analyze structural components and systems under blast loads. For most building use and occupancy, it is impractical to design the building to withstand a blast load without any damage. The objective of dynamic modeling is thus to predict the damage level of the building after the blast.

10.2.1 Resistance function

The dynamic response of CFS framing members or panels to blast pressure is a function of the force generated from the blast, the mass of the member or panel resisting the blast, and the system stiffness or resistance. It can be reasonably assumed that the dynamic response of CFS panels is dominated by flexural modes. However, panel end shear and end bearing are part of the flexural response and must be considered. A panel subjected to a uniform blast load can be simplified to a mass-spring system model, as illustrated in Fig. 10.2. The mass-spring model represents the panel as an equivalent single-degree-of-freedom (SDOF) system with a nonlinear spring. The mass, M, and the dynamic load, $F(t)$, of the equivalent system are based on the panel mass and blast load, respectively. The spring stiffness is based on the panel lateral load flexural capacity, and can be found as the ratio of force to displacement. Salim et al. (2005) explained that for complicated systems, the force–displacement relationship cannot be defined by a single stiffness value, k, and thus a static resistance function, R, is normally utilized. The equivalent SDOF system is chosen so that the displacement, y, of the concentrated mass is the same as that for a primary point on the panel, such as its midspan. The equation of motion of the system is written by applying the concept of dynamic equilibrium as:

$$M\ddot{y} + R - F(t) = 0 \qquad [10.1]$$

where $M\ddot{y}$ is the force of inertia; $\ddot{y}$ is the acceleration; R is the resistance function of the spring force; and $F(t)$ is the applied external blast force as a function of time t. To solve

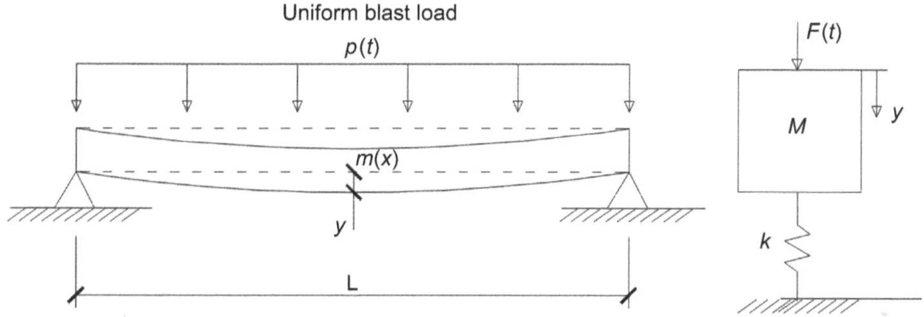

Figure 10.2 SDOF mass-spring system model.

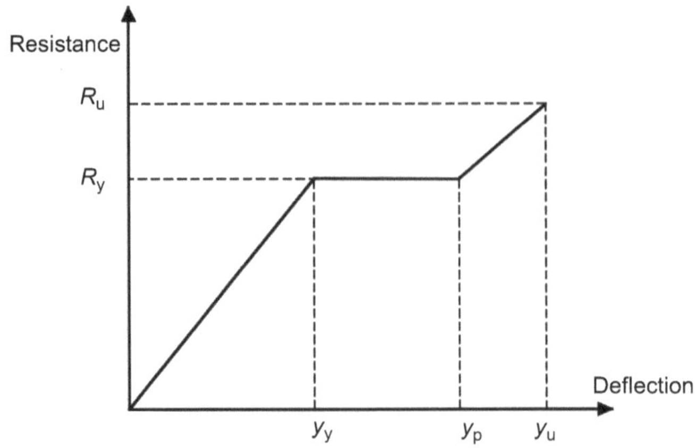

Figure 10.3 Idealized trilinear resistance function.

Eq. [10.1] for the displacement y, the resistance function R must be developed, which is typically completed using laboratory tests for the structural system under consideration. An idealized trilinear resistance function containing linear region, plastic region, and tension membrane region of behavior is shown in Fig. 10.3. Solution of Eq. [10.1] with resistance functions having multiple regions requires a numerical integration process.

10.2.2 Dynamic material properties

Since blast loading on buildings is not an ordinary design load, it is permitted under dynamic design procedures to consider available additional strength that can be quantified in the material. Multiple test data report that average yield strength of installed CFS material is typically greater than the minimum values specified by the building code. In addition, similar to hot-rolled steel, CFS exhibits higher strength under rapid applied load or rapid strain rate. As a result, the flexural capacity of CFS panels loaded dynamically is permitted to include the effects of the static increase factor (SIF) and the dynamic increase factor (DIF) applied to the steel minimum yield strength. The SIF accounts for the difference between the specified minimum and expected actual yield strength, while the DIF accounts for high strain rate effect when the member or panel is loaded dynamically. The recommended SIF and DIF values for CFS are summarized in Table 10.1, as reported in multiple blast design standards.

10.2.3 Performance criteria and response limits

Building components exposed to explosive threat such as exterior walls, roofs, openings, and skylights should be engineered to reduce the possibility of a breach. The performance criteria of the building as a whole or one of its components are a function of the building use and occupancy considerations. ASCE Standard 59 (ASCE, 2011) and the US Protective Design Center (PDC) TR-06-08 (DoD, 2008b) present the

Table 10.1 Static and dynamic increase factors for CFS components

	Static increase factor	Dynamic increase factor	
		Bending/ Shear	Tension/ Compression
ASCE Standard 59	1.1	1.1	1.1
CSA Standard S850 (far field for DIF)	1.1	Bending 1.1 Shear –	Tension 1.0 Compression 1.1
ASCE design of blast-resistant buildings in petrochemical facilities	1.21	1.1	1.1
DoD UFC 3-340-02	1.21	1.1	1.1

performance criteria in the form of a level of protection (LOP) and expected component damage. A very low LOP reflects extensive damage but no building collapse, with safe evacuation of occupants. A high LOP reflects a fully functional building with only superficial component damage and continuation of occupancy. Table 10.2 gives recommendations for the expected component damage for each LOP and whether the component is primary, secondary, or nonstructural.

For example, a gravity load-bearing CFS stud wall should be considered a primary structural component and be expected to suffer only superficial damage if the building is assigned a medium LOP. A lateral wind-bearing CFS stud wall can be treated as a secondary structural component and be expected to suffer heavy damage if the building is assigned a low LOP.

Response limits should be defined for various structural applications to accept or reject the results of a dynamic analysis. ASCE Standard 59 (ASCE, 2011) and PDC TR-06-08 (DoD, 2008b) define the response limits for individual components dominated by flexural mode in the form of maximum ductility ratio, μ (mu), and maximum support rotation, θ (theta). Ductility ratio μ is the ratio of the maximum deflection of a component to its maximum elastic deflection. Support rotation θ is the angle through

Table 10.2 Expected component damage with level of protection

Level of protection	Primary structural component	Secondary structural component	Nonstructural component
Very low	Heavy	Hazardous	Hazardous
Low	Moderate	Heavy	Heavy
Medium	Superficial	Moderate	Moderate
High	Superficial	Superficial	Superficial

Data from ASCE, 2011. Blast Protection of Buildings, ASCE 59–11 Standard, American Society of Civil Engineers, USA, with permission from ASCE.

Table 10.3 **Flexural response limits for CFS components**

Component		Superficial damage		Moderate damage		Heavy damage		Hazardous damage	
		μ	θ	μ	θ	μ	θ	μ	θ
Wall studs	Top slip track	0.5	–	0.8	–	0.9	–	1.0	–
	Connected top and bottom	0.5	–	1.0	–	2.0	–	3.0	–
	Ends anchored to develop full tensile membrane capacity	0.5	–	1.0	0.5°	2.0	2.0°	5.0	5.0°
Corrugated panel (one way)	Full tension membrane	1.0	–	3.0	3.0°	6.0	6.0°	10.0	12.0°
	Some tension membrane	1.0	–	–	1.0°	–	4.0°	–	8.0°
	Limited tension membrane	1.0	–	1.8	1.3°	3.0	2.0°	6.0	4.0°
Girt or purlin		1.0	–	–	3.0°	–	10.0°	–	20.0°

Data from ASCE, 2011. Blast Protection of Buildings, ASCE 59−11 Standard, American Society of Civil Engineers, USA, with permission from ASCE.

which a flexural component subject to blast loading has rotated at its supports when it achieves its maximum dynamic deflection. Table 10.3 shows the maximum allowed μ and θ for CFS wall studs, corrugated panels, wind girts, and purlins in flexure with various end connection conditions and various expected damage levels, as addressed above. It should be noted that some of the response limits in Table 10.3 are relatively stringent, primarily due to the limited shear and web-crippling capacity of the end connections of those CFS components. Higher response limits can be achieved when a tension membrane is utilized; however, end connections must be designed to deliver matching shear and crippling capacity.

10.3 Blast resistance test studies

There has been a flow of work in the last two decades to study and categorize the response of CFS components to blast loads. The bulk of this research has been supported by federal agencies of the US government, and driven by the need to retrofit exterior facades or curtain-wall cladding of existing reinforced concrete or structural steel-framed buildings. This section presents some notable test studies that provide primary knowledge about resistance of CFS components to blast loads.

10.3.1 Full-scale wall tests

The US Army Engineer Research and Development Center conducted multiple full-scale field tests to study blast response of steel stud exterior walls (Salim et al., 2003, 2005; Salim and Townsend, 2004). The objectives of the tests were to validate the performance of a wall stud anchoring systems in developing the full tension membrane capacity of the studs and to demonstrate the contribution of the wall mass to its dynamic response. In a field test designated BREW-1, two steel stud walls with blast-resistant anchor connections were tested, one with a brick finish and one with an exterior insulation finish system (EIFS). The walls were subjected to a blast loading corresponding to a medium vehicle bomb with reflected pressure and reflected impulse. Fig. 10.4 shows the posttest conditions of the two walls. The deflection measurements from the tests, as a measure of damage, demonstrated major difference in wall response due to inertial mass effects. The steel stud wall with the heavier brick finish showed less deflection and limited damage compared to the steel stud wall with EIFS finish, which showed extensive damage. In addition, the tests proved that steel stud walls with full tensile anchor connections are adequate to retrofit existing exterior facades or curtain walls. In a field test designated EWRP-8, one steel stud wall with a punched window opening and EIFS finish along with anchor connections was tested. The wall used double studs, in comparison to single studs in test BREW-1, to examine the reliance on stud strength and ductility only to absorb the energy from a blast. The wall was subjected to a blast loading corresponding to a large vehicle bomb with reflected pressure and impulse. Fig. 10.5 shows the posttest condition of the wall with

Figure 10.4 Blast posttest exterior views of test BREW-1: (a) EIFS façade; (b) brick façade. From Salim, H., Dinan, R., Townsend, P.T., 2005. Analysis and experimental evaluation of in-fill steel-stud wall systems under blast loading. Journal of Structural Engineering, ASCE, 131–8, 1216–1225, with permission from ASCE.

Figure 10.5 Blast posttest exterior view of test EWRP-8.
From Salim, H., Dinan, R., Townsend, P.T., 2005. Analysis and experimental evaluation of in-fill steel-stud wall systems under blast loading. Journal of Structural Engineering, ASCE, 131–8, 1216–1225, with permission from ASCE.

some recognized stud damage. However, the steel stud wall as a whole survived the test and provided adequate capacity to resist the imposed blast pressures on the wall with window openings.

The same researchers obtained the static resistance function of the steel studs by statically loading a pair of studs via a loading tree that distributed the load from hydraulic actuators to equally spaced points on the studs, as shown in Fig. 10.6. The

Figure 10.6 Loading tree system.
From Salim, H., Dinan, R., Townsend, P.T., 2005. Analysis and experimental evaluation of in-fill steel-stud wall systems under blast loading. Journal of Structural Engineering, ASCE, 131–8, 1216–1225, with permission from ASCE.

load-deflection equilibrium was used to calculate an equivalent pressure per unit width to draw the resistance function. A resistance chamber was another method used to obtain the static resistance function on a full-scale steel stud wall section. A vacuum pump was used to apply uniform pressure to failure. The steel stud wall went into tension membrane phase after local buckling and yielding of the studs, followed by a well-defined softening region.

10.3.2 Blast simulator wall tests

The blast simulator system was developed at the University of California San Diego in 2005 as an alternative to open-field blast testing. The system uses a combination of pressurized nitrogen and hydraulic oil in conjunction with actuators, known as blast generators, to reproduce the levels of pressures and impulses generated in an actual explosion. The blast generators in the system control a piston assembly which is rapidly forced out of the generators under a controlled motion. This motion can be programmed by specifying input parameters, including pressure levels and starting point of the piston before the test. A heavy metal plate is attached to each piston, and the mass of this plate can be adjusted to control the incoming energy applied to the test specimen. A urethane pad with a specific pyramidal geometry is mounted on the front face of the metal plate to control the loading duration of the blast, and consequently control the applied impulse. The system is equipped with pressure transducers in each blast generator to provide readings of the actuator pressures, and accelerometers placed on the back of the impacting metal plate and the test specimen to provide acceleration history throughout the test.

Multiple steel stud wall specimens with a composite gypsum board and steel sheet sheathing were tested using the blast simulator (Stewart et al., 2014; SGH, 2012). The objectives of the first test series were to optimize the wall design for typical low LOP blast requirements and to design the end connections to develop the response demand without premature shear failure. The second test series added additional objectives, such as examining the use of high-strength steel sheet sheathing, calibrating a nonlinear finite element numerical model, loading the stud walls to increasing dynamic force demands, and identifying limit states and failure modes for the steel stud, end connection, and sheathing under equivalent blast loads. A combination of reaction steel angles, stud hold-down clips, and wall track bearing washers was used at end connections of the steel studs in different wall specimens. The tests showed that a combination of stud hold-down clips and wall track bearing washers provide the needed stud anchorage against applied impact loads, and the use of high-strength steel sheet resulted in reduced lateral deformations and enhanced overall wall response (see Fig. 10.7). The developed finite element numerical model closely predicted the peak and residual response, as well as strength degradation characteristics of the steel stud walls. Design guidelines and construction details have been proposed to aid design engineers with a range of explosive threats and target performance criteria.

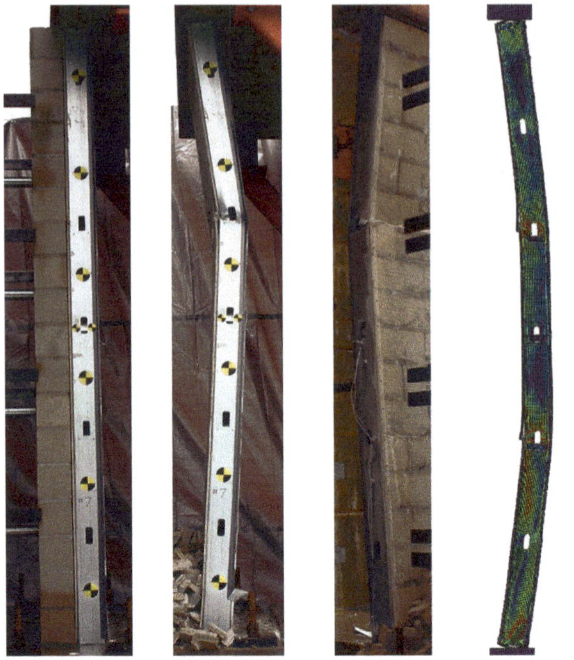

Figure 10.7 Blast simulator pretest and posttest conditions of steel stud wall.
From SGH, 2012. Use of HSLA-V Steel Studs for Blast-resistant Walls, Simpson Gumpertz & Heger Inc., Project 047079 for Vanadium Technology Program, USA, with permission from SGH.

10.3.3 Connection tests

The University of North Carolina at Charlotte tested multiple CFS stud walls subjected to blast loads in an open-field explosion (Whelan et al., 2015). The objective of the tests was to investigate the performance of the head-of-wall deflection connection and the midheight stud bracing (bridging) system. The blast loads applied to the specimens had reflected impulses and pressures consistent with the design values of a package bomb at a conventional standoff distance. Recorded data included reflected and incident pressures, acceleration, midheight displacement time series, and high-speed and regular-speed videos of each test. The observed damage patterns included local and distortional buckling of the stud flanges, web crippling of the stud webs, bridging connection pull-out of screws, and screw pull-out between the studs and the bottom wall track (see Fig. 10.8). However, the investigated head-of-wall deflection clip connection and the bridging system used generally kept the wall intact and allowed the steel studs to develop their end rotation and flexural ductility.

Figure 10.8 Blast posttest for head-of-wall connection and midheight bracing system. Courtesy of and with permission from the Steel Network, Inc.

10.4 Analysis and design methods

The UFC published by the US Department of Defense provide analysis and design approaches for CFS framing members subjected to blast loads. Standard 10 of UFC 4-010-01 (DoD, 2013) provides static and dynamic design approaches to size the supporting structural members of exterior window openings and roof skylights. Recommendations are given to evaluate the support reactions and determine the strength resistance factors for the design of end connections. Reaction forces from the exterior wall system into the primary structure do not have to be carried through the floor diaphragm as long as the mass of the floor can dissipate the reactions before they are transferred to the primary building frame.

10.4.1 Static design approach for framing around openings

The static design approach in UFC 4-010-01 (DoD, 2013) is a simplified method to determine the required moment and shear strengths of the jamb, header, and sill that frame an exterior wall opening. However, this approach is only allowed if the blast is expected at a distance equals to or greater than a conventional standoff distance where the wall around the opening does not need to be designed for blast, and it only can be used for punched openings. A punched window can be defined as an opening that is bound by a header, a sill, and two jambs, and has façade material on all four sides.

In the static design approach, a tributary area increase factor (C) is calculated as:

$$C = \frac{a_{\text{trib}}}{a_{\text{wall}}} \geq 1.0 \qquad [10.2]$$

where a_{wall} is the tributary area for the typical conventional wall section (the typical steel stud), and a_{trib} is the combined tributary area for the supported window and wall section. The required moment and shear strengths of the supporting structural member are calculated as:

$$M_{\text{SSE}} = C \cdot M_{\text{CW}}, \quad V_{\text{SSE}} = C \cdot V_{\text{CW}} \qquad [10.3]$$

where M_{SSE} and V_{SSE} are the required moment and shear strength of the supporting jamb, header, or sill, while M_{CW} and V_{CW} are the moment and shear strength of the conventional steel stud section in the wall. The static approach is limited to punched openings, since it uses the tributary area of the typical steel stud in the wall to calculate factor C. The end connections of the supporting structural member must be designed for a force equal to the increased member shear capacity V_{SSE}.

10.4.2 Dynamic design approach

In general, it is preferable to use dynamic analysis over static analysis in design for blast loads to utilize the extra material dynamic strength. The wall studs and supporting structural members of exterior wall openings can be designed for the appropriate blast pressure, impulse, and duration using dynamic analysis. These design parameters shall be determined from the applicable explosive weight at the actual standoff distance at which the member is sited. The blast loads should be applied over the areas tributary to the member being analyzed. UFC 4-010-01 (DoD, 2013) recommends a dynamic design approach using SDOF analysis and a strength design with load factors of 1.0 and strength reduction factors of 1.0 for the analysis and member sizing. The designer can take advantage of using the SIF and DIF applied to the steel minimum yield strength, as detailed in Section 10.2.2 above. The US PDC offers an Excel-based spreadsheet to design structural components using the SDOF approach. The tool is called the SDOF blast effects design spreadsheet, or SBEDS (DoD, 2008c), and is available at www.pdc.usace.army.mil/software/sbeds after obtaining approved access. An engineer with a basic knowledge of structural dynamics and SDOF mass-spring systems is expected to be able to write his/her own software tool to analyze CFS framing under blast loads.

The flexural response limits (ductility ratio, μ, and support rotation, θ) defined in Table 10.3 for CFS wall studs are to be used as the design limits in SDOF analysis. The "Top slip track" category in Table 10.3 refers to single- or double-slip track detail at the top end of the wall where there is no mechanical attachment between the wall track and the wall stud, and therefore it is assigned a relatively low ductility ratio response limit. The "Connected top and bottom" category refers to the presence of a mechanical attachment between the wall stud and the top and the bottom wall tracks using screws or weld, and therefore is assigned a higher ductility ratio response limit. This mechanical attachment is also permitted to have a fixed or a deflection clip angle. The "Ends anchored to develop full tensile membrane capacity" category in Table 10.3 refers to the wall having structural plate and angle bolted connection at the top and the bottom ends of the studs to ensure the wall studs reach their tensile membrane capacity at yield and allow large support rotation before any connection failure occurs.

10.4.3 Equivalent static loads for roof trusses

There is no recognized design approach for CFS roof truss systems under blast loads, since such systems have not been assessed through field blast tests or equivalent laboratory tests to develop their flexural response limits. A method to calculate an equivalent static design pressure to blast loads has been developed by independent practicing engineers (Warfield et al., 2015). The method considers both the positive and the negative pressure loading phases to act on the top chord of the truss, and idealizes the blast load as a uniform applied pressure to the top chord. A closed-form expression for midspan peak truss deflection based on an SDOF mass-spring model for a simply supported beam was derived, and used to develop the response spectra for the inbound and rebound deflection responses of the truss. The input parameters for the expression are the blast load pressure–time history, the ductility ratio limit of the truss, and the fundamental period of the truss, which is in turn a function of the mass and the inbound and rebound elastic stiffness of the truss. Since there are no recommended ductility ratio response limits for CFS trusses, the ductility ratio limits for open-web steel joists are used, assuming the two systems have similar flexural response. The equivalent static blast pressure can then be calculated as a truss properties constant multiplied by the peak dynamic pressure. In addition to the global response of the truss, attention should be given to the flexural response of the individual members of the top chord facing the blast load, the shear response of the web members near supports, and the maximum uplift forces at anchorage points to the structure below.

10.4.4 Design of connections

It is recommended that connections of CFS members are designed dynamically to the equivalent static reaction of that member subjected to a specific blast load, and the design should account for all potential failure modes. The equivalent static reaction is calculated based on the ultimate flexural strength of the CFS members, to ensure the connection has a strength equal to or greater than the full ultimate strength of the member. Unlike the design of members, a load and resistance factor design with load factors of 1.0 and strength reduction factors of ø (phi) less than 1.0 is recommended for sizing the components of the connection (DoD, 2012). A connection of CFS members typically consists of the connector (clip), the screw attachment between the connector and the member, and the anchor attachment between the connector and the supporting structure, as illustrated in Fig. 10.9.

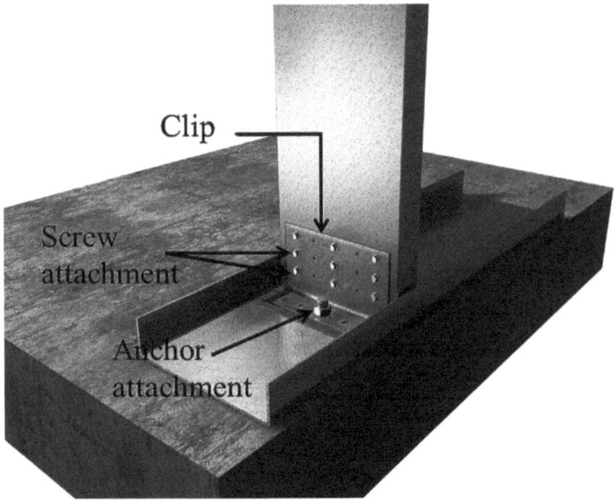

Figure 10.9 Components of CFS connection.

Appropriate strength reduction factors ø of all the components should be obtained from the related specifications and standards or approved test data. The reduced component design strength is obtained by applying the ø factors to the component test ultimate strength or calculated nominal strength, as depicted in Fig. 10.10 (Rahman and Kennedy, 2012). Determination of the reference strength is typically controlled by the desired LOP for the building. If the test ultimate strength of the connection component is not available, an equivalent dynamic strength for the component can be calculated, as illustrated in Fig. 10.10.

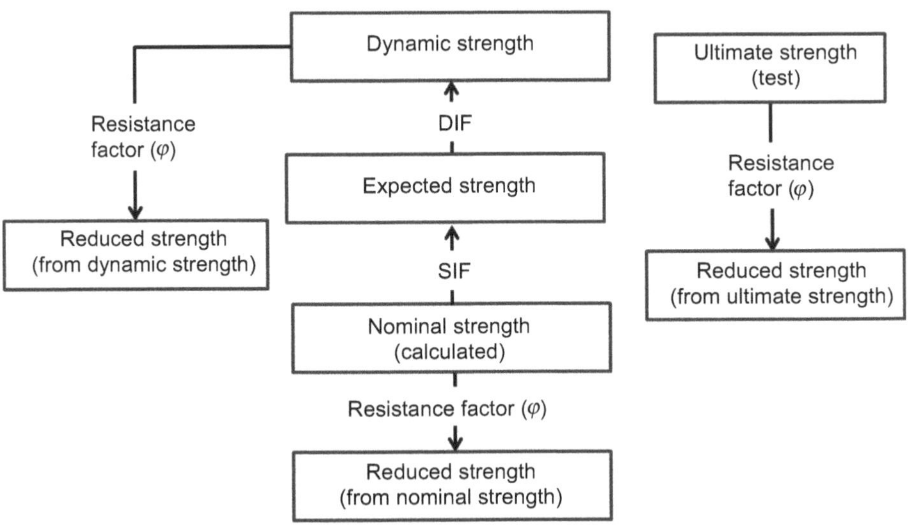

Figure 10.10 Strength relationship diagram.

References

ASCE, 2010. Design of Blast-resistant Buildings in Petrochemical Facilities, second ed. American Society of Civil Engineers, USA.

ASCE, 2011. Blast Protection of Buildings. ASCE 59−11 Standard. American Society of Civil Engineers, USA.

CSA, 2012. Design and Assessment of Buildings Subjected to Blast Loads. S850−12 Standard. Canadian Standards Association, Canada.

DoD, 2008a. Design of Structures to Resist Accidental Explosions. UFC 3-340-02. Department of Defense, USA.

DoD, 2008b. Single Degree of Freedom Structural Response Limits for Antiterrorism Design. PDC TR-06−08. Protective Design Center, Army Corps of Engineers, USA.

DoD, 2008c. User's Guide for the Single-degree-of-freedom Blast Effects Design Spreadsheets (SBEDS). PDC TR-06-02. Protective Design Center, Army Corps of Engineers, USA.

DoD, 2012. Blast Resistance Design Methodology for Window Systems Designed Statically and Dynamically. PDC TR-10−02. Protective Design Center, Army Corps of Engineers, USA.

DoD, 2013. DoD Minimum Antiterrorism Standards for Buildings. UFC 4-010-01. Department of Defense, USA.

FEMA, 2003. Risk Management Series: Reference Manual to Mitigate Potential Terrorist Attacks Against Buildings. FEMA 426. Federal Emergency Management Agency, USA.

FEMA, 2005. Risk Management Series: Risk Assessment − A How-to Guide to Mitigate Potential Terrorist Attacks Against Buildings. FEMA 452. Federal Emergency Management Agency, USA.

Rahman, N.A., Kennedy, K., 2012. Special seismic and blast design − cold-formed steel connection strength. Structure Magazine 7, 18−19.

Salim, H., Townsend, P.T., 2004. Explosion-resistant steel-stud wall systems. In: Proc., Structures 2004. ASCE, pp. 1−10.

Salim, H., Dinan, R., Kiger, S., Townsend, P.T., Shull, J., 2003. Blast-retrofit wall systems using cold-formed steel stud. In: Proc., 16th Engineering Mechanics Conf. ASCE.

Salim, H., Dinan, R., Townsend, P.T., 2005. Analysis and experimental evaluation of in-fill steel-stud wall systems under blast loading. Journal of Structural Engineering, ASCE 131−8, 1216−1225.

SCI, 1999. Design Guide for Stainless Steel Blast Wall. Technical Note 5. The Steel Construction Institute, UK.

SGH, 2012. Use of HSLA-V Steel Studs for Blast-resistant Walls. Simpson Gumpertz & Heger Inc., Project 047079 for Vanadium Technology Program, USA.

Stewart, L.K., Freidenberg, A., Hegemier, G., 2014. Design and Testing of Steel Stud Wall System for Blast Mitigation. In: Structures under Shock and Impact XIII, vol. 141, pp. 39−50.

Warfield, B., Weaver, M., Pellock, M., 2015. Analyzing cold-formed steel roof trusses for blast loading. Structure Magazine 8, 9−12.

Whelan, M., Ralston, A., Weggel, D., 2015. Blast testing of cold-formed steel-stud wall panels. Journal of Performance of Constructed Facilities. http://dx.doi.org/10.1061/(ASCE)CF.1943-5509.0000734, 04015008.

Part Four

Innovation and sustainability in cold-formed steel construction

New section shapes using high-strength steels in cold-formed steel structures in Australia

11

G.J. Hancock, C.H. Pham
University of Sydney, Sydney, NSW, Australia

11.1 Introduction to innovative shapes

11.1.1 High-strength steels and design standards and specifications

High-strength cold-formed lipped C- and Z-sections have been available in Australian and international markets for over 30 years (Pham and Hancock, 2009a). The high-strength steels are normally cold-reduced steel to Australian Standard AS1397 (Standards Australia, 2011) and have typical yield stress values in the range 450–550 MPa (65–80 ksi) depending upon the thickness and the degree of cold reduction. They are normally galvanized with zinc. A recent development is to include additional flange, web, and lip stiffeners to increase the buckling capacity in line with the increased yield strength. A further development is to decrease the gauge of the sections once additional stiffeners are included. This chapter describes the behavior of these sections, demonstrating their structural advantages.

The development of the direct strength method (DSM) for design of cold-formed sections (Schafer and Pekoz, 1998) has allowed the buckling behavior of innovative and complex shapes to be better understood. It is now incorporated in both the North American specification NAS:2012 (AISI, 2012) and the Australian/New Zealand standard AS/NZS 4600:2005 (Standards Australia, 2005). The signature curves of buckling stress versus buckle half-wavelength for innovative sections can be investigated to optimize the shapes for subsequent use with the DSM. The original DSM was developed by Schafer and Pekoz for sections in compression and bending. Pham and Hancock (2012a) extended it to shear and combined bending and shear, allowing the effects of web stiffeners to be better included in the design strengths.

11.1.2 Shapes in the Australian and international markets

Some of the new and innovative shapes using high-strength steels which have been introduced into the Australian and New Zealand marketplace over the last 20 years

Recent Trends in Cold-Formed Steel Construction. http://dx.doi.org/10.1016/B978-0-08-100160-8.00011-6

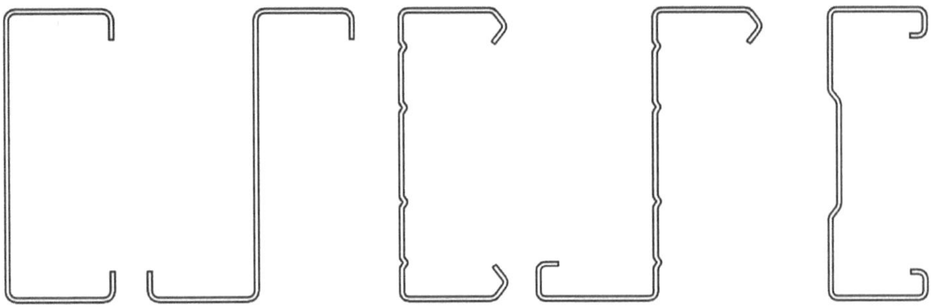

Figure 11.1 Profile shapes of manufactured purlins.

are shown in Fig. 11.1(a)–(c). This chapter concentrates on sections of the type in Fig. 11.1, including testing and finite element method (FEM) modeling to demonstrate their advantages. Signature curves in both flexure and shear are shown for use in DSM design equations.

11.1.3 Buckling analysis and the signature curves for innovative shapes

Thin-walled sections subjected to compression and bending undergo a range of elastic buckling modes, from local through distortional to overall flexural and flexural–torsional modes. The curves describing the buckling stress versus the half-wavelength of a buckle were first studied by Hancock (1978) for I-sections in flexure where distortional modes were demonstrated. The curve is commonly called a "signature curve." The signature curves for a range of channel, zed, and hollow flange beam sections in flexure are given in Hancock (2007). One of these curves for a plain C-section is shown in Fig. 11.2. The local, distortional, and flexural–torsional modes are clearly visible at short, intermediate, and long half-wavelengths respectively. The same curve for the section, but with four small intermediate stiffeners in the web and a sloping lip with return lip, is also shown in Fig. 11.2. This section has the trade name SupaCee (Bluescope Lysaght, 2003) and is also shown in Fig. 11.1(b). The increase in the local and distortional buckling stresses is clearly visible in Fig. 11.2, demonstrating the effect of these stiffeners on the elastic buckling stresses and hence their flexural strength according to the DSM.

Shear buckling modes were investigated by Hancock and Pham (2012) and Pham et al. (2012a,b, 2014) using a finite strip buckling analysis program (bfinst7.cpp) including shear stresses. The finite strip theory used is based on that in Plank and Wittrick (1974), and includes complex terms to allow for the phase shifts in the buckling modes across the sections. Several signature curves for channel sections in shear with an intermediate web stiffener are shown in Fig. 11.3. As the depth of the web stiffener increases, the buckling stress increases and the minimum point on the shear

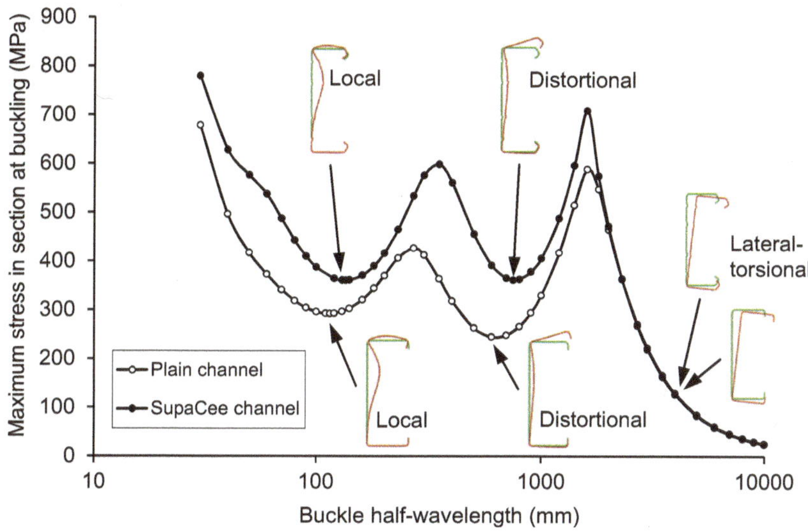

Figure 11.2 Flexural signature curve and buckling modes of cold-formed plain C-sections and SupaCee sections.

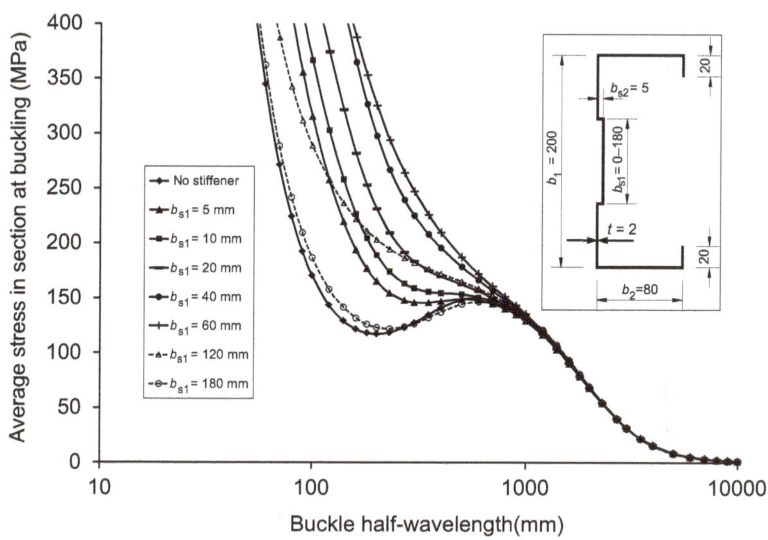

Figure 11.3 Shear signature curves for stiffened web C-section.

signature curve disappears until the web stiffener becomes very deep, at which point the minimum returns and the buckling stresses drop. The corresponding buckling modes are shown in Fig. 11.4. The bending and shear buckling stress curves are used in this chapter within the DSM for these types of stiffened sections.

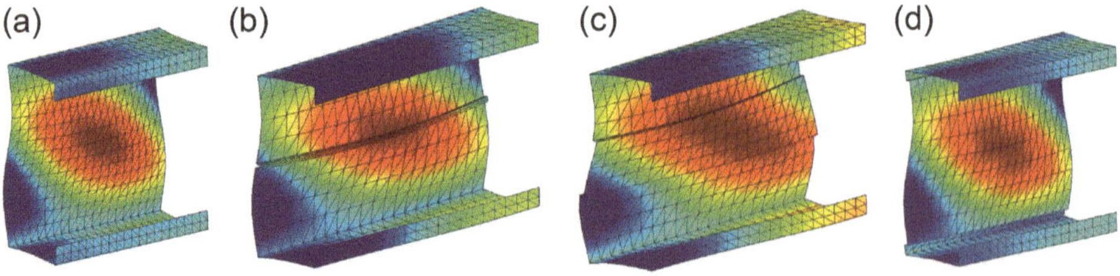

Figure 11.4 Shear buckling mode shapes for stiffened web C-section. (a) No stiffener; (b) $b_{s1} = 5$ mm; (c) $b_{s1} = 60$ mm; (d) $b_{s1} = 180$ mm.

11.2 Cold-reduced high-strength steel standards and specifications

11.2.1 Australian steel standard AS1397

The Australian standard AS1397 (Standards Australia, 2011) describes a range of coated steels including formability and structural grades. Steels manufactured according to AS1397 typically have a designation of the form G450-Z200, where the first symbol (G) indicates that mechanical properties have been achieved or modified by in-line heat treatment prior to hot dipping, the first three-digit number (eg, 450) denotes the minimum yield stress in MPa taken in the rolling direction, the second symbol (Z) denotes zinc coating, and the second three-digit number denotes the coating mass on both surfaces of the steel base measured in grams per square meter of sheet or strip. The structural grades range from G250 to G550: the G450 grade which has thicknesses above 1.5 mm is commonly used for purlins, and the G550 grade which has thicknesses below 1.0 mm is commonly used for studs, tracks, and roof trusses in steel-framed houses. Six coating types are specified: Type Z (zinc coating), Type ZF (zinc coating converted to zinc–iron alloy), Type ZA (zinc/aluminum coating where zinc prevails), Type ZM (zinc/aluminum/magnesium coating where zinc prevails but including aluminum and magnesium), Type AZ (aluminum/zinc coating where aluminum prevails), and Type AM (aluminum/zinc/magnesium where aluminum prevails but including zinc and magnesium).

The stress–strain characteristic of G450 steel to AS1397 is shown in Fig. 11.5, where the low tensile strength (f_u) to yield stress (f_y) ratio can be seen. As the stress–strain curve is rounded due to the cold reduction of the sheet steel, the yield stress f_y is normally based on a 0.2% proof stress, as shown in Fig. 11.5. The steel also has considerably reduced ductility, but still within an acceptable range for structural

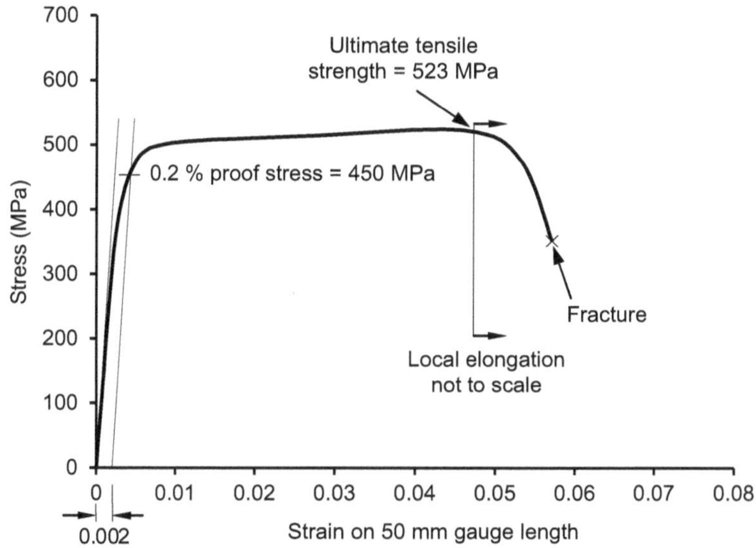

Figure 11.5 AS1397 G450 steel stress–strain curve.

design. G450 steel has a minimum ductility on a 50 mm gauge length of 10%, and G550 steel has a minimum ductility of 2% down to steels of 0.6 mm thickness.

11.2.2 The American Society for Testing Materials steel specifications for high-strength steel

The American Society for Testing Materials (ASTM) has a range of steel specifications which partly cover the structural steels in AS1397. These include the following.

ASTM A653 (SS Grades 60(410), 70(480), 80(550) Class 3) standard specification for steel sheet, zinc-coated (galvanized) or zinc–iron alloy coated (galvannealed) by the hot-dip process.

ASTM A792 (SS Grades 60(410), 70(480), 80(550) Class 3) standard specification for steel sheet, 55% aluminum–zinc alloy coated by the hot-dip process.

ASTM A1039 (SS Grades 55(380), 60(410), 70(480), 80(550)) standard specification for steel, sheet, hot-rolled, carbon, commercial, and structural, produced by the twin-roll casting process.

ASTM A1063 (SS Grades 55 (380), 60(410), 70(480), 80(550) Class 1) standard specification for steel sheet, twin-roll cast, zinc-coated (galvanized) by the hot-dip process.

Steels to ASTM A653 and ASTM A1063 in the classes specified have a minimum elongation of 3% and are very similar to AS1397 G450, G500, and G550 steels with Z and ZF coatings. Steels to ASTM A792 in the classes specified have a minimum elongation of 3% and are very similar to AS1397 G450, G500, and G550 steels with AZ coatings.

11.3 Bending behavior and design

The extension of the DSM of design to these innovative sections has required calibration against test data to include the sections in those prequalified according to the North American specification NAS:2012 (AISI, 2012) and the Australian/New Zealand standard AS/NZS 4600:2005 (Standards Australia, 2005). This section discusses the testing of both the innovative SupaCee section and the plain C-section in flexure, and compares the results with the DSM.

11.3.1 DSM design for flexure, including plastic behavior

11.3.1.1 Local buckling strength

The nominal flexural strength at local buckling (M_{nl}) of beams without holes is determined from NAS:2012, Appendix 1, Section 1.2.2.1.2 (AISI, 2012) as follows:

$$\text{For } \lambda_l \leq 0.776: \quad M_{nl} = M_{ne} \qquad [11.1]$$

$$\text{For } \lambda_l > 0.776: \quad M_{nl} = \left[1 - 0.15\left(\frac{M_{crl}}{M_{ne}}\right)^{0.4}\right]\left(\frac{M_{crl}}{M_{ne}}\right)^{0.4} M_{ne} \qquad [11.2]$$

where

λ_l is the nondimensional slenderness used to determine M_{nl} $\left(\lambda_l = \sqrt{M_{ne}/M_{crl}}\right)$
M_{ne} is the critical elastic lateral–torsional buckling moment
M_{crl} is the elastic local buckling moment of the section ($M_{crl} = S_f f_{crl}$)
f_{crl} is the elastic local buckling stress of the section in bending
S_f is the section modulus about a horizontal axis of the full section.

11.3.1.2 Distortional buckling strength

The nominal flexural strength at distortional buckling (M_{nd}) of beams without holes is determined from NAS:2012, Appendix 1, Section 1.2.2.1.3 (AISI, 2012) as follows:

$$\text{For } \lambda_d \leq 0.673: \quad M_{nd} = M_y \qquad [11.3]$$

$$\text{For } \lambda_d > 0.673: \quad M_{nd} = \left[1 - 0.22\left(\frac{M_{crd}}{M_y}\right)^{0.5}\right]\left(\frac{M_{crd}}{M_y}\right)^{0.5} M_y \qquad [11.4]$$

where

λ_d is the nondimensional slenderness used to determine M_{nd} $\left(\lambda_d = \sqrt{M_y/M_{crd}}\right)$
M_y is the yield moment of the full section ($M_y = S_f F_y$)
M_{crd} is the elastic distortional buckling moment of the section ($M_{crd} = S_f f_{crd}$)
f_{crd} is the elastic distortional buckling stress of the section in bending.

11.3.2 Inelastic reserve strength

In the 2012 edition of the North American specification, inelastic reserve for local and distortional buckling strengths has been included based on the work of Shifferaw and Schafer (2007). The resulting DSM equations are given by Eq. [11.5] as follows.

For $\lambda_l \leq 0.776$ and $\lambda_d \leq 0.673$, and sections symmetric about the axis of bending:

$$M_n = M_y + \left(1 - 1\Big/C_y^2\right)(M_P - M_y) \quad [11.5]$$

where $C_{yl} = \sqrt{0.776/\lambda_l} \leq 3$, $C_{yd} = \sqrt{0.673/\lambda_d} \leq 3$, and M_p is the plastic moment equal to $S_f f_y$ where S_f is the plastic section modulus of the full section.

An alternative extended inelastic local and distortional buckling strength was proposed in Pham and Hancock (2013) as follows.

For $\lambda_l \leq 1.55$ and $\lambda_d \leq 1.45$, and sections symmetric about the axis of bending:

$$M_{ny} = M_y + \left(1 - 1\Big/C_y^2\right)(M_P - M_y) \quad [11.6]$$

where $C_{yl} = \sqrt{1.55/\lambda_l} \leq 3$, $C_{yd} = \sqrt{1.45/\lambda_d} \leq 3$, and M_{ny} is an inelastic moment with extended slenderness limit.

11.3.3 Flexural testing of plain C-sections and SupaCee sections

The experimental program comprised a total of 24 tests (12 for plain C-lipped sections and 12 for SupaCee sections) conducted in the J. W. Roderick Laboratory for Materials and Structures at the University of Sydney. All tests were performed in the 2000 kN capacity DARTEC testing machine, using a servo-controlled hydraulic ram. A photo of the test setup for the common four-point loading configuration is shown in Fig. 11.6(a). Two different commercially available plain C and SupaCee sections of 150 and 200 mm depths, as shown in Fig. 11.1, were chosen, with three different thicknesses of 1.5, 1.9, and 2.4 mm for plain C-lipped sections and 1.2, 1.5, and 2.4 mm for SupaCee sections. The average measured dimensions and properties for the bending series of both plain C-lipped and SupaCee sections are given in Pham and Hancock (2010a, 2013) respectively.

Twelve tests (six of plain C-lipped sections and six of SupaCee sections) had eight 25 × 25 × 5EA (equal angle) straps connected by self-tapping screws on the top flanges which were uniformly distributed in the pure bending moment region between the two loading points, as shown in Fig. 11.6(b). The purpose of the straps was to force the channel members to buckle locally rather than by distortional buckling. The remaining tests were run without the six middle 25 × 25 × 5EA straps, as shown in Fig. 11.6(c). Only two straps adjacent to the loading points were attached to the channel members to prevent distortion at the loading points.

The comparisons of the DSM design rules with the test data are shown graphically in Figs. 11.7–11.10. In Fig. 11.7 the yield moment (M_y) is used in both the horizontal axis for nondimensional slenderness and the vertical axis for testing to predicted

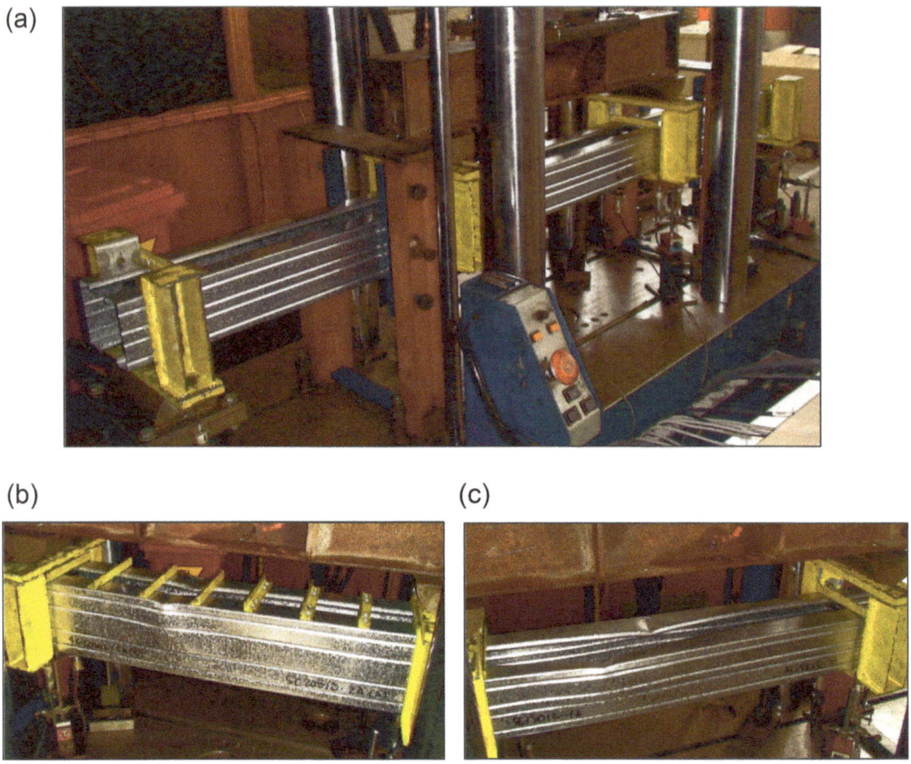

Figure 11.6 Bending test setup configuration for tests with and without straps. (a) Four-point bending rig; (b) Supa sections with straps; (c) Supa sections without straps.

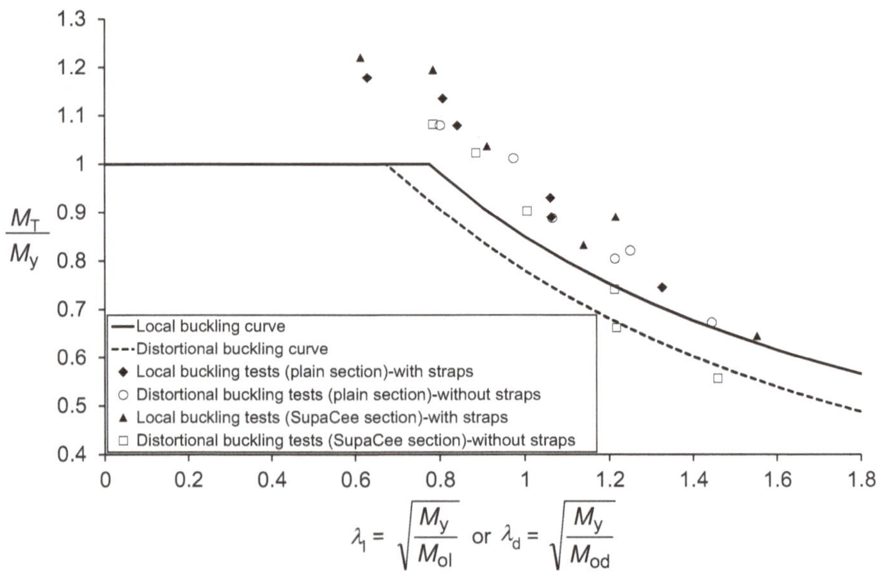

Figure 11.7 Comparison of DSM predictor curves with test data for beam: yield moment case.

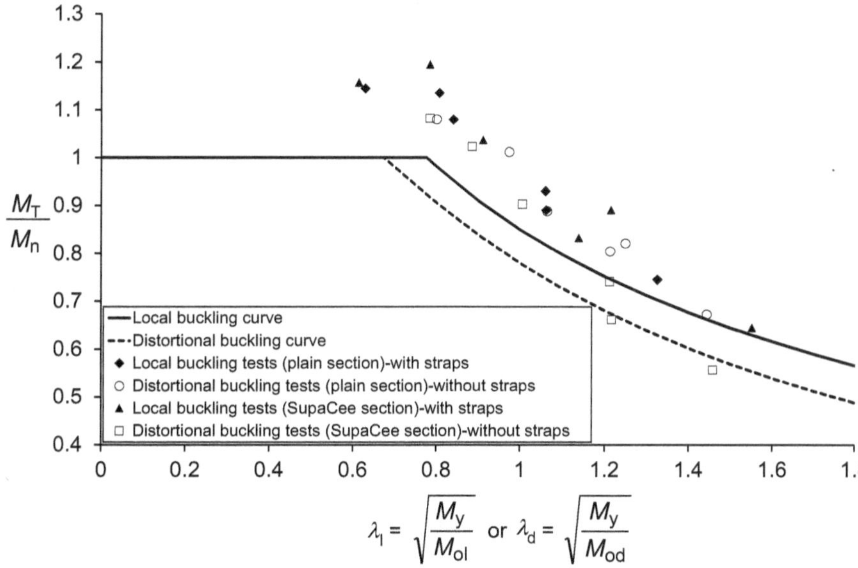

Figure 11.8 Comparison of DSM predictor curves with test data for beam: inelastic moment case.

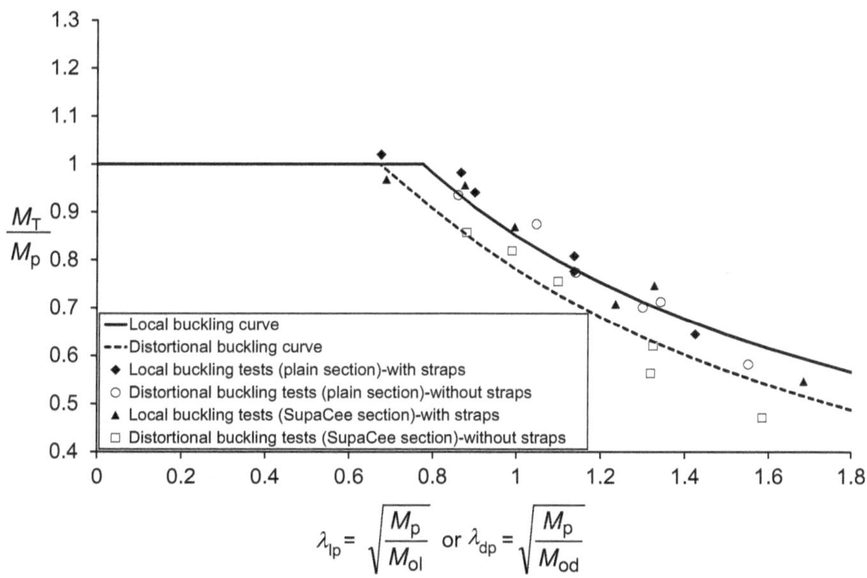

Figure 11.9 Comparison of DSM predictor curves with test data for beam: plastic moment case.

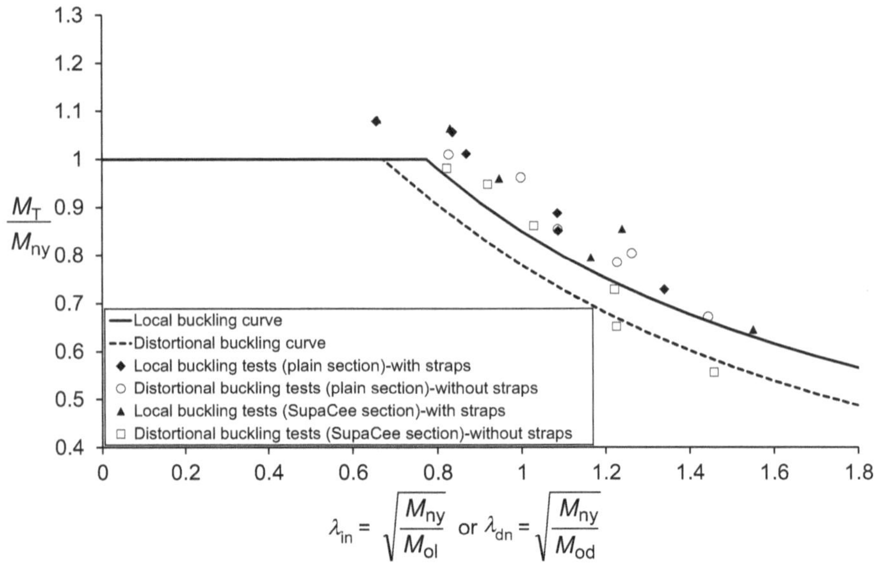

Figure 11.10 Comparison of DSM predictor curves with test data for beam: extended inelastic moment case.

capacities (M_T/M_y), as defined in existing DSM rules (Eqs. [11.1]–[11.4]). In Fig. 11.8, while the nondimensional slenderness is unchanged in the horizontal axis, the yield moment (M_y) in the vertical axis is replaced by the inelastic bending moment (M_n) as given by Eq. [11.5] to take into account the effect of inelastic reserve capacity. In Fig. 11.9 the yield moment (M_y) is replaced by the plastic moment (M_p) in both axes so that $\lambda_{lp} = \sqrt{M_p/M_{ol}}$ and $\lambda_{dp} = \sqrt{M_p/M_{od}}$ are the plastic slenderness values. In Fig. 11.10 the inelastic nondimensional slenderness is calculated for both local and distortional buckling respectively as $\lambda_{ln} = \sqrt{M_{ny}/M_{ol}}$ and $\lambda_{dn} = \sqrt{M_{ny}/M_{od}}$, where the inelastic moment (M_{ny}) from Eq. [11.6] is used for both axes.

For the local buckling tests, it can be seen in Figs. 11.7 and 11.8 that the test results for both plain and SupaCee sections are high compared with the DSM curves, so the mean of the test results divided by the DSM local buckling predicted capacities are conservative at 1.135 and 1.127 for the yield and inelastic moment cases respectively. The corresponding coefficients of variations are 0.062 and 0.055. The slightly different results for these two cases are due to the two stocky section tests (C15024 and SC15024) which have nondimensional slenderness $\left(\lambda_l = \sqrt{M_y/M_{od}}\right)$ in the inelastic local buckling reserve range ($\lambda_l \leq 0.776$) so that Eq. [11.5] can be used. For the plastic moment case in Fig. 11.9, the results are better predicted, with a mean equal to 1.005. The standard deviation for this case is quite low at 0.044. However, there are some low values at high slenderness, which may be unacceptable for design.

For the distortional buckling tests, the means of the test results divided by the DSM distortional buckling predicted capacities are the same at 1.152 for both the yield moment and inelastic moment cases. This is due to the fact that no sections have nondimensional slenderness $\left(\lambda_d = \sqrt{M_y/M_{od}}\right)$ in the inelastic distortional buckling reserve range ($\lambda_d \leq 0.673$). For the plastic moment case in Fig. 11.9, the DSM also provides better strength predictions for the distortional buckling failures, with a mean equal to 1.038. The corresponding standard deviation for the plastic moment case is reasonably low at 0.090, compared with 0.096 for both the yield and inelastic moment cases. However, there are some low values at high slenderness, especially for the SupaCee sections, which may be unacceptable for design.

For the extended inelastic moment case in Fig. 11.10, the means of the test results divided by the DSM local and distortional buckling predicted capacities are 1.082 and 1.123, which are more conservative than the plastic moment case in Fig. 11.9 but well predicted in comparison with those in Figs. 11.7 and 11.8. It is interesting to note that the corresponding standard deviations for the inelastic moment case are 0.038 and 0.086 for the local and distortional buckling tests respectively, which are the lowest of all cases. It is recommended that the extended inelastic moment case according to Eq. [11.6] can be used for plain C and SupaCee sections in place of that in the NAS S100:2012 specification based on Eq. [11.5].

11.4 Shear behavior and design

This section demonstrates the testing and FEM modeling of stiffened web sections in shear compared with the plain C-section. The strengths are compared with the DSM equations for both shear and combined bending and shear.

11.4.1 DSM design rules in shear without tension field action

The nominal shear strength (V_n) of beams without holes in the web and without web stiffeners is determined from NAS:2012, Appendix 1, Section 1.2.2.2.1 (AISI, 2012) as follows:

$$\text{For } \lambda_v \leq 0.815: \quad V_n = V_y \tag{11.7}$$

$$\text{For } 0.815 < \lambda_v \leq 1.227: \quad V_n = 0.815\sqrt{V_{cr}V_y} \tag{11.8}$$

$$\text{For } \lambda_v > 1.227: \quad V_n = V_{cr} \tag{11.9}$$

where

$$V_y = 0.6A_wF_y \tag{11.10}$$

and

$$V_{cr} = \frac{k_v \pi^2 EA_w}{12(1 - \nu^2)(d_1/t_w)^2} \quad [11.11]$$

whereV_y is the yield load of web based on an average shear yield stress of $0.6F_y V_{cr}$ is the elastic shear buckling force of the whole section derived by integration of the shear stress distribution at buckling over the whole section; $\lambda_v = \sqrt{V_y/V_{cr}}$

k_v is the shear buckling coefficient of the whole section based on the spline finite strip method (SFSM) (Pham and Hancock, 2009b, 2012b) or the semianalytical finite strip method (Hancock and Pham, 2012, 2013; Pham et al., 2012a,b, 2014).

11.4.2 DSM design rules in shear with tension field action

The nominal shear strength (V_n) of beams without holes in the web including tension field action (TFA) is determined from NAS:2012, Appendix 1, Section 1.2.2.2.1 (AISI, 2012) as follows:

$$V_n = \left[1 - 0.15\left(\frac{V_{cr}}{V_y}\right)^{0.4}\right]\left(\frac{V_{cr}}{V_y}\right)^{0.4} V_y \quad [11.12]$$

11.4.3 DSM design rules for combined bending and shear

In limit states design standards, the interaction is expressed in terms of bending moment and shear force. The upper limit interaction formula for combined bending and shear of a section with a vertically unstiffened web is given in NAS:2012, Section C3.3.2 (AISI, 2012):

$$\sqrt{\left(\frac{M^*}{M_{nxo}}\right)^2 + \left(\frac{V^*}{V_n}\right)^2} = 1 \quad [11.13]$$

where M^* is bending action, M_{nxo} is the bending section capacity (strength) in pure bending based on the lesser of Eqs. [11.1] and [11.2] with $M_{ne} = M_y$, V^* is the shear action, and V_n is the shear capacity (strength) in pure shear. The upper limit equation for combined bending and shear of vertically stiffened webs is also given in NAS: 2012, Section C3.3.2 (AISI, 2012):

$$0.6\left(\frac{M^*}{M_{nxo}}\right) + \frac{V^*}{V_n} = 1.3 \quad [11.14]$$

where M_{nxo} is the bending section capacity (strength) in pure bending based on the lesser of Eqs. [11.1] and [11.2] with $M_{ne} = M_y$.

11.4.4 Shear testing of stiffened web channel sections

The experimental program comprised a total of 14 tests conducted in the J. W. Roderick Laboratory for Materials and Structures at the University of Sydney. All tests were performed in the 2000 kN capacity DARTEC testing machine, using a servo-controlled hydraulic ram. Photos of the test setup configuration are shown in Figs. 11.11 and 11.12. This is the same rig as used by Pham and Hancock (2010a), and is reported in detail in Pham et al. (2015). The load was transferred to two stiffened channel sections measuring 250 × 90 × 6CC (cold-formed channel) which were connected to the test beam specimens by two vertical rows of M12 high-tensile bolts. The distance between these two vertical rows of bolts was 50 mm. At the supports, the two test specimens were bolted through the webs by vertical rows of M12 high-tensile bolts. These rows of bolts were connected through the webs of two stiffened channel sections of 250 × 90 × 6CC. The load was subsequently transferred to steel load transfer plates of 20 mm thickness bolted through the flanges of the stiffened channel sections of 250 × 90 × 6CC. These load-bearing plates eventually rested on the half rounds of the DARTEC supports to simulate a set of simple supports. At each bolt, a nut was located between the test beam specimen and CC section so the specimen did not attach directly to the CC, thus minimizing restraint to the web. Five bolts were used at each support and 10 bolts at the loading point. The test beam specimens were also connected by four 25 × 25 × 5EA steel straps on each top and bottom flange

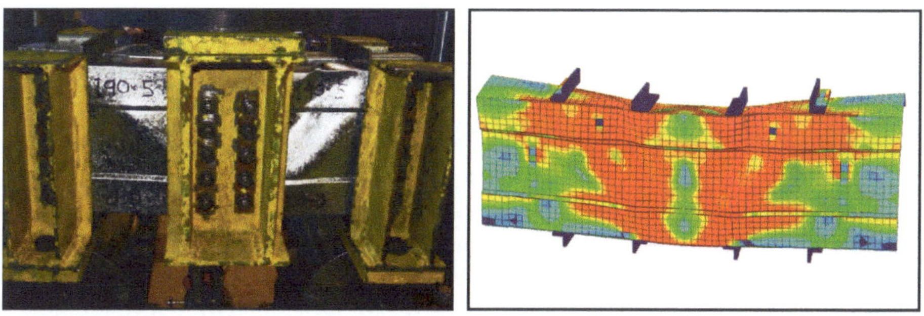

Figure 11.11 Shear failure mode shapes of the test and ABAQUS model SWC20012-I90*5.

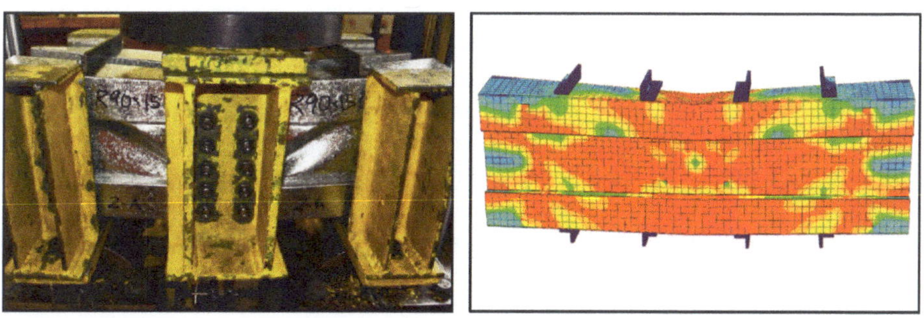

Figure 11.12 Shear failure mode shapes of the test and ABAQUS model SWC20012-R90*15.

adjacent to the loading point and supports, as shown in Figs. 11.11 and 11.12. Self-tapping screws were used to attach these straps to the test beam specimens. The objective of the straps was to prevent section distortion at the loading point and supports. The 250 × 90 × 6CC stiffened channel sections were used to prevent a bearing failure at the loading point and supports which could be caused by using conventional bearing plates.

Based on a standard plain lipped C-section with a web depth of 200 mm, a flange width of 80 mm and a lip size of 20 mm as a preferred section, six different stiffened web channel (SWC) sections were designed by adding a single longitudinal web stiffener of various sizes. The sections are similar to those in Fig. 11.1(c) but without the return lip. The nominal geometries of the seven SWCs including the plain C-section are given in Pham et al. (2015). SWCs with two web indents of 5 and 15 mm and three web stiffener depths of 20, 40, and 90 mm were chosen for the longitudinal web stiffener dimensions. The two segments between the inner and outer web portions of 5 mm indent are inclined 45 degree due to the minimum required gap between two bends, whereas those of 15 mm indent are folded perpendicularly to form a rectangular web stiffener. All sections have the same thickness of $t = 1.2$ mm in G450 steel. Fourteen tests of the seven types of SWC including a plain C-section were conducted. Each type of section was tested twice to ensure accuracy. In each test, two identical specimens were tested in pairs with top flanges facing inward and a gap between them to ensure that the inside assembly was possible.

11.4.5 Finite element modeling of test rig and SWC

The test data only contained specimens with straps on the top flanges at the support points and loading points, as shown in Figs. 11.11 and 11.12. To extend the range of test data to sections without straps where distortional buckling may occur, an FEM study was performed. Pham and Hancock (2010b) presented the modeling and analysis of experimental specimens in a shear and combined bending and shear test series on cold-formed C-sections using the FEM program ABAQUS (2008). A detailed FEM based on that of Pham and Hancock (2010b) was developed to study the structural behavior and validate the test results of SWCs in shear. The test rig was generated in ABAQUS using three-dimensional deformable solid elements and was assigned as normal steel properties. The SWC sections were modeled by using the four-node shell elements with reduced integration, type S4R. Quadrilateral element mesh was used for both test rig and channel sections. While a coarse mesh size of 20 mm was used for the test rig, a finer 10 mm element mesh was selected for the channel sections. In this study, the first method for imperfection modeling as used in Pham and Hancock (2010b) is chosen, where scaled buckling modes are separately superimposed on the initial geometry. An initial analysis is carried out on a perfect mesh using elastic buckling analysis to generate the possible buckling modes and nodal displacements of these modes. The imperfections are introduced to the perfect mesh by means of linearly superimposing the elastic buckling modes on to the mesh. The lowest

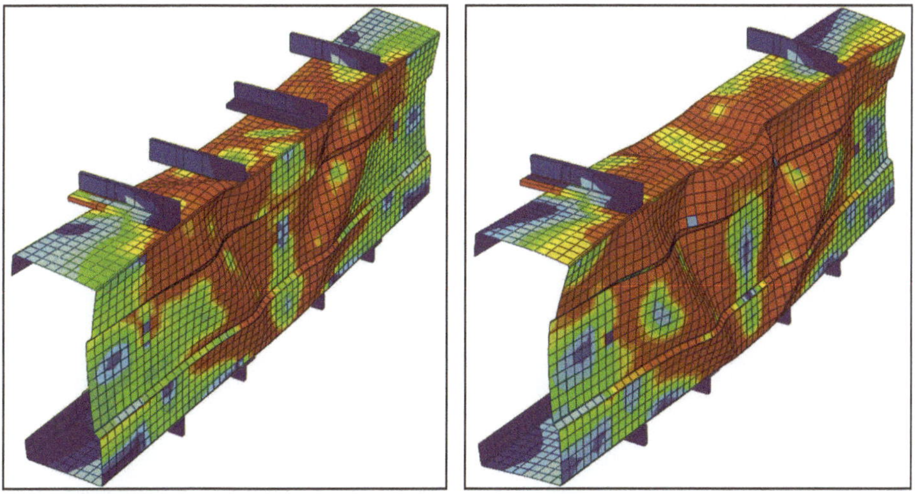

Figure 11.13 Shear FEM failure modes of sections with and without straps.

buckling modes are usually the critical modes, and these are used to generate the imperfections. Typical FEM failure models of the sections with and without straps at the loading points are shown in Fig. 11.13.

For the test observations, typical failure modes for the predominantly shear tests and ABAQUS models of the SWC sections are shown in Figs. 11.11 and 11.12. For SWC-I90 × 5-1 ($b_{s1} = 5$ mm, $d_{s1} = 90$ mm) (see Fig. 11.11), the shear failure mode appears to be influenced by the bending moment. It can also be seen in Fig. 11.11 that the failure mode of the channel members occurs adjacent to the middle of the shear span in the combined bending and shear buckling mode. For the tests with the larger stiffener width ($b_{s1} = 15$ mm), SWC-R90 × 15-2 ($b_{s1} = 15$ mm, $d_{s1} = 90$ mm) (see Fig. 11.12), the failures in local buckling modes were first observed in the top flanges between the loading points. The shear buckles subsequently occurred in the middle web portion (or web stiffener depth, d_{s1}), as shown in Fig. 11.12 for both test and ABAQUS model.

11.4.6 Comparison with DSM design equations for shear

The test result points are plotted in Fig. 11.14 against both DSM design curves for shear with TFA (Eqs. [11.7]−[11.9]) and without TFA (Eq. [11.12]). The TFA curve and elastic buckling curve (V_{cr}) are also graphically reproduced in Fig. 11.12. While all the SWC channels were tested with four 25 × 25 × 5EA straps attached on each top and bottom flange adjacent to the loading point and supports, as shown in Fig. 11.11, the ABAQUS-FEM results shown in Fig. 11.14 were performed without two straps adjacent to the loading points screwed on the top flanges of the SWC sections. Without

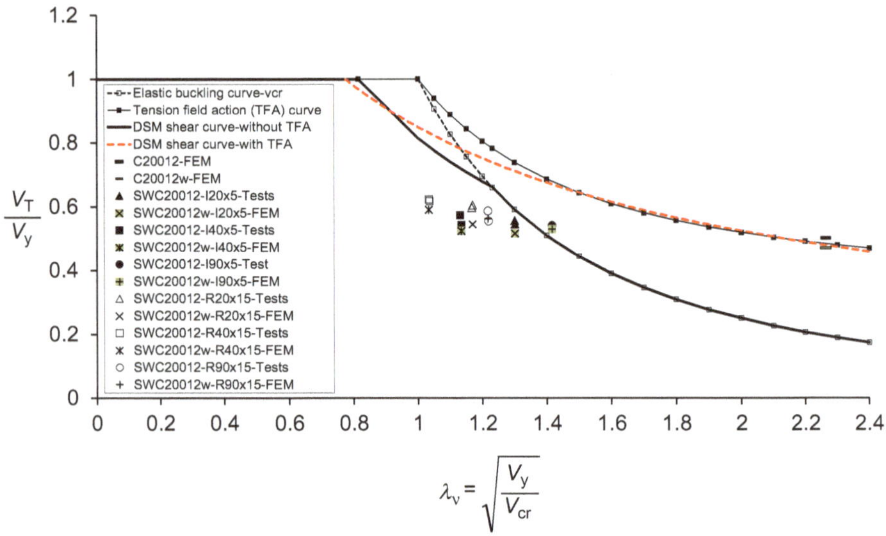

Figure 11.14 Shear test data (with straps) and FEM results (without straps) versus DSM shear design curves.

these straps, section distortion at the loading point may occur, as shown in Fig. 11.13(b), and reduce the bending and shear capacity. The specimen nomenclature for these ABAQUS models without straps in Fig. 11.14 has "*w*" and "FEM" added. For instance, "SWC20012w-I20 × 5-FEM" expresses that the SWC20012-I20 × 5 test was modeled using FEM "without" straps on the top flanges.

For channel sections with stiffeners of different sizes, the elastic shear buckling loads (V_{cr}) based on the SFSM (Pham and Hancock, 2009a, 2012a) increase significantly as the stiffeners become larger. The test points are therefore shifted horizontally to the left, as shown in Fig. 11.14. However, it is interesting to note that the shear strengths (V_T) of the swc sections only increase slightly on the vertical axis of Fig. 11.14 in association with the increased stiffener sizes. As a result, the test points of the SWC lie below the DSM shear curve with TFA. The reason for this may be a significant increase in the shear capacity (V_n) relative to the bending capacity (M_{nl} or M_{nd}), so that combined bending and shear now becomes important. For the ABAQUS model of the SWC sections without straps, the FEM results lie slightly below those of the tests with straps.

11.4.7 Comparison with the existing DSM design specification for combined bending and shear

Fig. 11.15 shows the interaction between (M_T/M_{nl}) and (V_T/V_n) where M_{nl} is based on the local buckling moment and V_v is based on Eq. [11.12] with TFA. M_T is the maximum moment in the shear span, and has been used in Fig. 11.15 because calibration of Eqs. [11.13] and [11.14] in the design standards assumes the maximum value in the span. It can be seen in Fig. 11.15 that the tests for SWC sections are no longer pure shear as for the plain sections. They lie around the bending and shear limits given by Eq. [11.14], but above Eq. [11.13] unit circular. The interaction between bending and shear for the SWC sections is therefore significant. The failure mode of the SWC20012-I90 × 5 tests under combined bending and shear can be observed in Fig. 11.11 with a local buckle in the flange as well as a shear buckle in the web. Fig. 11.16 shows the interaction between (M_{FEM}/M_{nd}) and (V_{FEM}/V_n) where M_{FEM}

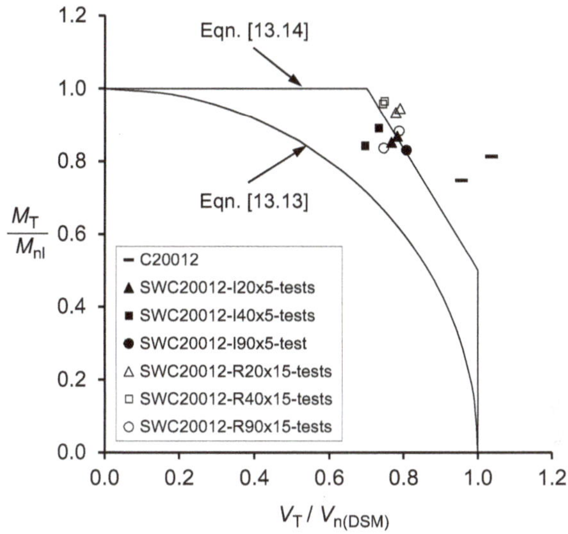

Figure 11.15 Interaction between (M_T/M_{nl}) and ($V_T/V_{n(DSM)}$) with straps.

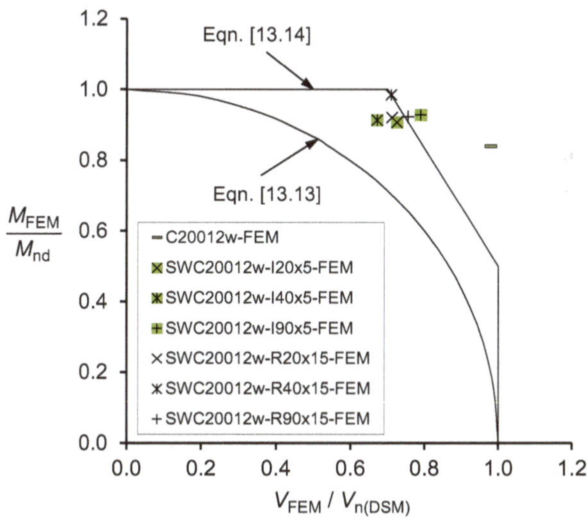

Figure 11.16 Interaction between (M_{FEM}/M_{nd}) and ($V_{FEM}/V_{n(DSM)}$) without straps.

and V_{FEM} are the ultimate bending moment and the ultimate shear load based on ABAQUS-FEM models without the straps, respectively. Also in Fig. 11.16, M_{nd} is based on the distortional buckling moment and V_v is based on Eq. [11.12] with TFA. The FEM results still lie around the bending and shear limits given by Eq. [11.14], but above Eq. [11.13] unit circular. Eq. [11.14] still provides a mean fit for the case without straps where M_{nd} is used instead of M_{nl}

11.5 Conclusions

This chapter discusses the new and innovative cold-formed shapes manufactured in Australia and New Zealand using high-strength AS1397 steel. These shapes have additional web stiffeners and return lips, and allow thinner gauges to be used as a result of the significant increases in their local and distortional buckling stresses. The main purpose of the chapter is to validate these shapes for use in DSM design as prequalified sections in NAS S100:2012 and AS/NZS 4600:2005.

From the flexural tests, it has been demonstrated that the inelastic reserve capacity can be taken into account when using the DSM. The tests also show that with four small additional multiple longitudinal web stiffeners and return lips, as designed in the SupaCee sections, the ultimate bending capacities of the complex channel sections were improved in the range of 4.5–22.4% in comparison with the plain C-sections. From the shear tests, it has been demonstrated that the addition of web stiffeners may lead to failure by combined bending and shear, which can be adequately predicted using the DSM in NAS S100:2012 and AS/NZS 4600:2005. Their superior structural capability is a result of the high-strength steels combined with lighter gauge sections and more complex shapes with intermediate web stiffeners and lips with return lips on the sections.

References

ABAQUS/Standard Version 6. 8–2, 2008. Abaqus/CAE User's Manual. Dassault Systèmes Simulia Corp, Providence, RI, USA.

American Iron and Steel Institute (AISI), 2012. North American Specification for the Design of Cold-formed Steel Structural Members, 2012 Edition, NAS S100–2012.

Bluescope Lysaght, 2003. Lysaght Purlins and Girts User's Manual (NSW, Australia).

Hancock, G.J., Pham, C.H., 2012. Direct strength method of design for shear of cold-formed channels based on a shear signature curve. In: Proceedings, the 21st International Specialty Conference on Cold-formed Steel Structures. Missouri University of Science & Technology, St Louis, Missouri, pp. 207–221.

Hancock, G.J., Pham, C.H., 2013. Shear buckling of channel sections with simply supported ends using the semi-analytical finite strip method. Thin-walled Structures 71, 72–80.

Hancock, G.J., 1978. Local, distortional and lateral buckling of I-beams. Journal of Structural Engineering, American Society of Civil Engineers 104 (11), 1787–1798.

Hancock, G.J., 2007. Design of Cold-formed Steel Structures to AS/NZS 4600:2005. Australian Steel Institute, North Sydney, Australia.

Pham, C.H., Hancock, G.J., 2009a. Direct strength design of cold-formed purlins. Journal of Structural Engineering, American Society of Civil Engineers 135 (3), 229–238.

Pham, C.H., Hancock, G.J., 2009b. Shear buckling of thin-walled channel sections. Journal of Constructional Steel Research 65 (3), 578–585.

Pham, C.H., Hancock, G.J., 2010a. Experimental investigation of high strength cold-formed C-sections in combined bending and shear. Journal of Structural Engineering, American Society of Civil Engineers 136 (7), 866–878.

Pham, C.H., Hancock, G.J., 2010b. Numerical simulation of high strength cold-formed purlins in combined bending and shear. Journal of Constructional Steel Research 66 (10), 1205–1217.

Pham, C.H., Hancock, G.J., 2012a. Direct strength design of cold-formed C-section for shear and combined actions. Journal of Structural Engineering, American Society of Civil Engineers 138 (6), 759–768.

Pham, C.H., Hancock, G.J., 2012b. Elastic buckling of cold-formed channel sections in shear. Thin-Walled Structures 61, 22–26.

Pham, C.H., Hancock, G.J., 2013. Experimental investigation and direct strength design of high strength complex C-sections in pure bending. Journal of Structural Engineering, American Society of Civil Engineers 139 (11), 1842–1852.

Pham, S.H., Pham, C.H., Hancock, G.J., 2012a. Shear buckling of thin-walled channel sections with complex stiffened webs. In: Proceedings, the 21st International Specialty Conference on Cold-formed Steel Structures. Missouri University of Science & Technology, St Louis, Missouri, pp. 262–281.

Pham, S.H., Pham, C.H., Hancock, G.J., 2012b. Direct strength method of design for shear – the third dimension. In: Proceedings, the 6th International Conference on Coupled Instabilities in Metal Structures. Strathclyde University, Glasgow, Scotland (Keynote Paper), pp. 17–32.

Pham, S.H., Pham, C.H., Hancock, G.J., 2014. Direct strength method of design for shear including sections with longitudinal web stiffeners. Thin-walled Structures 81, 19–28.

Pham, C.H., Bruneau, L.A., Hancock, G.J., 2015. Experimental study of longitudinally stiffened web channels subjected to combined bending and shear. Journal of Structural Engineering, American Society of Civil Engineers. http://dx.doi.org/10.1061/(ASCE)ST.1943-541X.0001259.

Plank, R.J., Wittrick, W.H., 1974. Buckling under combined loading of thin, flat-walled structures by a complex finite strip method. International Journal for Numerical Methods in Engineering 8 (2), 323–329.

Schafer, B.W., Pekoz, T., 1998. Direct strength predictions of cold-formed steel members using numerical elastic buckling solutions. In: Proceedings, Second International Conference on Thin-walled Structures. National University of Singapore, pp. 137–144 (also Thin-Walled Structures, Research and Development, Elsevier, Oxford, Eds Shanmugam, N. E., Liew, R. J. Y and Thevendran, V.).

Shifferaw, Y., Schafer, B.W., 2007. Inelastic bending capacity in cold-formed steel members. In: Structural Stability Research Council – Proceedings of the 2007 Annual Stability Conference, pp. 279–300.

Standards Australia, 2005. AS/NZS 4600:2005, Cold-formed Steel Structures. Standards Australia/Standards New Zealand, Sydney Australia.

Standards Australia, 2011. AS 1397:2011, Continuous Hot-dip Metallic Coated Steel Sheet and Strip – Coatings of Zinc and Zinc Alloyed with Aluminium and Magnesium. Standards Australia.

Sustainable applications of cold-formed steel structures: connections and joints

12

W. Lu
Aalto University, Espoo, Finland

12.1 Principles and methods

12.1.1 Integrated life-cycle design of sustainable cold-formed steel structures

12.1.1.1 Integrated life-cycle design process

The fundamental stages in the life of a structure as a product can be defined as life initiation, design, construction, operation, and end of life, as shown in Fig. 12.1. The life of a structure is initiated by a client's needs, and a design team is formed. Design activities are carried out according to the main aim at this stage: transform the functional descriptions of the structure according to the client's need to physical instructions that can then be used to ensure that the structure can be economically built; reliably, safely, and economically operated; and safely deconstructed or demolished in the end of its life. This methodology applies to any type of structure, including cold-formed steel (CFS) structures.

In the design stage, design teams first analyze available resources, clarify clients' needs, identify the technical requirements and constraints of the task, and transform these requirements into functional specifications. Then they create conceptual drawings

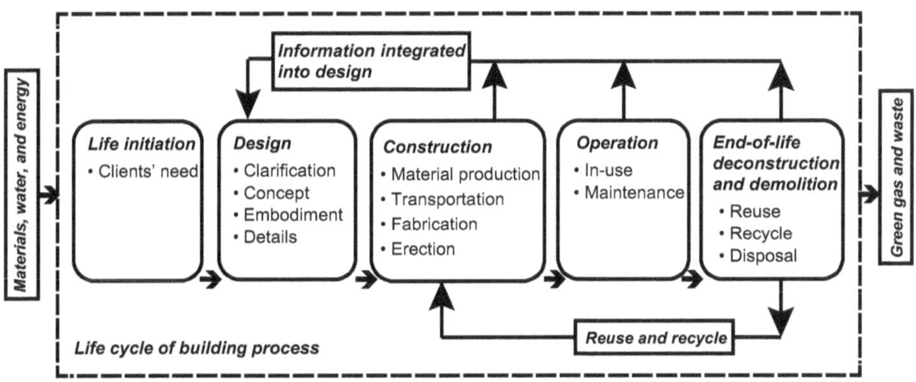

Figure 12.1 Integrated life-cycle design of a building process.

Recent Trends in Cold-Formed Steel Construction. http://dx.doi.org/10.1016/B978-0-08-100160-8.00012-8

to fulfill the requirements in the specifications, and produce preliminary drawings for further detailed design. In the process of detailed design, design teams select materials, determine precise dimensions of members, and propose methods for manufacturing and construction. The end results of the design stage are the detailed production instructions. In the production stage, the structural elements are manufactured in shop, and assembled into components or modules according to the production instructions. The structural elements, assemblies, and modules are transported to the construction site. Resource extractions can be included in the production stage. Technical systems inside a building such as heating, water supply and sewerage, air-conditioning, electrical, information, and solid waste management are installed on site. After construction, the building is ready to use and operation starts. The operational stage is the longest part of the working life of a structure, and regular maintenance is necessary. At the end of its life the structure is dismantled or demolished: materials are recycled, structural components are reused, and the rest is disposed for landfill. (Wallace and Clarkson, 1999).

The integrated design process incorporates other stages of the building process at the design stage, especially in conceptual and detailed design. The life cycle refers to the reuse of structural components and recycling of materials after deconstruction and demolition. As shown in Fig. 12.1, the inputs to the life-cycle building process are materials, water, and energy; the outputs are greenhouse gases and waste. The objectives of integrated life-cycle design are to reduce both consumption of resources and energy on the input side and pollution and waste on the output side.

12.1.1.2 Sustainable requirements and functional descriptions

A sustainable structure is a structure satisfying three-dimensional requirements: economic, societal, and environmental. In the economic dimension, besides the costs of material, construction, and energy other elements such as quality, flexibility, adaptability, and reusability of building and components affect the economic performance of a building. In the societal dimension, sustainable attributes can enhance the safety and the service life of the structure, and benefit the health and the convenience of the users by improving the qualities such as indoor air quality, thermal performance, acoustic performance, moisture, fire safety etc. In the environmental dimension, sustainable attributes can have environmental impacts by reducing resource consumptions, pollution, and wastes; by increasing the usage of reusable materials and energies Couchman and Sansom (2012). The performance of sustainable structures can be treated as additional technical requirements together with other requirements when design teams try to identify clients' needs during the design stage in the process of integrated life-cycle design.

12.1.1.3 Service life of CFS building products

Modular building products can be divided into building systems and technical and equipment systems. The hierarchy of building systems can be defined in levels: building, subbuilding, module, component, and basic element. Typical modules inside a building are load-bearing frames, façades, roofing systems, partition walls, floor

Table 12.1 Cold-formed steel modules, their service life, and performance requirements

Modules	Service life	Performance requirements
Load-bearing frames	Long term	Flexibility, adaptability
Wall panels	Long term	Load-bearing, flexibility
Roofing systems	Long term	Load-bearing, flexibility
Partition walls (nonload-bearing)	Short term	Easily removed for adaptability
Floor systems	Long term	Load-bearing, compatibility with building installations

systems, and other systems such as staircases, etc. (Sarja, 2002). The typical applications of CFS at building level include industrial and residential buildings, and at the subbuilding level may include modular construction and storage racking systems. The design working lifetimes of various types of buildings are different, so the lifetimes of building modules vary considerably. The CFS modules, their service life, and the corresponding performance requirements are summarized in Table 12.1.

12.1.1.4 Integrated life-cycle design methods for sustainable CFS structures

When sustainable performance is integrated into the design process of CFS structures, strength and serviceability are not the only parameters, as in traditional design, but other factors such as safety, flexibility, adaptability, quality, comfort, and reusability through life cycles of structures are considered. As pointed out by Sarja (2002), the methods in integrated life-cycle design include design for adaptability, design for durability, design for safety, design for deconstruction (DfD), and so on. Other stages of a building process are implicitly embedded in these methods during the design stage. Both conceptual and detailed designs are made for structural systems, which are divided into submodules taking the service life of each module into consideration. The output from the design process is detailed drawings for further manufacture, assembly, and on-site construction. Details of the integrated life-cycle design method for CFS structures are illustrated in Fig. 12.2.

12.1.2 Performance properties and design of sustainable connections and joints

12.1.2.1 DfD and modular construction

Connectors are used in the production stage to join structural elements by connections and joints in load-bearing frames, wall panels, or other building modules in building systems. When sustainable performance is integrated in the design stage,

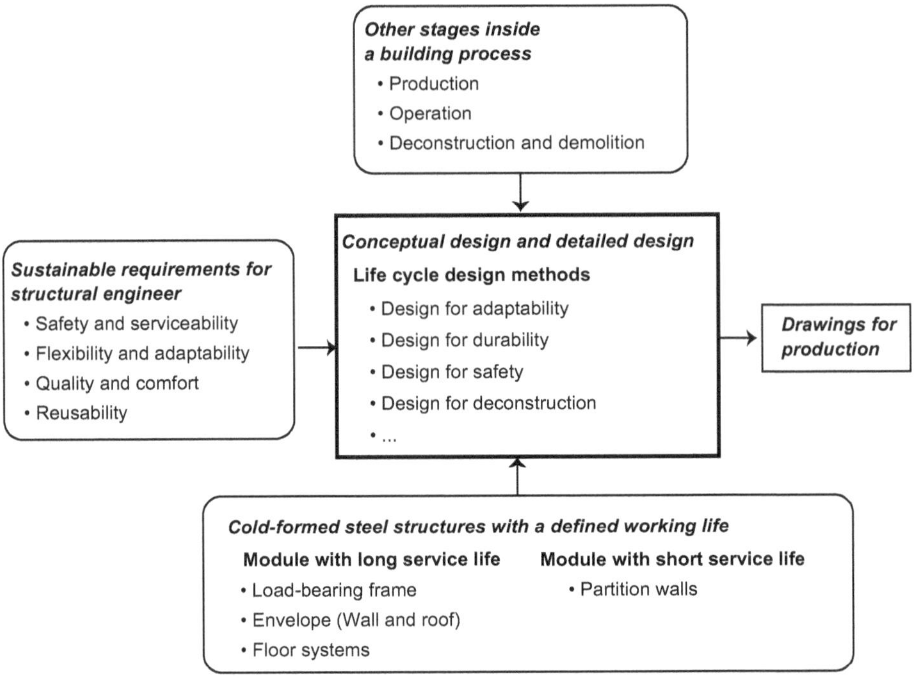

Figure 12.2 Integrated life-cycle design methods for sustainable cold-formed steel structures.

connections and joints also play key roles in life-endings of building systems. For modules with a short service life, connections and joints are designed for easy deconstruction so the structural components can be frequently taken apart for further reuse and recycling. Even for modules with longer service life, designing connections and joints should take into account future deconstruction. DfD and modular construction are two innovations making the construction industry more sustainable (Olson, 2010). Hechler et al. (2011) broke down the share of total costs for steel structural work (without transportation) as follows: 48% materials costs, 20% fabrication, 17% erection, 10% design, and 5% painting. Since connections are responsible for the lion's share of design, fabrication, and erection, the need for connection design in constructability and deconstructability is obvious. Thus the performance and design of sustainable connections and joints are discussed here for DfD and modular construction.

Olson (2010) distinguished the terminologies of disassembly, deconstruction, and demolition: "Disassembly represents the components being taken part without damaging, but not being reused elsewhere; Deconstruction is similar to disassembly but with the thought of reusing the components; Demolition refers to both the name of the industry and the process of intentional destruction." In this chapter, the terminology of deconstruction is used when discussing DfD. The contributions of DfD to sustainable performance, according to Olson (2010), are summarized in three categories.

- Societal aspects: creating more job opportunities with special skills and methods for deconstruction; salvaging materials in low-income areas for community revitalization.
- Environmental aspects: reducing landfill waste by reusing and recycling materials; reducing demand for new materials; reducing fabrication and transportation energy.
- Economic aspect: lowering the cost by reusing and recycling.

According to Olson (2010), modular construction can be thought of as DfD on a larger scale. Instead of dismantling to individual elements, modularity entails preserving whole modules in deconstruction, and reuse at module level. Olson (2010) proposed a theory of hybridizing DfD and modular construction. The innovation of this theory is the combination of conventional stick-built construction, panelized construction, and modular construction taking DfD into consideration. Its main benefit is to increase ease of deconstruction of a building, and at the same time make construction more customer or client oriented. Beside the contribution to DfD, modular construction provides possibilities for design considering flexibility and adaptability, and greater control over quality. In addition, the vast majority of buildings are constructed with some level of prefabrication of components, panelized structures, and modular structures (Garrison and Tweedie, 2008).

12.1.2.2 General principles for ease of deconstruction

Deconstruction is a reverse of construction. The general principles for guiding ease of construction can be used to guide deconstruction. Olson (2010) cited three elements for ease of construction: simplicity, standardization, and clear communication. These guidelines can also be applied to DfD. More principles related to DfD are discussed in the literature (Chiodo, 2005; Hechler et al., 2011; Edmonds and Gorgolewski, 2000; Morgan and Stevenson, 2005; Olson, 2010; Pulaski et al., 2004), mainly in relation to materials, components, and buildings, and design of connections and joints. Two other design methods, design for environment and design for recycling, were included in the discussions of the principles. Based on this literature, the general principles for deconstruction related to materials and buildings are summarized as follows.

- Using materials with reuse potential.
- Using materials that are nonhazardous, nontoxic, and durable.
- Considering the actual process of deconstruction at design stage.
- Considering a building as a series of layers with different lifespans.
- Avoiding multiple types of structural systems.
- Using dry construction processes where possible.
- Seeking alternatives to spray-on fire-proofing.

As steel is 100% reusable material, CFS has sustainable potential. According to the layering concept defined by Brand (1994), a CFS building can be layered into different lifespan modules.

- Site, including foundations: over 300 years.
- Structure, including load-bearing elements, frames, walls, floor systems, and roof trusses: 30–300 years.
- Skin, including cladding for weather protection: 20 years.

- Services, including service systems: 7–15 years.
- Space plan, including internal partitions: 3–30 years.
- Stuff, including furniture and equipment: daily.

The lifespan of a structure is normally longer than that of claddings and services. As far as DfD is concerned, the layers with shorter lifespans should be easily accessible without affecting the layers with longer lifespans. Components like internal partitions, equipment, and fire protection can be removed for reuse (Brand, 1994). This requires that the connections and joints inside these components can be easily removed with only minor damage, if any.

12.1.2.3 Special principles in designing connections and joints for ease of deconstruction

The principles of DfD for connections and joints in the literature mentioned above can be summarized as follows.

- Considering connections that can be reversed and avoiding irreversible processes.
- Using removable fasteners, and avoiding adhesives.
- Being easy to locate and access with power tools during deconstruction.
- Using common, standard shapes.
- Simplifying and standardizing connection details.
- Being easily accommodated by few connecting elements rather than many smaller elements.
- Minimizing the number of different member sizes.

These principles are further applied in design and assessment of connections and joints commonly used in CFS constructions in the next section.

12.2 Sustainable design and assessment of commonly used connections and joints

12.2.1 Classifications of commonly used CFS connections and joints

In EN 1993-1-8 (2005) a connection is defined as the location where connected elements meet; and a joint is defined as the zone where connected members are interconnected. For instance, a single-side beam-to-column joint is composed of a connection and a web panel in shear. The same definitions of joint and connection are used here.

12.2.1.1 Classification of connections and joints for sustainable performance

Connections can be classified according to different criteria, such as working principle, physical effect, geometrical characteristics, assembly-related properties, disassembly-related properties, and necessity of additives. Klett (2009) proposed systematic classification criteria for connections in three aspects: geometrical interface, locking mechanism, and combinations of geometrical interface and locking mechanism.

- According to the geometrical connecting interfaces of the connected components, three joining concepts are defined: planar, concave or convex (or profiled), and inserting. The interfaces of components in planar joining are flat, whereas in the other two cases both interfaces are profiled: one component is convex and the other is concave. The only difference is that in inserting joining one component can be inserted into the other component.
- Further, three types of locking mechanisms are defined: by force field, by added materials, or by additional mechanical fasteners. The locking technologies corresponding to each mechanism are vacuum or magnetism for force-field locking; welding or gluing for material locking; and bearing bolt, friction bolt, or screw-clamp for mechanical fasteners.
- By combining both locking mechanism and geometrical interfaces, various connection principles can be created. For different connection principles, ease of disassembly is different. For instance, two actions are necessary for mechanically locked connections, ie, unlocking and separating; while one action of separating is enough for both force-field locked and materially locked connections.

Klett (2009) reviewed available literature on the classification of disassembly processes and proposed a new classification based on principles such as degree of damages, tool and level of automation, and disassembly rules.

- The first classification is made according to the degree of damage to both connected components and additives: in a partially destructive process only additives are damaged, and reassembly is possible when additives are replaced; in a fully destructive process both additives and components are damaged, and reassembly is not possible; and in a nondestructive process neither additives nor components are damaged, and reassembly is possible.
- The second classification of a disassembly process is made based on disassembly tools and their levels of automation: disassembly can be performed automatically, manually without tools, manually with manual tools, or manually with mechanized tools.
- The criteria of the third classification concern the disassembly process: contact-based disassembly through movement and forces; or contactless disassembly through other mechanisms such as electric current or magnetic field.

Based on the discussions above, classification criteria of connections and joints for further evaluation of their sustainable performance in applications in CFS buildings are listed in Table 12.2. The criteria include connection solutions, joining mechanism, locking mechanism, disassembly process, and ease of deconstruction. Ease of deconstruction can be evaluated by both degree of damage and tools and level of automation. These criteria can be further used in innovation of sustainable connections and joints.

12.2.1.2 Classification of commonly used CFS connections

Typical connecting methods used in CFS applications include screwing (self-tapping and self-drilling screws), bolting, riveting, joining with shot-fired pins, nailing, and welding (Yu and LaBoube, 2010; ESDEP; Lee et al., 2014). Different connecting solutions can have different applications. ESDEP summarizes the typical applications of mechanical connectors: when connecting thin to thick steel, bolts, self-tapping screws,

Table 12.2 Classification criteria of connection and joints

	Principles for assembly and disassembly			Ease of deconstruction	
Connecting solutions	**Joining mechanism**	**Locking mechanism**	**Disassembly mechanism**	**Degree of damage**	**Tool and level of automation**
Screwed, bolted	Planar, profiled, or inserting	Force field, added materials, or mechanical fasteners	Unlocking, separating, or unlocking and separating	Fully destructive, partially destructive, or nondestructive	Manually without tools, manually with hand tools, manually with mechanized tools, or automation

Table 12.3 Application of classifications criteria to typical CFS connecting solutions

	Principles for assembly and disassembly			Ease of deconstruction	
Connecting solutions	**Joining mechanism**	**Locking mechanism**	**Disassembly process**	**Degree of damage**	**Tool and level of automation**
Screwed	Planar	Mechanical	Unlocking and separating	Partial	Manually with tools
Bolted	Planar	Mechanical	Unlocking and separating	None	Manually with tools
Riveted	Planar	Mechanical	Unlocking and separating	Partial	Manually with tools
Nailed	Planar	Mechanical	Unlocking and separating	Partial	Manually with tools
Shot-fired	Planar	Mechanical	Unlocking and separating	Partial	Manually with tools
Welded	Planar	Material	Separating	Partial	Manually with tools

self-drilling screws, thread-cutting screws, and shot-fired pins can be used; when connecting thin to thin steel, hexagon-head screws, self-drilling screws, blind rivets, and seam locking can be used. The classification of typical CFS connections according to the criteria mentioned above is shown in Table 12.3. The classification table can be further used in the sustainable design and assessment of connections and joints.

12.2.2 Design and assessment in accordance with performance of sustainability

12.2.2.1 Design and assessment of connections

In integrated life-cycle design, designers receive information from other stages of the design process, as shown in Fig. 12.2. Designers can cooperate with assembly engineers using guidelines, and qualitative and quantitative methods. Guidelines are important for design, since they provide the designers with background information, although it is not sufficient to provide principles when various design solutions are evaluated. Qualitative evaluation methods can rate various design solutions without measurable values. Quantitative methods can calculate the time or cost savings for redesign. (Jürisoo and Staaf, 2007). The first two methods are used for designing and assessing sustainability of connections and joints in CFS applications. Quantitative methods applied in connection design for easy assembly on site can be found in, for instance, Jürisoo and Staaf (2007).

For the purpose of demonstration, the qualitative evaluation method given in Chiodo (2005) is extended to evaluate the sustainable performance of CFS connections and joints. The evaluation is carried out for strength, ease of disassembly, recycling, and economic aspects. Both static and fatigue strength are included in the evaluation of strength performance. Ease of deconstruction is assessed from the two aspects described in Table 12.3. Material recycling and product reuse are taken into account in reusable performance. The economy aspect mainly includes costs. According to the detailed requirements, each performance is evaluated on various levels such as good, average, or poor; easy, average, or difficult; non, partially, or fully; and more, average, or less. Each performance is rated as grade 1, 2, or 3, with better performance being given a higher grade. The detailed grading corresponding to each performance is listed in Table 12.4. The rating table for evaluating sustainable performance of connections and joints (Table 12.5) shows the results when applying this qualitative evaluation method for designing and assessing CFS connections with respect to sustainable performance. The final decision can be made according to the total grade obtained by summing the grades of each requirement.

12.2.2.2 Design and assessment of joints

The design and assessment of joints in DfD can be carried out for both the fastener and the parts to be fastened. The detailed requirements related to fasteners for DfD include using detachable fasteners, minimizing the number of fasteners, minimizing types of fasteners for selection of removal tools, using snap fits that are easy to locate and

Table 12.4 Rating table for evaluating sustainable performance of connections and joints

Performance	Detailed requirements	Rating	
		Performance	**Grade**
Strength	Static and fatigue	Good	3
		Average	2
		Poor	1
Ease of deconstruction	Manually without tools, manually with hand tools, manually with mechanized tools, or automation	Easy	3
		Average	2
		Difficult	1
	Fully destructive, partially destructive, or nondestructive	Non	3
		Partially	2
		Fully	1
Reusable	Recycling of materials Reuse of fasteners and connected parts	Good	3
		Average	2
		Poor	1
Economy	Cost	Less	3
		Average	2
		More	1

easy to open with standard tools, using connections that can be partially destructive, using connections with easy-to-access points, and using fasteners with materials compatible with parts' materials. The detailed requirements related to connected parts for DfD include selecting parts with materials that are easy to disassembly, positioning joints so the parts does not need to be turned or moved for dismantling, indicating on the parts how a connection should be opened nondestructively, placing parts that are likely wear out at the same time in close proximity so they can be replaced simultaneously, choosing the same tool for assembly and disassembly, designing for multiple detachments with one operation, and decreasing numbers of parts to limit the joints (Manner, 2015). This section demonstrates the method of writing guidelines for two example joints using the listed, detailed requirements in combination with adopting the layering concept and the corresponding service life of each layer.

The portal frame as a load-bearing structure has the longest service lifespan at 30–50 years. Possible reuse and recycling after deconstruction will come into consideration at the end of the building life. Wrzesein and Lim (2008) reviewed typical CFS portal-frame joints for 1982–2006. Three eaves joints, as shown in Fig. 12.3 after Mills and LaBoube (2004), are taken from a review paper for writing guidelines for DfD. Fig. 12.3(a) is a bolted end plate joint with four components: beam (1), column

Table 12.5 Grading of sustainable performance of cold-formed steel connections

Criteria and detailed requirements		Screws	Bearing bolts	Rivets	Shot-fired pins	Welds
Strength	Static	3	3	3	3	3
	Fatigue	2	2	2	2	2
Ease of disassembly	Tool and level of automation	2	2	1	1	1
	Degree of damage	2	2	2	1	1
Reusable aspects	Reuse of fasteners	3	3	2	1	1
	Reuse of connected parts	2	2	2	2	1
	Material recycling	3	3	3	3	3
Economic aspect	Costs	3	3	2	2	1
Total grade		**20**	**20**	**17**	**15**	**13**

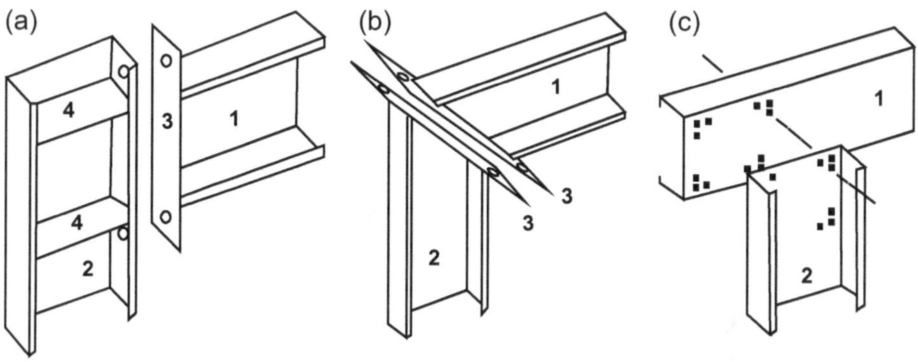

Figure 12.3 Eaves joints in portable frame: (a) bolted end plate joint; (b) miter joint; (c) self-drilling screw joint.
Redraw from Wrzesein, A.M., Lim, J.B., 2008. Cold-formed steel portal frame joints a review. In: 19th International Specialty Conference on Cold-formed Steel Structures, Paper 4, 16 pages. Available from: http://scholarsmine.mst.edu/isccss/19iccfss/19iccfss-session9/4 (accessed 09.04.15.).

(2), end plate (3), and web-stiffening plate (4). Both end plate and two web-stiffening plates are welded to the beam end and column web, respectively. Fig. 12.3(b) is a miter joint with three components: beam (1), column (2), and end plates (3). Two end plates are welded to either beam end or column end, separately. In addition, both beam end and column end are cut at 45 degree beforehand. Fig. 12.3(c) is a self-drilling screwed joint with two components: beam (1) and column (2). The beam and column are connected through webs with six self-drilling screws. Since all these are moment joints, the fasteners are located close to both tension and compression flanges of beams so the moment arms can be as large as possible.

The strengths of the bolted joints are 9.91 and 10.04 kN m, respectively. The predicted joint capacity for the screwed joint was 10.4 kN m, but the capacity in tests was 7.8 kN m (Mills and LaBoube, 2002). Table 12.6 shows that with a similar level of

Table 12.6 Guidelines for three types of portal-frame joints for DfD

	Criteria	Bolted end plate joint	Miter joint	Self-drilling screw joint
Fastener related	Moment resistance under gravity load (kNm)	9.91	10.04	7.80
	Detachable fasteners	Bolts Welds	Bolts Welds	Screws
	Connection can be partially destructive	Welds	Welds	Yes
	Number of fasteners	2	2	12
	Minimize type of fasteners for selection of removal tool	1 type	1 type	1 type
	Snap fits easy to locate and easy to open with standard tool	–	–	–
	Fastening points easy to access	Easy	Easy	Easy
	Fasteners' materials compatible with parts' materials	Yes	Yes	Yes
Connected parts related	Parts with materials easy to disassembly	Yes	Yes	Yes
	Indicate on parts how joint should be opened nondestructively	Yes	Yes	No
	Place parts that are likely wear out at the same time in close proximity	Yes	Yes	Yes
	Use the same tool for assembly and disassembly	Yes	Yes	Yes
	Design for multiple detachments with one operation	No	No	Yes
	Decrease number of parts to limit the joints	No	No	Partly yes

capacity, each type of joint has its own advantages and disadvantages for deconstruction according to the listed guidelines. Further quantitative evaluations are necessary when other criteria such as time or cost of deconstruction are integrated into the design.

Another typical application of CFS members is in constructing storage rack structures. Zaharia and Franssen (2000) reviewed various types of storage racks, such as pallet racks, drive-in racks, drive-through racks, cantilever racks, selective pallet racks, narrow aisle racks, double-deep pallet racks, push-back racks, and gravity flow and pallet flow racks. Lee et al. (2014) reviewed the connections in various types of rack systems. The main frame of a storage rack normally consists of two parallel beams and two upright framed columns, which are similar to laced built-up columns. In upright framed columns, CFS members and bracings are either bolted or welded together. The pallets are connected to the beams to hold the goods if drive-in and drive-through are not necessary. The joints between upright framed columns and beams are similar to beam-to-column joints in steel frames.

Two examples of beam-to-column joints in rack systems, taken from Gilbert and Rasmussen (2010), are shown in Fig. 12.4. In Fig. 12.4(a) a T-shaped connector tap is welded to the end of a beam. The beam is connected to the perforated column by hooking built-in taps of the beam to the perforations of the upright column. Extra safety-locking pins or clips are necessary to prevent the beam being accidentally dislodged from its connections. In Fig. 12.4(b) the beams are connected to the perforated column by bolts.

In the layering concept, the storage rack can be considered as a stuff subsystem inside the building. The service lifespan of a storage rack ranges from daily to some years. To satisfy users' daily special needs, the storage rack structure should be easily disassembled, especially the joints between beams and upright framed columns. Similar guidelines to those in Table 12.6 for DfD are given in Table 12.7. Compared

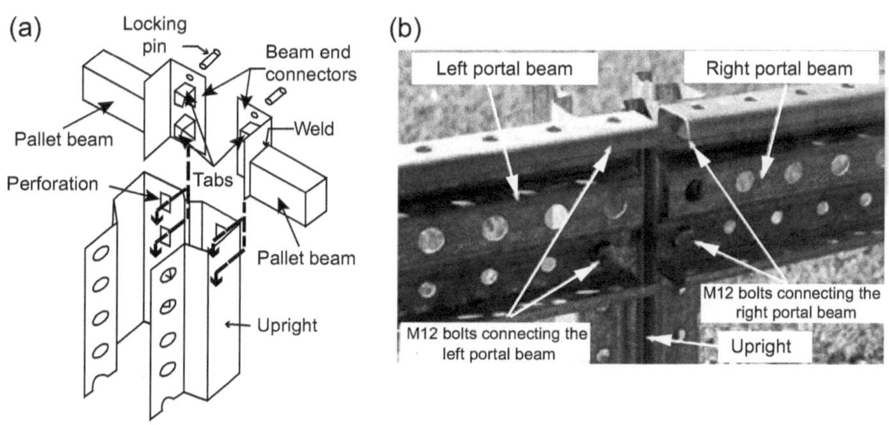

Figure 12.4 Examples of beam-to-column joint in pallet rack: (a) speed lock joint; (b) bolted joint.
Redraw from Gilbert, B.P. Rasmussen, K.J.R., 2010. Bolted moment connections in drive-in and drive-through steel storage racks. Journal of Constructional Steel Research 66 (6), 755–766, Available from: http://dx.doi.org/10.1016/j.jcsr.2010.01.013.

Table 12.7 **Evaluation of example beam-to-column joints in rack systems for DfD**

	Criteria	Locked joint	Bolted joint
Fastener related	Detachable fasteners	Safety pins	Bolts
	Connection can be partially destructive	–	–
	Number of fasteners	1 or 2	2
	Minimize type of fasteners for selection of removal tool	1 type	1 type
	Snap fits easy to locate and easy to open with standard tool	Tabs on beam end connectors	–
	Fastening points easy to access	Easy	Easy
	Fasteners' materials compatible with parts' materials	Yes	Yes
Connected parts related	Parts with materials easy to disassemble	Yes	Yes
	Indicate on parts how joint should be opened nondestructively	Yes	Yes
	Place parts that are likely wear out at the same time in close proximity	Yes	Yes
	Use the same tool for assembly and disassembly	Yes	Yes
	Design for multiple detachments with one operation	No	No
	Decrease number of parts to limit the joints	Yes	Yes

to the joints used in portal frames shown in Table 12.6, the beam-to-column joints in pallet rack systems shown in Table 12.7 give better performance in both easy assembly and easy disassembly; thus similar beam-to-column joints to those in rack systems may be extended to use in beam-to-column joints in steel frames for DfD.

12.3 Innovations in sustainable connections and joints

12.3.1 Connections and joints for deconstruction

12.3.1.1 General considerations for creating novel connections for deconstruction

Creating novel sustainable connections and joints for DfD is also an integrated design process. Other design methods such as design for environment and design for recycling can be integrated into the DfD (Manner, 2015). Consideration of service lifespan of the connected components is also necessary. The general criteria for creating sustainable connections for DfD and their functional requirements are listed in Table 12.8. The criteria include easy of disassembly, reusability, and safety. These criteria are further applied as guidelines in creating novel sustainable connections and joints.

12.3.1.2 Future trends of innovative connections and joints for CFS construction for DfD

The concept of active disassembly was put forward based on unaligned trigger paths for locking and unlocking. Several types of unaligned triggers for locking and unlocking are available: temperature, magnetism, radio waves, pressure, and liquids. (Klett, 2009). The most common trigger in active disassembly is temperature. Instead of using a traditional fastener-specific tool, self-disassembly is triggered by temperature. The materials used for fasteners are so-called smart materials, ie., shape memory alloys (SMAs). A fastener can change its shape when exposed to a specific temperature in

Table 12.8 Performance and functional requirements for creating novel connections for DfD

Performance		Functional requirements
Easy to disassemble	Additives	Avoid added materials or fasteners
	Directions	Separate in multiple directions: planar or profiled geometries
	Space	Restrict geometries due to available space: planar or profiled
	Disassembly	Unlock easily by hand
	Frequency	Short service lifespan Fasteners or added materials can withstand frequent unlocking Geometry and material should not be changed Avoid fasteners with plastic deformations
Reusability	Destructive or not	Both components and connections are nondestructive or can be reused with minor reparation Unlock through movement or deformation
	Trigger path	One trigger path (separating and unlocking at the same time) Unaligned trigger path for locking and unlocking
	Recycling	Materials can be recycled
Safety		Movements should be restricted in multiple directions

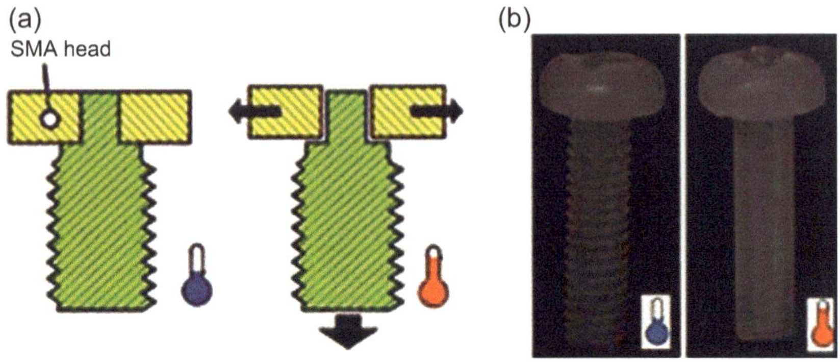

Figure 12.5 Fasteners in active disassembly: (a) bolt with smart head; (b) smart screw. Modified from Chiodo, J., Jones, N., 2012. Smart materials use in active disassembly. Assembly Automation 32, 8–24. Available from: http://dx.doi.org/10.1108/01445151211198683 (accessed 15.04.15.).

the range of 65–120°C (Chiodo, 2005). In the examples shown in Fig. 12.5, either the bolt head or the screw thread is made from SMAs. When exposed to certain temperatures, the screw threads disappear and a screw can be naturally disassembled from its connected parts. Similarly, the smart bolt head can be unattached from its shank, and the bolt shank without head can be self-disassembled from its connected parts. The connected parts are nondestructive, and similar new bolts or screws can be reshaped to fit new connections. Applicability in CFS connections, especially when exposed to fire, needs further investigation. Klett (2009) provided a detailed list of other types of nondestructive or partially destructive connections, some of which can be applied in CFS connections.

According to the guidelines given in Table 12.8, one option to promote a novel connection is to avoid added materials or fasteners. Fig. 12.6(a) and (b) shows examples of an inserting type of connection: Quicon®. The Quicon® connection was developed by the Steel Construction Institute in 2002, and can be used as both beam-to-column and beam-to-beam connections in steel frames. In a beam-to-column connection, the flange of a T-plate is bolted to the flange of a connected column. The studs welded to the web of the connected beam can slot into the keyholes in the web of the T-plate during assembly. With this connection, the beam can also be easily disassembled later. Other types of novel connections can be found in Cooper (2010). These innovation ideas can be also applied to CFS applications for DfD.

12.3.2 Connections and joints for prefabricated production

Modular construction can be thought of as DfD on a larger scale, with reusability accessed at module level. Modular units are often prefabricated in shop. Rrefabrication can be classified at one-dimensional (1D), two-dimensional (2D), and three-dimensional (3D) levels: line-like components, such as built-up columns or beams, are at the 1D level; panel-like components, such as wall panels, roof panels, roof

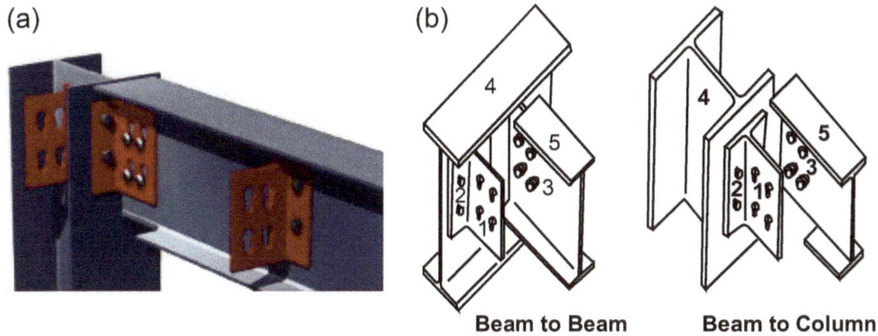

Figure 12.6 Quicon® connections: (a) example of a beam-to-column connection; (b) details in both beam-to-beam and beam-to-column connections.
From NSC: New Steel Construction, 2005. Quicon Wins Contract at IKEA, Available from: http://www.newsteelconstruction.com/wp/quicon-wins-contract-at-ikea/ (accessed 25.09.15.); Redraw from Ogden, R., Plank, R., Lawson, M., Atree, J., 2003, Architecture Teaching Resource Studio Guide, SCI publication P167, 28 pages. Available from: http://www.tatasteelconstruction.com/file_source/StaticFiles/Construction/studio%20guide.pdf (accessed 24.09.15.). Courtesy of Steel Construction Institute.

trusses, or frames, are at the 2D level; and volume-like modules are at the 3D level. A modular construction is normally a volume-like module. CFS has wide application in both prefabrications and modular constructions. In this section, some joining techniques in the prefabrication of CFS structural components in shop are presented first, and then the connections that are used in assembling or disassembling the modules both in shop and on site are discussed. Easy manufacture and fast assembly are the main factors to be considered.

12.3.2.1 Joining methods used in prefabrication of 1D and 2D components

Laser beam welding (LBW) technologies have been widely used in the automobile industry, but are relatively new in the use of CFS structures. LBW is a technology used for joining components with thin, small weld seams. Lasers generate high energy to heat the materials into a molten state, and fuse them together as shown in Fig. 12.7(a). The benefits of LBW are high automation, productivity, precision, and high quality. An LBW welded CFS built-up beam, shown in Fig. 12.7(b), is manufactured by welding two C-shaped beams with flange edge stiffeners and two reinforcing plates together. The reinforcing plates are placed inside the hollow flanges. The beam can be used in load-bearing frames to replace a single CFS C-shape beam to enhance the buckling resistance so as to carry a higher applied load. (Landolfo et al., 2009).

The ROSETTE technology was developed in the late 1990s for fully automated integrated joining of strip coil material to light-gauge steel frame components such as stud wall panels or roof trusses. A ROSETTE joint is formed in pairs between two steel sheets (Fig. 12.8(a)): one with a prefabricated hole and one with a collared hole. Firstly

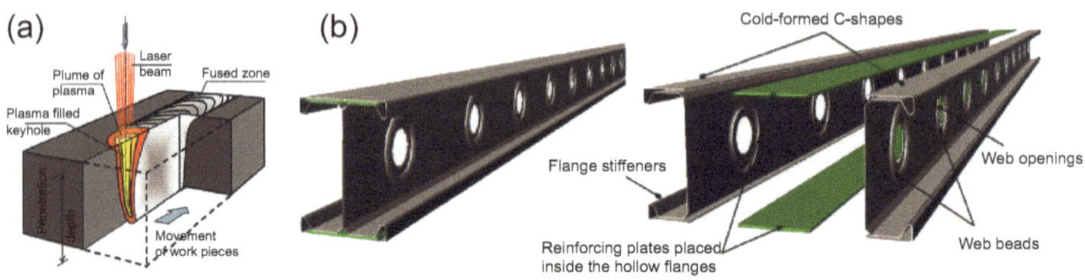

Figure 12.7 Application of LBW in CFS structures: (a) laser welding process; (b) CFS built-up beam by laser welding.
From Landolfo, R., Mammanaa, O., Portioli, F., Di Lorenzo, G., Guerrieri, M.R., 2009. Experimental investigation on laser welded connections for built-up cold-formed steel beams. Journal of Constructional Steel Research 65, 196–208, Available from: http://dx.doi.org/10.1016/j.jcsr.2008.03.015.

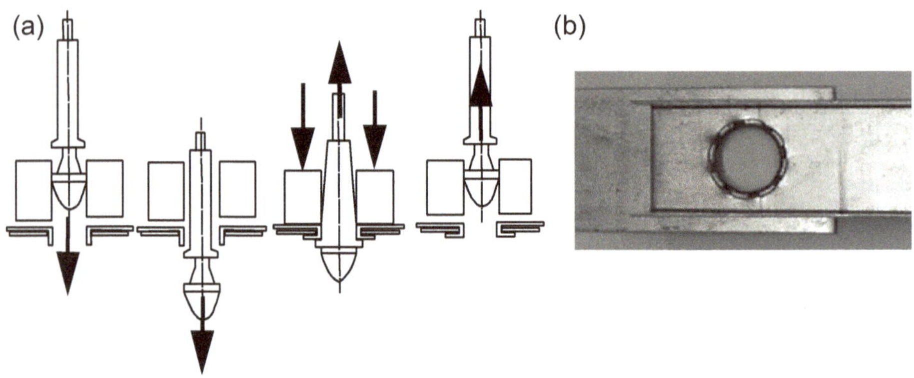

Figure 12.8 ROSTTE joint: (a) joining process; (b) final product.
Redraw from Lu, W., Segaro, P., Kesti, J., Mäkeläinen, P., 1998a. Study on the Shear Strength of a Single-Lap ROSETTE-Joint. Research report, 40 pages. TKK-TER-08, Publication 8. Laboratory of Steel Structures, Helsinki University of Technology.

the collar is snapped through into the prefabricated hole. Then the tool head penetrates the hole at the joint point. The head expands, and is pulled back with hydraulic force. The expanded tool head crimps the collar against the plate around the hole. Torque is enhanced by using multiple teeth in the joint perimeter, as shown in Fig. 12.8(b). (Kaitila et al., 2001; Lu et al., 1998a). Descriptions of various production lines for producing stud-wall frames, roof trusses, and floor frames can be found in FUSA Tech. Inc. (2003). In terms of DfD, ROSETTE belongs to plastically deformed connections, and is disassembled with destruction. However, roof trusses, stud-wall frames, and floor frames fabricated with ROSETTE joints are layers of structures with long service life, and can be disassembled as whole elements for further use.

Press joining forms a connection by deforming the parent metal of sheets to be connected; its benefits are fast production, high quality, long durability by preserving the original coating, efficient on-site construction, and waste reduction. Press joining is a single-step process with two phases of shearing and deforming, which occur

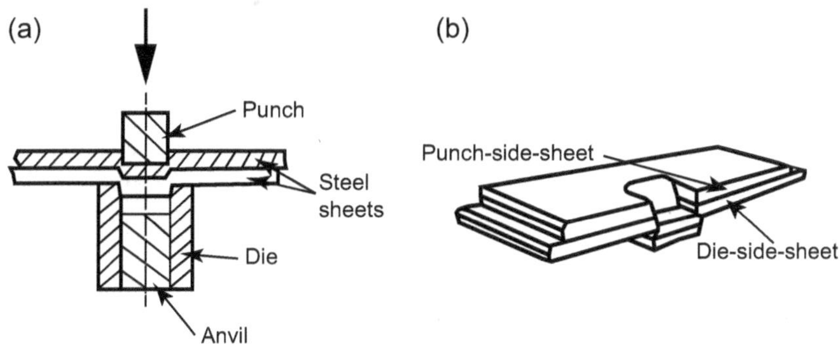

Figure 12.9 (a) Joining process of press joins; (b) final product of press joins.
Redraw from Lu, W., Kesti, J., Mäkeläinen, P., 1998b. Shear and Cross-Tension Tests for Press-Joins. Research reports, 55 pages, Report 7. Laboratory of Steel Structures, Helsinki University of Technology.

simultaneously (Fig. 12.9(a)). The shearing phase occurs when the steel sheets are cut along two parallel edges by the action of a punch. A bulged area is created out of the sheets. The deforming phase occurs as the punch continues its motion until the layers of steel are pressed between the anvil and the punch. The pressing causes the layers of steel to spread laterally against the spring plates. The steel layers become wider and this prevents the sheets from coming apart, as shown in Fig. 12.9(b). Depending on the shape of the punch, press joins can be circular or rectangular. (Pedreschi and Sinha, 1996; Lu et al., 1998b; Varis, 2006; Mucha et al., 2014). Press joining can be used in prefabricated CFS frames. Similar to ROSETTE joints, prefabricated elements using press joins can be disassembled as whole elements for further use.

Beside the development of joining technologies, one possible innovative solution taking sustainability into account is prefabricating novel truss systems using high-strength steel. Kadak (2014) studied the behavior of the joint inside a novel truss developed in the Nordic region. The truss is composed of a hat-beam (WQ-beam) as the upper chord, rectangular hollow sections as bracings, and steel plates as lower chords. Using a WQ-beam as top chord, slim floor systems can be easily integrated with the truss, and long-span open spaces in building structures can be created. The steel grades for both bracings and lower chord are up to S550. (Kadak, 2014).

12.3.2.2 Joining methods used in assembling CFS modules in shop and on site

Based on structural behavior, modular units can be classified as corner supported, continuously supported, and nonload-bearing modules. In a corner-supported module, the unit is designed with longitudinal beams being supported by four corner columns. The sides of the unit are braced. The profiles for corner columns are often hollow sections. In a continuously supported module, the unit is supported by longitudinal edge beams. The sides of the unit are braced only for transportation. The profiles for corner posts are often angle plate or hollow sections. A nonload-bearing module, often called a pod, sits on floors of framed buildings and is used as service unit. (Gorgolewski et al., 2009). Examples of corner-supported and continuously supported modules are shown in Fig. 12.10(a) and (b) (Lawson and Ogden, 2008). Connections and joints within modules vary not only with different forms of modules but also with different manufacturers. The typical connection solutions within modules, either corner supported or continuously supported, are bolted, welded, or screwed connections, as in the modular units produced by several UK manufacturers (Lawson, 1999). More practical applications of modular constructions can be found in Lawson (2007).

Modular-unit-assembled buildings can be classified as nonattached, semiattached, and attached modules based on how the modules are connected on site. In nonattached buildings each module is transported to the site and built separately; in semiattached buildings the pod within other frames is attached to pods in the current frame, or two modules are attached with separate field joints; and in attached buildings the modules are either horizontally attached or stacked in multiple stories. (Garrison and Tweedie, 2008). Structural connections between modules are mainly responsible for the integrity and robustness of the entire structure. The connection details vary with both the form of the module and the particular application. For instance, when joining modules in a stack, the connections should ensure that the load will be transferred at the locations of the supports, ie., at corner supports or continuous supports. Fig. 12.11 shows an example of how four modules are connected at corners. Modules 1 and 2 are stacked on top of Modules 3 and 4. Vertically, Modules 1 and 2 are bolted through their base plates to the end plates of Modules 3 and 4, respectively. Horizontally, Modules 1 and 2 are tied by a bolted tie plate. (Gorgolewski et al., 2009).

Another example of joining two modules vertically at corners is given by Garrison and Tweedie (2008: 68). Four pins are welded to the base of each corner column in the upper module (Module 1). Module 1 is then inserted into the module underneath (Module 2) through these pins. One of the four pins is taper-shaped, one is diamond-shaped, and the others are cylinder-shaped. The tapered pin is used to find the column of Module 2, the diamond pin is for vertical alignment, and the two cylinder-shaped floating pins allow for tolerances. A blind-riveted end plate is used to connect Modules 1 and 2 horizontally with two other adjacent modules on the surfaces.

(a) (b)

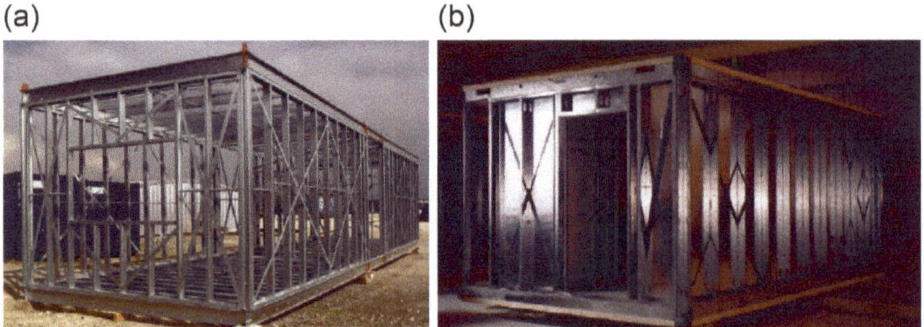

Figure 12.10 Load-bearing module forms: (a) corner supported; (b) continuously supported. From Lawson, R.M., Ogden, R.G., 2008. "Hybrid" light steel panel and modular systems. Thin-walled Structures 46, 720–730.

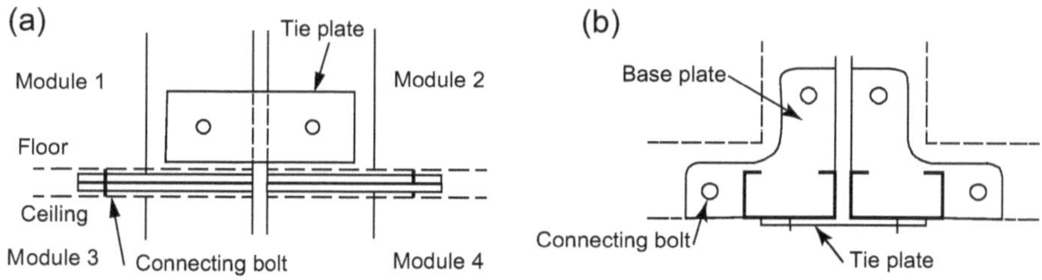

Figure 12.11 Corner connection details between modules: (a) elevation on post; (b) plan on corner.
From Gorgolewski, M.T., Grubb, P.J., Lawson, R.M., 2009. Modular Construction Using Light Steel Framing: Design of Residential Buildings', SCI Publications P302, 102 pages. Available from: http://www.steelconstruction.info/index.php?title=Special:ImagePage&t=SCI+P302.pdf (accessed 16.04.15.). Courtesy of Steel Construction Institute.

Because of limitations of roads and the dimensions of vehicles used for transportation, the dimensions of prefabricated modules are also limited and vary in different countries. For the purpose of simple and cost-effective transportation, varieties of portable or folded, flat-packed modules are available all over the world. One example is given on the Baühu website (2015): packs of folded, flat modules are stacked, and can be easily transported to a construction site; then the folded modules are unpacked and assembled on site one by one. The on-site-assembled modules are now ready for further assembly to a modular-unit-assembled building. The connections used for assembling each module and joining separate modules are mainly bolted.

Connections and joints make great contributions to structural integrity in fire. The fire resistance time for a modular building varies from 30 to 60 min according to the height and use of the building. Modular construction generally achieves fire requirements by the use of plasterboard, cement particle board, or gypsum fiber board. Ma et al. (2012) investigated the temperature development inside a joint of a roofing system. The joint is composed of the connected steel roof sheeting and the rectangular steel top chord of a roof truss. The temperature development of each element inside the joint without any fire protection is compared to that with fire protection of intumescent, Rockwool board, or both, respectively. Further research on the fire resistance of modular constructions is necessary, especially when modular construction is applied in highrise buildings with both walls and floors being load-bearing elements (Lawson et al., 2012).

Sources of further information and advice

Jürisoo, E., Staaf, R., 2007. Connection Design for Easy Assembly On-site. Master's thesis, 173 pages. Available from: http://documents.vsect.chalmers.se/CPL/exjobb2007/ex2007-099.pdf (accessed 08.04.15.).

Klett, J., 2009. Systematic Design of Connections under Consideration Assembly and Disassembly Related Properties. Doctoral dissertation, 193 pages. Available from: https://opus4.kobv.de/opus4-tuberlin/files/2253/klett_jan.pdf (accessed 31.01.15.).
Pahl, G., Beitz, W., Feldhusen, J., Grote, K.H., 2007. Engineering Design: A Systematic Approach, third ed. Springs-Verlag, London.
Sarja, A., 2002. Integrated Life Cycle Design of Structures. Spon Press, Taylor & Francis Group, London and New York.

References

Baühu Prefabricated Construction Solutions, 2015. Case Study – Formula E Race Control Buildings. Available from: http://www.bauhu.com/cubes.html (accessed 17.04.15.).
Brand, S., 1994. How Buildings Learn: What Happens After They're Built. Viking, New York, NY.
Chiodo, J., 2005. Design for Disassembly Guidelines. Active Disassembly Research, 8 pages. Available from: http://www.activedisassembly.com/guidelines/ADR_050202_DFD-guidelines.pdf (accessed 31.03.15.).
Chiodo, J., Jones, N., 2012. Smart materials use in active disassembly. Assembly Automation 32, 8–24. Available from: http://dx.doi.org/10.1108/01445151211198683 (accessed 15.04.15.).
Cooper, D., 2010. Novel Joining Techniques to Promote Deconstruction of Buildings. Working paper 7 for WellMet 2050, 8 pages. Available from: http://www.lcmp.eng.cam.ac.uk/wellmet2/publications#Theme1 (accessed 16.04.15.).
Couchman, G., Sansom, M., 2012. Integrated design for successful steel construction. In: Davison, C., Owens, G.W. (Eds.), Steel Designers' Manual, seventh ed. Steel Construction Institute. Pubished by Blackwell Publishing Ltd.
Edmonds, J., Gorgolewski, M., 2000. Steel Component Design for Deconstruction. Steel reuse information paper No.2, 4 pages. Available from: http://www.reuse-steel.org/files/Information%20papers/Deconstruction%20IP%2010-5.pdf (accessed 30.03.15.).
EN 1993-1-8, 2005. Eurocode 3: Design of Steel Structures. Part 1-8: Design of Joints. European Committee for Standardization, Management Centre: rue de Stassart, 36 B-1050 Brussels.
ESDEP: European Steel Design Education Programme, WG 9.6 Connections in Thin-walled Construction. Available from: http://www.fgg.uni-lj.si/~/pmoze/ESDEP/master/wg09/t0600.htm (accessed 08.04.15.).
FUSA Tech, Inc., 2003. ROSETTE automated steel collar fastening system: alternative wall panel and truss lines, and their product applications. Bulletin News, 4 pages. Available from: http://www.rosettesystems.com/acrobat/Rosette_Alternatives_2003.pdf (accessed 15.04.15.).
Garrison, J., Tweedie, A., 2008. Modular Architecture Manual: Prepared by Garrison Architects for Kullman Buildings Corp, 104 pages. Available from: http://www.garrisonarchitects.com/uploads/1000010/1394717065681/GA_ModularArchitectureManual.pdf (accessed 08.04.15.).
Gilbert, B.P., Rasmussen, K.J.R., 2010. Bolted moment connections in drive-in and drive-through steel storage racks. Journal of Constructional Steel Research 66 (6), 755–766. Available from: http://dx.doi.org/10.1016/j.jcsr.2010.01.013.
Gorgolewski, M.T., Grubb, P.J., Lawson, R.M., 2009. Modular Construction Using Light Steel Framing: Design of Residential Buildings. SCI Publications P302, 102 pages. Available from: http://www.steelconstruction.info/index.php?title=Special:ImagePage&t=SCI+P302.pdf (accessed 16.04.15.).
Hechler, O., Larsen, O.P., Nielsen, S., 2011. Design for Deconstruction, 17 pages. Available from: research.kadk.dk/files/31555965/DfD_16082010.doc (accessed 30.03.15.).

Jürisoo, E., Staaf, R., 2007. Connection Design for Easy Assembly On-site. Master's thesis, 173 pages. Available from: http://documents.vsect.chalmers.se/CPL/exjobb2007/ex2007-099.pdf (accessed 08.04.15.).

Kadak, J., 2014. Effect of Steel Strength on the Welded Joint between a Plate and Two Tubular Cross-sections. Master thesis, 100 pages. Aalto University. Available from: https://aaltodoc.aalto.fi/handle/123456789/13489 (accessed 16.04.15.).

Kaitila, O., Kesti, J., Mäkeläinen, P., 2001. ROSETTE-joints and ROSETTE Steel Trusses: Research Report and Design Recommendations. Research report, 124 pages. Tkk-TER-17, Publication 17. Laboratory of Steel Structures, Helsinki University of Technology.

Klett, J., 2009. Systematic Design of Connections under Consideration Assembly and Disassembly Related Properties. Doctoral dissertation, 193 pages. Available from: https://opus4.kobv.de/opus4-tuberlin/files/2253/klett_jan.pdf (accessed 31.03.15.).

Landolfo, R., Mammanaa, O., Portioli, F., Di Lorenzo, G., Guerrieri, M.R., 2009. Experimental investigation on laser welded connections for built-up cold-formed steel beams. Journal of Constructional Steel Research 65, 196–208. Available from: http://dx.doi.org/10.1016/j.jcsr.2008.03.015.

Lawson, R.M., 1999. Modular Construction Using Light Steel Framing: An Architect's Design Guide. SCI Publications P272. Available from: http://www.steelconstruction.info/index.php?title=Special:ImagePage&t=SCI+P272.pdf (accessed 17.04.15.).

Lawson, R.M., 2007. Building Design Using Modules. SCI Publications P348, 16 pages. Available from: http://www.steelconstruction.info/index.php?title=Special:ImagePage&t=SCI+P348.pdf (accessed 16.04.15.).

Lawson, R.M., Ogden, R.G., 2008. "Hybrid" light steel panel and modular systems. Thin-walled Structures 46, 720–730.

Lawson, R.M., Ogden, R.G., Bergin, R., 2012. Application of modular construction in high-rise buildings. Journal of Architecture Engineering, ASCE 18, 148–154.

Lee, Y.H., Tan, C.S., Mohammad, S., Md Tahir, M., Shek, P.N., 2014. Review on cold-formed steel connections. The Scientific World Journal, 11 pages. Available from: http://www.hindawi.com/journals/tswj/2014/951216/ (accessed 12.02.15.).

Lu, W., Segaro, P., Kesti, J., Mäkeläinen, P., 1998a. Study on the Shear Strength of a Single-lap ROSETTE-joint. Research report, 40 pages. TKK-TER-08, Publication 8. Laboratory of Steel Structures, Helsinki University of Technology.

Lu, W., Kesti, J., Mäkeläinen, P., 1998b. Shear and Cross-tension Tests for Press-joins. Research report, 55 pages, Report 7. Laboratory of Steel Structures, Helsinki University of Technology.

Ma, Z., Lu, W., Mäkeläinen, P., Outinen, J., 2012. Prediction of joint temperatures in shot-nailed cold-formed steel sheeting with finite element modelling. In: Proceeding of 21st International Specialty Conference on Cold-formed Steel Structures, St. Louis, Missouri, USA, pp. 599–609.

Manner, K.J., 2015. Design for Disassembly. Lecture notes on Engineering Design Projects, 7 pages. Available from: http://homepages.cae.wisc.edu/~me349/lecture_notes/me349_dfd.pdf (accessed 31.03.15.).

Mills, J., Laboube, R.A., 2004. Self-drilling screw joints for cold-formed channel portal frames. In: 16th International Specialty Conferences on Cold-formed Steel Structures. Paper 4, 14 pages. Available from: http://scholarsmine.mst.edu/cgi/viewcontent.cgi?article=1443&context=isccss (accessed 10.04.15.).

Morgan, C., Stevenson, F., 2005. Design for Deconstruction. SEDA design guides for Scotland: No.1, 70 pages. Available from: http://www.seda.uk.net/assets/files/guides/dfd.pdf (accessed 31.03.15.).

Mucha, J., Witkowski, W., 2014. The clinching joints strength analysis in the aspects of changes in the forming technology and load conditions. Thin-Walled Structures 82, 55–66.

NSC: New Steel Construction, 2005. Quicon Wins Contract at IKEA. Available from: http://www.newsteelconstruction.com/wp/quicon-wins-contract-at-ikea/ (accessed 25.09.15.).

Olson, T.P., 2010. Design for Deconstruction and Modularity in a Sustainable Built Environment, 37 pages. Available from: http://cmec.wsu.edu/publications/timothyolsonreport.pdf (accessed 30.03.15.).

Ogden, R., Plank, R., Lawson, M., Atree, J., 2003. Architecture Teaching Resource Studio Guide. SCI Publication P167, 28 pages. Available from: http://www.tatasteelconstruction.com/file_source/StaticFiles/Construction/studio%20guide.pdf (accessed 24.09.15.).

Pedreschi, R.F., Sinha, B.P., 1996. The potential of press-joining in cold-formed steel structures. Construction and Building Materials 10 (4), 243–250. Available from: http://dx.doi.org/10.1016/0950-0618(96)00006-2.

Pulaski, M., Hewitt, C., Horman, M., 2004. Design for Deconstruction. Modern Steel Construction, 4 pages. Available from: http://www.tboake.com/pdf/deconstruction.pdf (accessed 31.03.15.).

Sarja, A., 2002. Integrated Life Cycle Design of Structures. Spon Press, Taylor & Francis Group, London and New York.

Varis, J., 2006. Ensuring the integrity in clinching process. Journal of Materials Processing Technology 174, 277–285. Available from: http://dx.doi.org/10.1016/j.jmatprotec.2006.02.001.

Wallace, K., Clarkson, J., 1999. An Introduction to the Design Process. Course materials, 20 pages. Available from: http://www-mdp.eng.cam.ac.uk/web/library/enginfo/design/design_process.pdf (accessed 27.02.15.).

Wrzesein, A.M., Lim, J.B., 2008. Cold-formed steel portal frame joints a review. In: 19th International Specialty Conference on Cold-formed Steel Structures. Paper 4, 16 pages. Available from: http://scholarsmine.mst.edu/isccss/19iccfss/19iccfss-session9/4 (accessed 09.04.15.).

Yu, W.W., LaBoube, R.A., 2010. Cold-formed Steel Design, fourth ed. USA: John Wiley & Sons, Hoboken, NJ.

Zaharia, R., Franssen, J.M., 2000. Behaviour of Steel Pallet Rack Structures under Exceptional Loading Conditions. Partial Report, 69 pages. Available from. http://hdl.handle.net/2268/99377 (accessed 11.04.15.).

Sustainable applications of cold-formed steel structures: portal frames

13

J.B.P. Lim[1], *A.M. Wrzesien*[2], *D.A. Nethercot*[3]
[1]The University of Auckland, Auckland, New Zealand; [2]University of Strathclyde, Glasgow, United Kingdom; [3]Imperial College London, London, United Kingdom

Notation

a_B	length of bolt group
b_B	breadth of bolt group
b_{ec}	b_B pertaining to eaves-to-column connection
b_{er}	b_B pertaining to eaves-to-rafter connection
d_b	diameter of bolt
d_h	diameter of bolt hole
EA	axial rigidity of back-to-back channel used for column and rafter members of portal frame
EI	flexural rigidity of back-to-back channel used for column and rafter members of portal frame
f_f	height to apex less height to eaves
h_f	height to eaves of portal frame
k_{ab}	rotational stiffness of apex bracket
k_{ar}	rotational stiffness of apex to rafter connection
k_b	bolt-hole elongation stiffness determined from the initial gradient of load versus bolt-hole elongation curve
k_B	rotational stiffness of bolt group
k_{eb}	rotational stiffness of eaves bracket
k_{ec}	rotational stiffness of eaves-to-column connection
k_{er}	rotational stiffness of eaves-to-rafter connection
l'_{ar}	effective length of bracket pertaining to apex-to-rafter connection
l'_{ec}	effective length of bracket pertaining to eaves-to-column connection
l'_{er}	effective length of bracket pertaining to eaves-to-rafter connection
L_f	span of portal frame
n	flexibility factor after Zadanfarrokh and Bryan (1992)
b_f	bay spacing
s_f	length of rafter of portal frame
t	thickness of plate
t_1, t_2	thicknesses of metal sheets
θ_f	pitch of portal frame

Recent Trends in Cold-Formed Steel Construction. http://dx.doi.org/10.1016/B978-0-08-100160-8.00013-X

13.1 Introduction

Cold-formed steel (CFS) portal frames (Fig. 13.1) can be a viable alternative to conventional hot-rolled steel portal frames for lowrise commercial, light industrial, and agricultural buildings. Fig. 13.2 shows the parameters used to define a building in this chapter. Under loading conditions in the UK, spans of up to 18 m can be

Figure 13.1 CFS portal framing system.
Wrzesien, A.M., Lim, J.B.P., Nethercot, D.A., 2012. Optimum joint detail for a general cold-formed steel portal frame. Advances in Structural Engineering 15, 1623–1639.

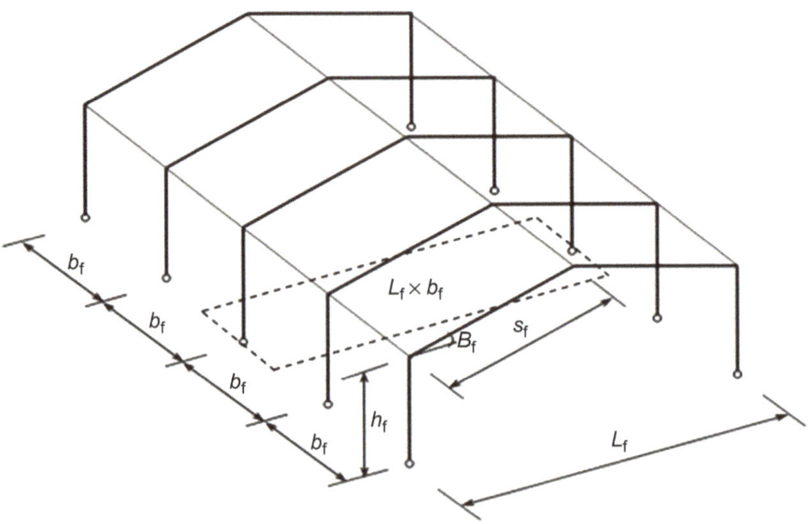

Figure 13.2 Parameters describing a portal frame geometry.
Phan, D.T., Lim, J.B.P., Tanyimboh, T.T., Wrzesien, A.M., Sha, W., Lawson, R.M., 2015. Optimal design of cold-formed steel portal frames for stressed-skin action using genetic algorithm. Engineering Structures 93, 36–49.

economical; in Australia, where in most areas snow loading is lower, this increases to spans of around 25 m. However, the majority of CFS portal frames erected are small, and spans of 10 m (or less) are typical.

The advantages of constructing a portal framing system using CFS sections instead of conventional hot-rolled sections include the following.

1. Sections delivered to site have been marked, cut to length, and punched at the factory to design specifications.
2. Maintenance-free pregalvanized CFS sections that do not require painting to prevent rusting.
3. Reduced transportation costs due to efficient stacking of CFS sections.
4. Reduced acquisition costs as the CFS sections used for secondary members can be purchased from the same manufacturer/supplier.
5. Frames erected manually by semiskilled labor without the need for an on-site crane.

Over the past three decades various researchers have undertaken tests on joint arrangements that can be used for the eaves and apex joints of CFS portal framing systems. Table 13.1 summarizes the joint tests reported in the literature, including the moment capacity of the CFS sections being connected and the number of components and fasteners required to form the joints. In chronological order, a brief description of each joint arrangement is given in Section 13.1.1.

It should be noted that while the joints tests reported in Section 13.1.1 are mainly concerned with the moment capacities of the joints, the majority of the joints were found to be semirigid when tested. It will be demonstrated that to obtain optimum frame performance, the joint detail for the eaves and apex should in fact not be designed independently of the frame. As such, taking joint effects explicitly into account in the design process provides an opportunity to devise an appropriate balance between joints and member properties, and thus reduces material use and construction costs.

However, when CFS frames are considered as part of a building system, a bare frame analysis may not be appropriate. As demonstrated for small-span buildings, research by Wrzesien et al. (2015) and Phan et al. (2015) highlights the importance of accurate predictions of bare frame stiffness. If the stiffness is overpredicted, loads transferred to gable frames due to stressed-skin action may be largely underestimated. As the end gable frames are often designed as braced frames, or act accordingly due to the bracing effect of the gable sheeting, their horizontal stiffness affects load redistribution. In some design cases the bare gable frame will experience much higher forces due to the effects of diaphragm action, and forces acting on it will be underestimated even if deflection limits, such as those recommended by the Steel Construction Institute (SCI), are adopted (see Table 13.2).

13.1.1 Literature review

The earliest tests reported in the literature on CFS portal frame joints are those by Baigent and Hancock (1982). Details of this joint are given in Fig. 13.3. As can be seen, the joints were formed through the web of the channel sections used for column and rafter members. The moment capacity of the channel sections being connected was

Table 13.1 **Joints reported in the literature**

	Sections				Brackets				Fasteners		Joint stiffness ratio	Joint resistance ratio
		Depth × breadth × t	σ_y	M_{cy}	l_e	t_b	σ_{yb}					
Author(s)		(mm)	(N/mm^2)	(kNm)	(mm)	(mm)	(N/mm^2)	No. of brackets	Description	No. of fixings/joint	K_j	$M_{j,exp}/M_{cy}$
Baigent and Hancock (1982)		153 × 79 × 1.86	325.8	9.2	260	12	–	1	M19 mm H.T. bolts	8	rigid	0.78
Kirk (1986)		220 × 65 × 2.4	280*	32.0	620	3	280	2	M16 G.8.8	8	rigid	1.30
Mäkeläinen and Kankaanpää (1996)		300 × 75** × 3.0	350*	77.0	–	12	355	1	M16 G.8.8	20	–	0.95
Chung and Lau (1999)		150 × 64 × 1.6	450*	17.9	460^h	6	343	1	M16 G.8.8	8	1.57	0.83
					460^{hs}	2.5	475	2				
Lim and Nethercot (2002)		340 × 90 × 3	280*	82.8	746^{hs}	3	209	2	M16 G.8.8	16	0.77	0.90
Mills and LaBoube (2004)		200 × 76 × 1.5	450*	10.8	End plate joint				Bolts	2	rigid	0.72
					Miterd joint				Bolts	2		
					Sections screwed back to back				Screws	12		
Dubina et al. (2004)		350 × 100 × 3	452	117.8	940	10	235	1	M20 G.6.6	32	0.55	0.99

Dundu and Kemp (2006)		300 × 75 × 3	468.9	51.6	Sections bolted back to back				M20 G.8.8	4	0.54	0.83
									M20 G.8.8	8		
Kwon et al. (2006)		PRY 150 × 40 × 0.8	570*	3.6	261.6	2.3	240	1	Screws ø 4.8 mm	16	0.13	1.97
Rhodes and Burns (2006)		342 × 97 × 2.5 rafter	343	76.7	Connection angles		275*	2	M16 G.8.8	32	1.06	1.10
		402 × 97 × 3.2 column	352	128.5	C 202 × 69 × 2 knee-brace		350*	2			–	1.06
Wrzesien et al. (2012)		152 × 64 × 2	395	21.2	400 (joint 1)	3.0	350	2	M16 G.8.8	30	1.36	0.91
					520 (joint 2)	3.0						

*Design yield strength; **Dimensions are not reported but standard sizes are assumed; *h* denotes a haunch; *hs* denotes a haunch with stiffeners, stiffness ratio $K_j = k_B/(EI_b/L_b)$ where k_B = rotational stiffness of the joint, E = modulus of elasticity, I_b = moment of inertia of the jointed beam member, L_b = length of the jointed beam member equal to 1 m, l_e = length of the connection bracket, $M_{j,exp}$ = experimental moment resistance of the joint, M_{cy} = moment resistance of the connected member calculated to BS 5950-5 (1998), t = thickness of the section, t_b = thickness of the connection bracket, σ_y = yield strength of the section, σ_{yb} = yield strength of the bracket.

Table 13.2 Recommended serviceability deflection limits for portal frames under both live and wind loads after SCI P313 (2002)

Deflection category	Reason for limit	Deflection limit
Lateral eaves deflection	Damage to side cladding	$h_f/100$
	Damage to roof cladding	$b_f/200$
Vertical apex deflection	Ponding of water	$\sqrt{b_f^2 + s_f^2}/125$
		$b_f/100$

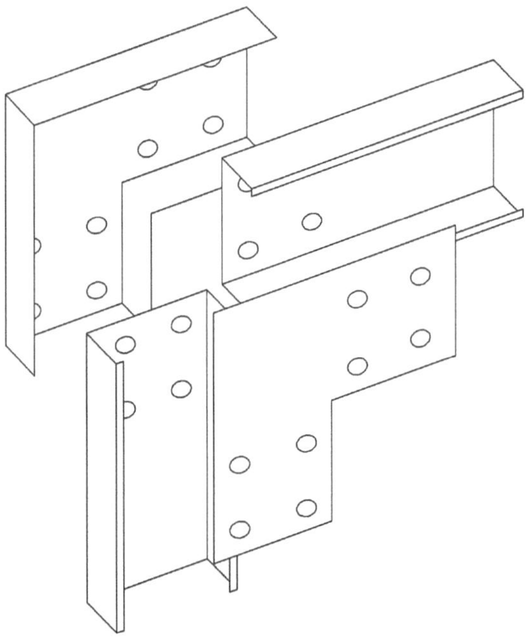

Figure 13.3 Eaves joint after Baigent and Hancock (1982).
Wrzesien, A.M., Lim, J.B.P., Nethercot, D.A., 2012. Optimum joint detail for a general cold-formed steel portal frame. Advances in Structural Engineering 15, 1623–1639.

9.19 kN m. The thickness of the channel sections was 1.86 mm, while the thickness of the plate used to connect the joints was 12 mm. Due to the use of high-tensile friction grip bolts, the joints could be considered as being rigid.

The next tests reported were those by Kirk (1986) on the Swagebeam portal framing system. These tests were undertaken at Salford University. Fig. 13.4 shows details of the joints. As can be seen, back-to-back channel sections were used for the column and rafter members. The joints were formed through back-to-back brackets bolted between the webs of the channel sections. The moment capacity of the back-to-back channel

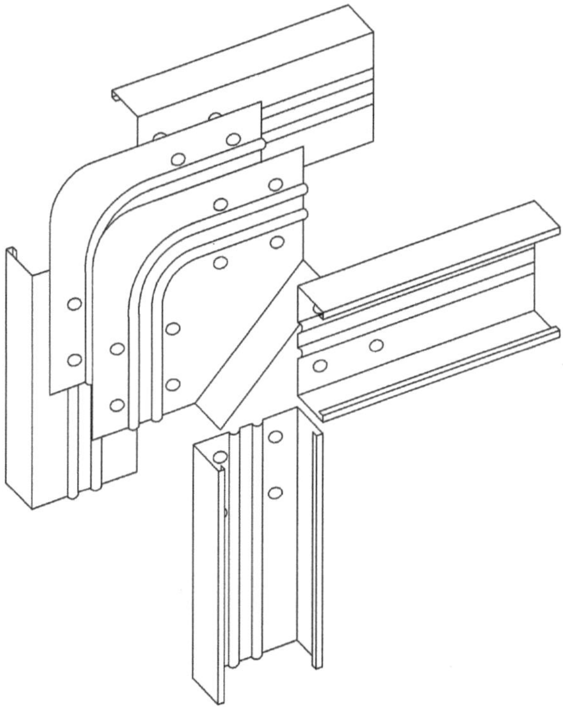

Figure 13.4 Swagebeam eaves joint after Kirk (1986).
Wrzesien, A.M., Lim, J.B.P., Nethercot, D.A., 2012. Optimum joint detail for a general cold-formed steel portal frame. Advances in Structural Engineering 15, 1623–1639.

sections was 32 kN m, the thickness of the channel sections was 2.4 mm, and the thickness of each bracket was 3.0 mm. The primary innovation was that the joints could be formed through the swages rolled in the brackets which connected with matching swages in the webs of the channel sections. This arrangement produced stiff connections.

Mäkeläinen and Kankaanpää (1996) described tests on a portal framing system constructed from back-to-back sigma sections connected through the web via brackets. To provide additional stiffness to the frame, a tie bar (double angle 50 × 50 × 2.5 mm) was bolted to both eaves brackets (Fig. 13.5(a)). The depths of the sections used for the tests were 250, 300, and 400 mm, and thicknesses of 2.5 and 3.0 mm were considered. Fig. 13.5 shows details of the joint brackets. These included a single plate of thicknesses of 8, 10, and 12 mm (see Fig. 13.5(a)), four CFS plates with a thickness of 2.5 mm each (see Fig. 13.5(b)), and four CFS plates with two outer plates outwardly lipped (see Fig. 13.5(c)). Although the moment capacities of the sections were not provided, similar compound members made from back-to-back standard sigma sections of 300 mm deep, 75 mm wide, and 3.0 mm thick were calculated to have a moment capacity of 77 kN m.

Chung et al. (Chung and Lau, 1999; Wong and Chung, 2002; Yu et al., 2005) and Lim and Nethercot (2002, 2003a,b, 2004a,b) independently reported tests on an

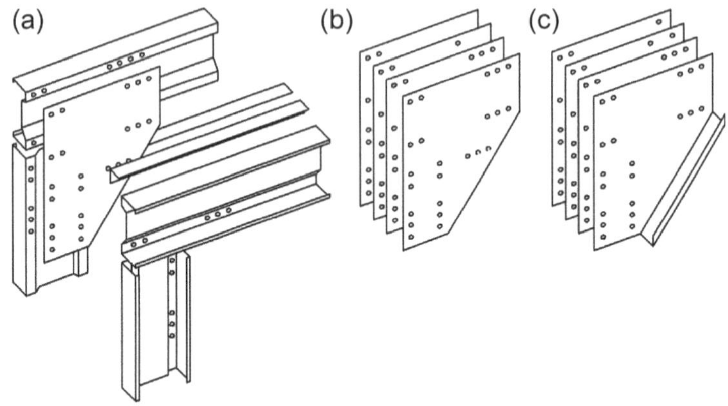

Figure 13.5 Eaves joint with different configurations after Mäkeläinen and Kankaanpää (1996): (a) single layer; (b) four layers; (c) four layers with lip stiffeners.
Wrzesien, A.M., Lim, J.B.P., Nethercot, D.A., 2012. Optimum joint detail for a general cold-formed steel portal frame. Advances in Structural Engineering 15, 1623–1639.

arrangement where the joint was formed through back-to-back brackets bolted between the webs of the channel sections being connected. In the tests described by Chung and Lau (1999), the moment capacity of the sections tested was 17.88 kN m, while that of Lim and Nethercot was 82.8 kN m. Fig. 13.6(a)–(d) shows the different shapes of the brackets studied by Chung. In the case of the joint details shown in Fig. 13.6(c) and (d), the joints were tested twice. In the first stage the joints were formed through a hot-rolled steel single-gusset plate of 6 mm thickness. In the second phase the joints were formed through two back-to-back CFS brackets, each 2.5 mm thick and with lip stiffeners along the cathetus and hypotenuse of the bracket

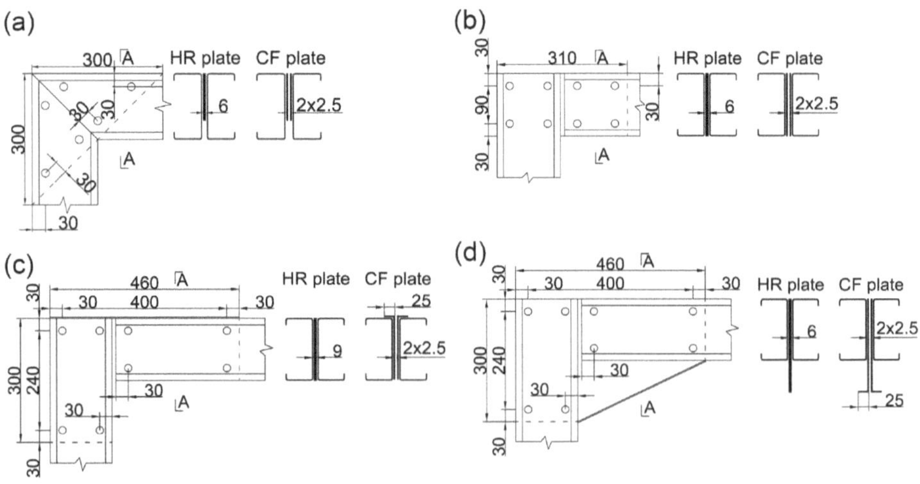

Figure 13.6 Eaves joint brackets after Chung and Lau (1999): (a) triangular; (b) rectangular; (c) L-shape with stiffener; (d) haunched with stiffener.

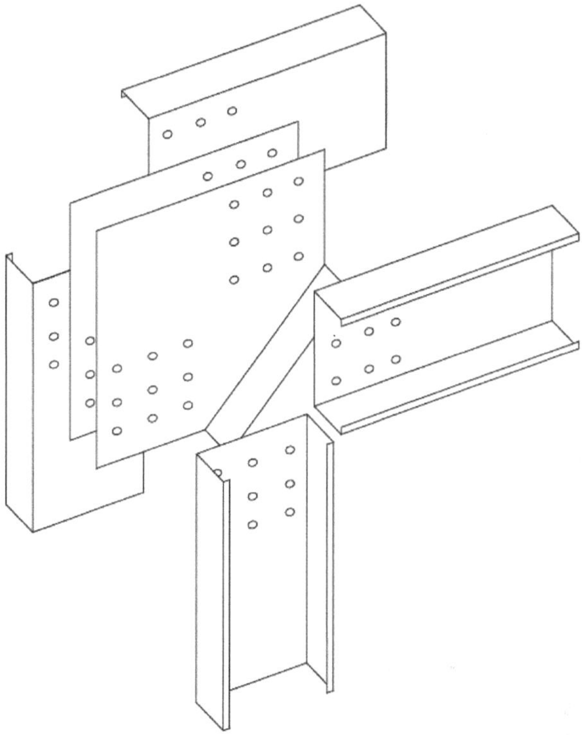

Figure 13.7 Eaves joint after Lim and Nethercot (2002).
Wrzesien, A.M., Lim, J.B.P., Nethercot, D.A., 2012. Optimum joint detail for a general cold-formed steel portal frame. Advances in Structural Engineering 15, 1623–1639.

respectively (Fig. 13.6(c) and (d)). Unlike the joints tested by Chung and Lau (1999), the joints tested by Lim and Nethercot (Fig. 13.7) isolated failure of the brackets from that of the channel sections. Ensuring that the brackets themselves would not fail, research was focused on the strength and stiffness of the channel sections, as influenced by the bolt group size.

Mills and LaBoube (2004) conducted experimental studies on joints currently used in Australia for CFS portal frame sheds. The joints were constructed from single channel sections with a moment capacity of 10.84 kN m. Popular joints included an end plate connection bolted to the column and welded to the rafter (Fig. 13.8(a)), and a miterd joint (Fig. 13.8(b)). Self-drilling screws were used as an alternative to conventional bolting (Fig. 13.8(c)). A similar arrangement for the apex joint was also studied, in which double-lipped channel sections were used as the gusset plate and screwed back to back to the rafters.

Dubina et al. (2004) described three different types of joints (Fig. 13.9). As can be seen from Fig. 13.9(a) and (c), the channel sections were bolted only through the web of welded I-section brackets (KIS and KIP in Fig. 13.9) and spaced gusset

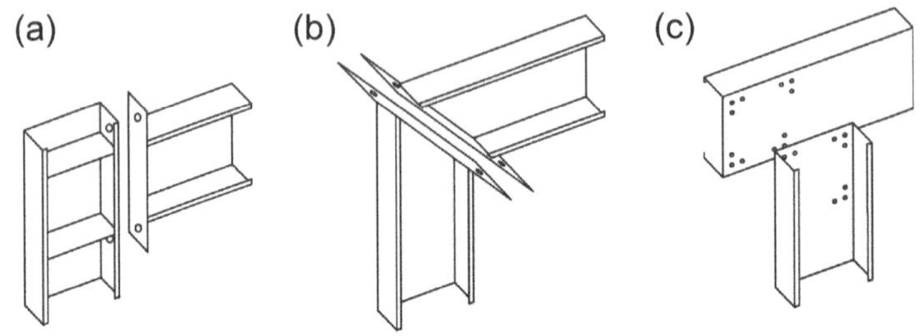

Figure 13.8 Eaves joints after Mills and LaBoube (2004): (a) bolted end plate joint; (b) miterd joint; (c) self-drilling screw joint.
Wrzesien, A.M., Lim, J.B.P., Nethercot, D.A., 2012. Optimum joint detail for a general cold-formed steel portal frame. Advances in Structural Engineering 15, 1623–1639.

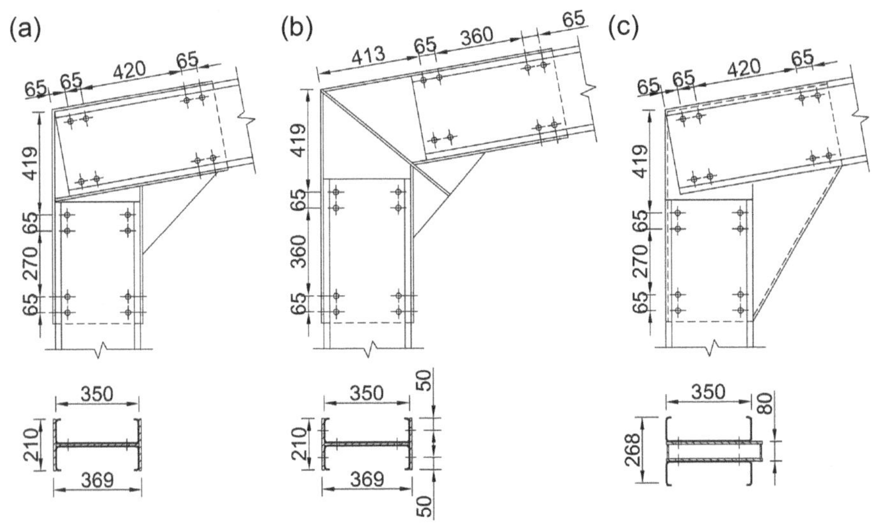

Figure 13.9 Eaves joints after Dubina et al. (2004): (a) bolts on the web (KIS, KIP); (b) bolts on the web and on the flange (KIS, KIP); (c) bolts on the web (KSG).

brackets (KSG). In a second variant, bolts were located on both the web and the flange (Fig. 13.9(b)) of I-section bracket (KIS) and I-section bracket with plate bisector (KIP). The moment capacity of the channel sections being connected was 117.8 kN m. It should be noted that unlike the joints tested by Chung and Lau (1999), the joint was formed through hot-rolled steel sections instead of back-to-back brackets. However, as the strength of the hot-rolled steel sections is much greater than that of the channel sections, the behavior of the joints is dominated by that of the channel sections.

Dundu and Kemp (2006) conducted research on single channels connected back to back (Fig. 13.10). Such an arrangement is similar to that of Mills and LaBoube (2004). Dundu and Kemp proposed the use of a plastic hinge analysis and concentrated on the ductility of the joints. A novel method for providing lateral restraint was introduced through an angle connection between the web of the rafter and that of the purlin. It

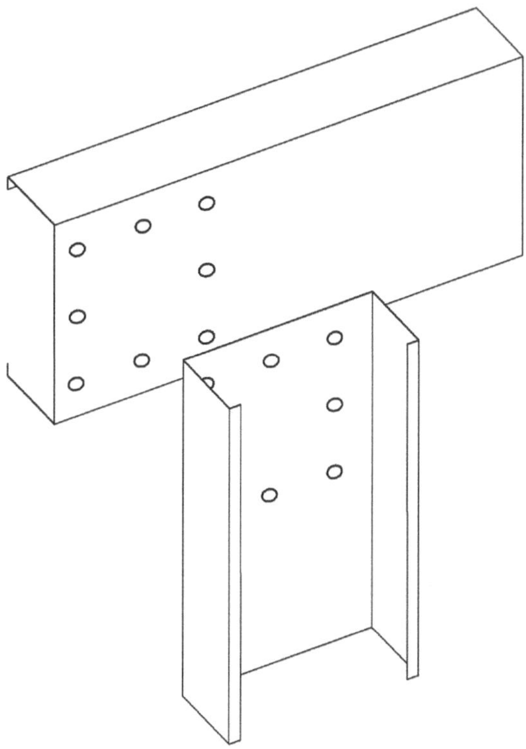

Figure 13.10 Bolted joints after Dundu and Kemp (2006).
Wrzesien, A.M., Lim, J.B.P., Nethercot, D.A., 2012. Optimum joint detail for a general cold-formed steel portal frame. Advances in Structural Engineering 15, 1623–1639.

was demonstrated that this arrangement eliminated the lateral–torsional buckling failure mode, since both the top and bottom flanges were effectively restrained.

Kwon et al. (2006) reported research on applications of closed sections produced by a combination of cold-rolling and clinching techniques. The sections used for the tests were 150 mm deep, 40 mm wide, and 0.8 mm thick. The moment capacity calculated from the gross section modulus was 3.55 kN m. Connection brackets for the eaves and apex joints were constructed from 2.3 mm thick mild steel plates through a combination of folding and welding, with four different connection types. The brackets of connection types 1 and 2 were produced by cutting the bottom flange of a C-shape bracket and welding a lipped plate to build a haunch stiffener, with and without a lip on the flange respectively. However, the bracket was made by press-breaking 3 mm thick CFS strips. Fig. 13.11 shows the general joint arrangement of connection type 3 with a lip on the flange. In connection type 4, the bracket of the same shape lip on the flange was not provided.

Rhodes and Burns (2006) conducted extensive component tests on the eaves joints of a CFS portal framing system that used a knee-brace arrangement, formed through back-to-back channel sections bolted to the flanges of the column and the rafter through a welded bracket. Fig. 13.12 shows details of the joint. The columns and

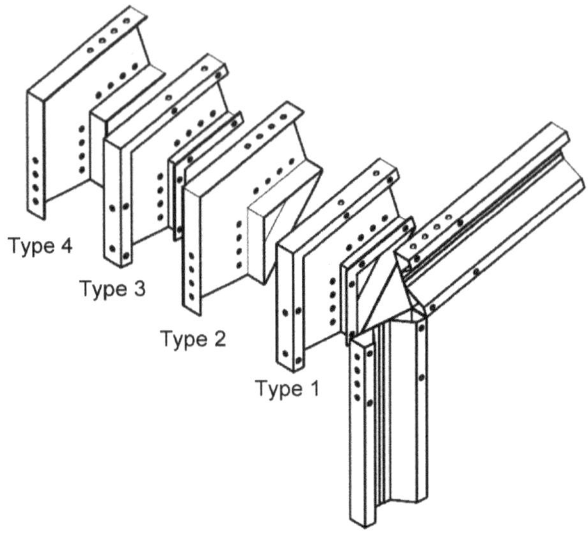

Figure 13.11 Eaves joints after Kwon et al. (2006).

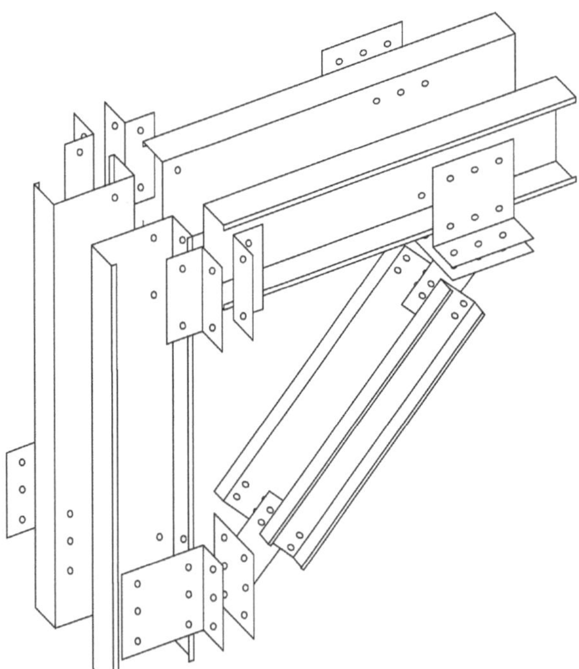

Figure 13.12 Eaves joint after Rhodes and Burns (2006).

Wrzesien, A.M., Lim, J.B.P., Nethercot, D.A., 2012. Optimum joint detail for a general cold-formed steel portal frame. Advances in Structural Engineering 15, 1623–1639.

rafters were formed from back-to-back channel sections with moment capacities of 128.54 and 76.68 kN m, respectively.

Wrzesien et al. (2012) reported six full-scale tests on CFS portal frame buildings. Unlike the other tests reported in the literature, the purpose here was to investigate the effects of stressed-skin action. Details of the eaves and apex joints considered are shown in Fig. 13.13; such joints are typical of those used for CFS portal frames in practice. Two different bolt group sizes were considered for the joints, with each bolt group size (and therefore bracket size) having a different rotational stiffness. Tests on frames without roof sheeting were described. Vertical loading was applied in two tests, and horizontal loading in four tests. For horizontal loading only, the frame tests

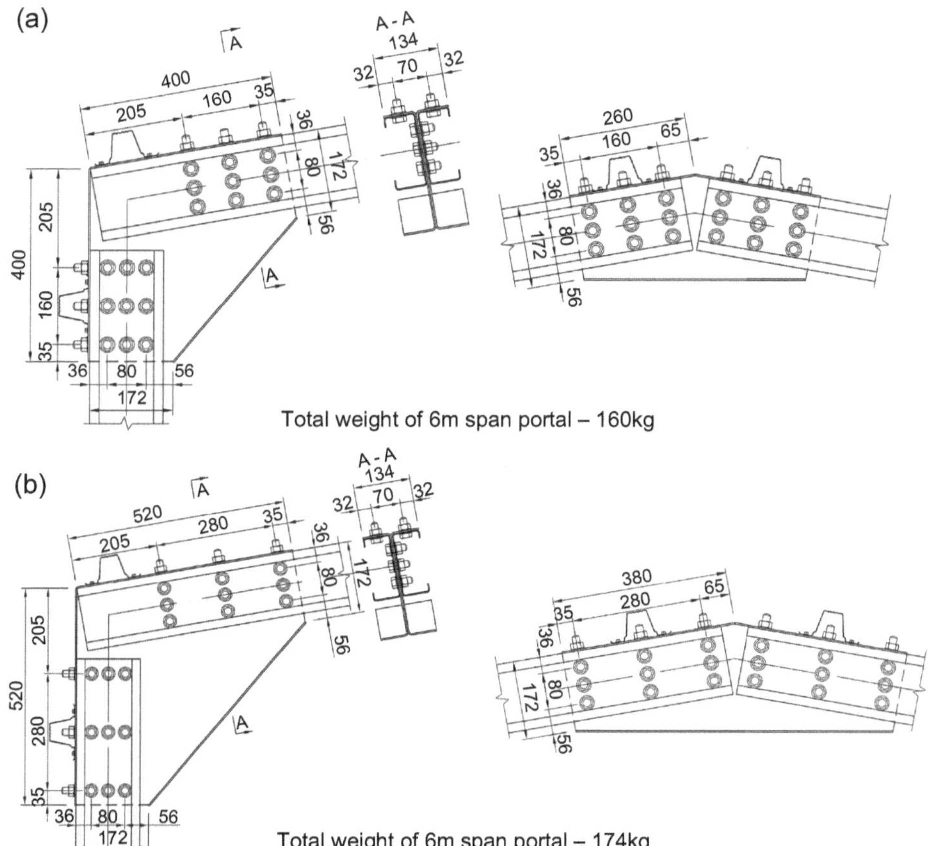

Figure 13.13 Details of joint eaves and apex joints after Wrzesien et al. (2012). (a) Joint 1; (b) Joint 2.
Wrzesien, A.M., Lim, J.B.P., Xu, Y., Macleod, I.A., Lawson, R.M., 2015. Effect of stressed skin action on the behaviour of cold-formed steel portal frames. Engineering Structures 105, 123–136.

were carried out with and without roof sheeting. As mentioned earlier, it was shown that careful attention should be paid to the load returned to the gable through stressed-skin action, and that for certain geometries of frames the SCI deflection limits are inadequate to prevent damage to the sheeting or gable frames.

13.1.2 Summary of literature

In the joints described in the literature, the general joint arrangement comprises back-to-back channel sections for the members, connected through gusset plates bolted to the web of the channel sections. Table 13.1 summarizes the moment capacities of the sections used for the joints. The ratio of the experimental moment resistance of the joint to the moment resistance of the cross-section is also shown, to indicate if the joint is full or partial strength according to BS EN 1993-1-8 (2005). The values of experimental rotational stiffness are expressed in the form of stiffness ratio K_j, (shown in Table 13.1 for a member length of 1 m).

It should be noted that in the case of the joints described by Dubina et al. (2004), the gusset plates are replaced by hot-rolled steel sections. However, the connection of the back-to-back channel sections to the hot-rolled steel sections remains through bolts in the webs of the channel sections. The main difference between the joint described by Dubina et al. and the general joint arrangement is that there are also bolts through the flanges of the channel sections to the hot-rolled steel sections; such bolts increase the moment capacity and rotational stiffness of the bolt groups. On the other hand, in the joint described by Dundu and Kemp (2006), while there is no gusset plate, the connection is still through bolts in the webs of the channel sections. However, the lack of gusset plates means that the moment capacity and stiffness of the bolt groups in the webs are limited by the depth of the channel sections.

13.1.3 Outline of chapter

In the remainder of this chapter, the requirement for a beam idealization to take into account both the semirigidity and finite connection length is shown through comparisons against full-scale test results. In the case of joints formed through the web, a reduced strength as a result of the bimoment in the section is demonstrated. The effect of stressed-skin action, for portal frames with small spans is shown, highlighting the need for serviceability limits which, if ignored, would lead to tearing of the cladding or the gable end frames. Finally, the performance in fire conditions is described, suggesting the potential need for fire protection to the side rails.

13.2 Behavior of bare frames

For application to portal frames, virtually all practical forms of CFS portal framing joints are semirigid and have finite connection lengths, and both these effects should

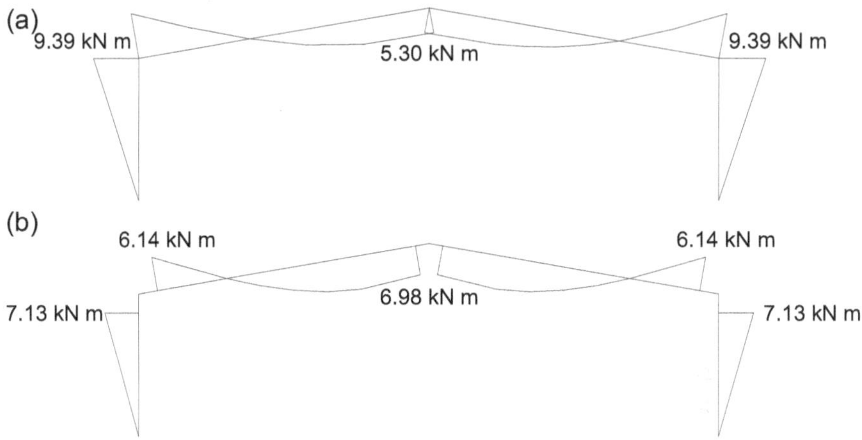

Figure 13.14 Bending moment distribution of a cold-formed steel portal frame: (a) rigid frame; (b) semirigid frame.

be taken into account in analysis of portal frames since the semirigidity and finite connection length of the joints influence the bending moment distribution as well as the deflected profile of the frame. Treating the joints as rigid will result in a very different pattern of moments and underestimate frame deflections.

As an example, Fig. 13.14 shows the changes in bending moment for a 12 m span portal frame, contrasting calculated bending moments based on a rigid joint analysis with realistic bending moments that take into account the semirigidity and finite connection lengths of the joints.

As can be seen, the rafter members would be designed to sustain a bending moment around 50% larger in a rigid joint analysis, as opposed a frame analysis that takes into account the semirigidity and finite connection length of the joints. More importantly, a rigid joint analysis would result in an underestimation of the bending moment in the apex by 32%. In this section, analysis and design guidance applicable to CFS portal frames are presented.

In terms of structural design, two joint effects should be taken into account. Firstly, as mentioned previously, owing to localized bolt-hole elongation, the joints are semirigid. To illustrate this, Fig. 13.15 shows a bolt group rotating under moment. As the bolt holes elongate, the bolt group rotates around the center of rotation. The effects of joint flexibility are increased frame deflections, so serviceability deflections may be more important than for the case of a rigid-jointed frame. Bolt-hole elongation stiffness has been investigated by Zadanfarrokh and Bryan (1992) and Zaharia and Dubina (2006), and formulae are given in Table 13.3.

Secondly, unlike rigid joints, moment is resisted by bolt forces developed in the web over the length of the bolt group (see Fig. 13.16), which in turn results in a bimoment (see Fig. 13.17) (Lim et al., 2016). As a result of this bimoment, the moment capacity of the joints can be expected to be lower than that of the channel sections; the strength of the joints can therefore be seen to be influenced by the size of the bolt

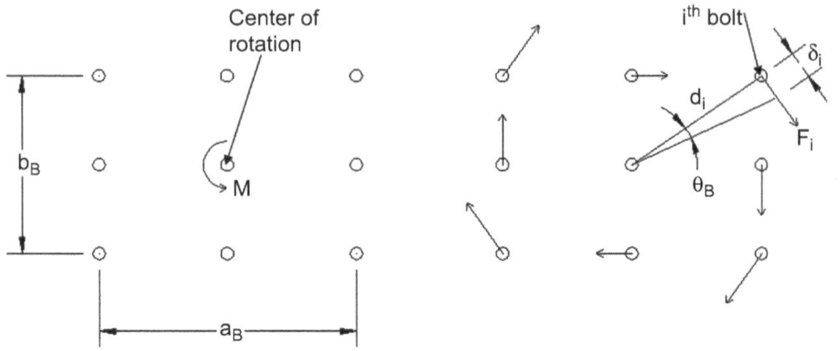

Figure 13.15 Details of bolt group resisting moment.
Wrzesien, A.M., Lim, J.B.P., Nethercot, D.A., 2012. Optimum joint detail for a general cold-formed steel portal frame. Advances in Structural Engineering 15, 1623–1639.

Table 13.3 Analytical formulae for predicting elongation stiffness (k_b) of bolted joint

Elongation stiffness	Authors
$k_{Zad} = \dfrac{1}{5n\left(\frac{10}{t_1} + \frac{10}{t_2} - 2\right)10^{-3}}$ (kN/mm)	Zadanfarrokh and Bryan (1992)
$k_{Zah} = 6.8\dfrac{\sqrt{d_b}}{\left(\frac{5}{t_1} + \frac{5}{t_2} - 1\right)}$ (kN/mm)	Zaharia and Dubina (2006)

Figure 13.16 Free body diagram of channel section when joint is in pure bending: (a) details of typical channel section and bolt group; (b) free body diagram.
Lim, J.B.P., Nethercot, D.A., 2003. Ultimate strength of bolted moment-connections between cold-formed steel members. Thin-walled Structures 41, 1019–1039.

Figure 13.17 Failure due to combined bending and bimoment stresses.

group. In more general CFS portal framing systems, larger bolt group sizes are required to increase both the rigidity and the strength of the joints.

13.2.1 Beam idealization and semirigid joints

Fig. 13.18 shows details of a beam idealization of a CFS portal frame. The semirigidity of the joints is idealized by a rotational spring at the center of rotation of the bolt group. The size of the bolt groups (and brackets) is shown in Fig. 13.19. The rotational stiffnesses at the eaves of the bolt group connecting the column and rafter to the eaves bracket are defined as k_{ec} and k_{er}, respectively. Similarly, the rotational stiffness at the apex of the bolt group connecting the rafter to the apex bracket is k_{ar}. It is worth noting that the rotational stiffness of the bolt group in semirigid joints is based on the bolt-hole elongation stiffness, k_b (Lim and Nethercot, 2003a). The value of the bolt-hole elongation stiffness can either be measured experimentally in the laboratory or determined from finite element analysis.

Also, the bolt-hole elongation stiffness depends on the thickness of the CFS plates, and can be calculated based on equations proposed in Table 13.3. The distance from the intersection of the members to the center of rotation of each bolt group is referred to herein as the effective length of the bolt group. For the eaves bracket, the effective

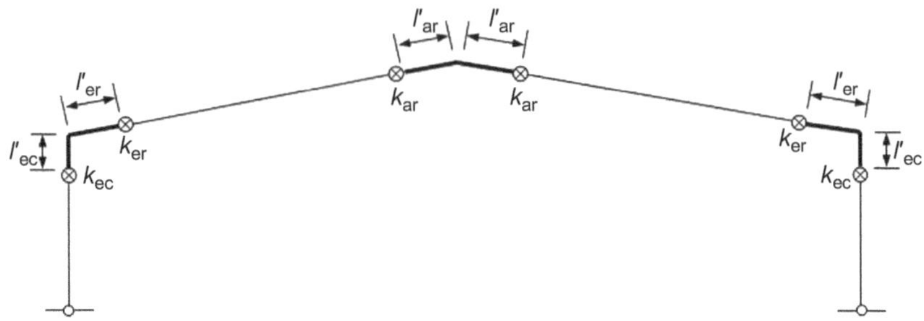

Figure 13.18 Parameters used to define semirigid joints of internal frames.
Phan, D.T., Lim, J.B.P., Tanyimboh, T.T., Wrzesien, A.M., Sha, W., Lawson, R.M., 2015. Optimal design of cold-formed steel portal frames for stressed-skin action using genetic algorithm. Engineering Structures 93, 36–49.

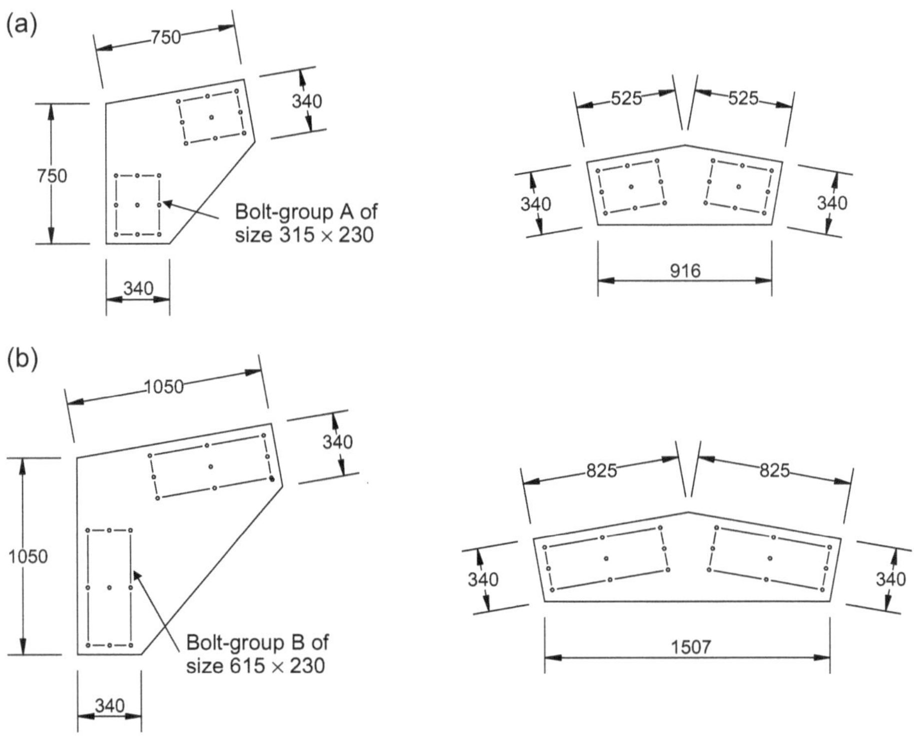

Figure 13.19 Details of the bolt-groups and the corresponding sizes of eaves and apex brackets: (a) frame A; (b) frame B.
Lim, J.B.P., Hancock, G.J., Charles Clifton, G., Pham, C.H., Das, R., 2016. DSM for ultimate strength of bolted moment-connections between cold-formed steel channel members. Journal of Constructional Steel Research 117, 196–203.

lengths of the bolt group connecting the column and rafter to the eaves bracket are l'_{ec} and l'_{er}, respectively. Similarly, the effective length of the bolt group connecting the rafter to the apex bracket is l'_{ar} (see Fig. 13.18). It is worth noting that the influence of imperfections and elongation of bolt holes has an effect on the center of rotation. However, owing to the deflection limits for stressed-skin design being small, the elongation of bolt holes is considered to be large, so the effect on the rotational center can be neglected.

13.2.2 Effect of bimoment on moment capacity of joint

Baigent and Hancock (1978) described full-scale portal frame tests in which single CFS channel sections were used for the column and rafter members. The frames tested had rigid joints. Failure was observed at a strength less than that of the in-plane major axis moment capacity of the channel sections, which was attributed to the presence of a bimoment (Vlasov, 1961; Zbirohowski-Koscia, 1967) (resulting from the eccentricity from the shear center of the channel section of the major axis moment generated in the web of the channel section by the bolt group). The bimoment B is equal to the product of the major axis moment, M^*x, and eccentricity of the web centerline from the shear center. Further detail is given in Hancock (1985) and Lim et al. (2016). The stress distribution of this combined effect is shown in Fig. 13.20.

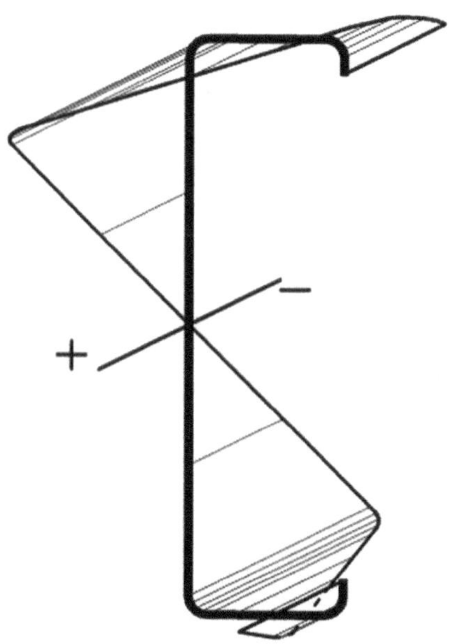

Figure 13.20 Stress distribution due to combined bending and bimoment (compression is negative and tension is positive).
Lim, J.B.P., Hancock, G.J., Charles Clifton, G., Pham, C.H., Das, R., 2016. DSM for ultimate strength of bolted moment-connections between cold-formed steel channel members. Journal of Constructional Steel Research 117, 196–203.

The bimoment, generated by the eccentricity as described above, puts each flange into bending about its own (horizontal) plane. For a channel subject to negative major axis moment (ie, bottom flange in compression), the bimoment generates compression stress at the bottom flange/web junction, which adds to the major axis bending-induced compression stress. At the bottom flange/lip junction the bimoment generates tensile stress, which opposes the major axis bending-induced compression stress. The result is increased compression in the bottom half of the web and the inside part of the bottom flange, but with the outside part of the bottom flange in tension, as shown in Fig. 13.20. This effectively anchors the bottom flange against distortional buckling, but reduces the local buckling critical stress in the web.

This bimoment effect on the strength of the channel sections as used in the column and rafter members of CFS portal frames was seen in apex joint tests conducted by Lim and Nethercot (2003a,b). These tests are briefly described here.

The parameters used to define the geometry of the back-to-back apex brackets used in each test are shown in Fig. 13.21. The lip stiffener along the compression edge of the apex brackets prevents buckling of the free edge. In all four tests, no buckling of the apex brackets was observed.

The dimensions of the apex bracket used for each joint are summarized in Table 13.4. Each joint used a different length of bolt group (and therefore a different size of apex bracket); all bolt groups were formed from an array of nine bolts. The nominal thickness of each bracket was 4 mm, and the nominal diameters of the bolts and bolt holes were 16 mm and 18 mm, respectively. The average yield and ultimate strengths of the brackets, measured from three tensile coupons taken from each bracket, were 341 N/mm^2 and 511 N/mm^2, respectively. The average dimensions of the channel sections used in the tests are shown in Fig. 13.22; their average yield and ultimate strengths were determined from tensile testing to be 358 N/mm^2 and 425 N/mm^2, respectively.

A photograph of the laboratory test arrangement is shown in Fig. 13.23. The apex joint was tested horizontally on the laboratory floor, and was loaded under pure

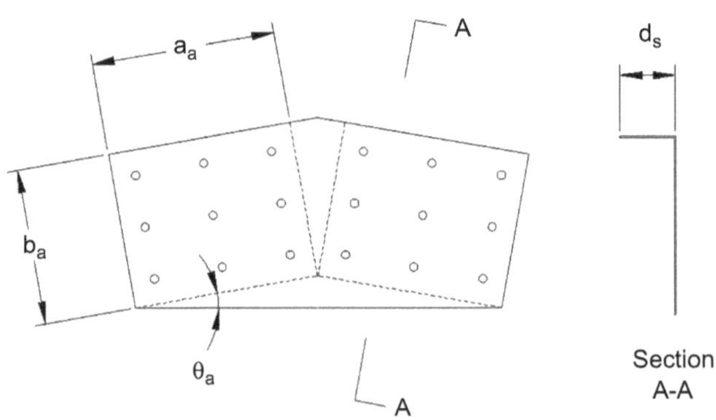

Figure 13.21 Diagram showing parameters of apex bracket and bolt group size.
Lim, J.B.P., Nethercot, D.A., 2003. Ultimate strength of bolted moment-connections between cold-formed steel members. Thin-walled Structures 41, 1019–1039.

Table 13.4 **Dimension of apex brackets tested**

Test	Cold-formed steel sections			Bolt group	
	a (mm)	*b* (mm)	*t* (mm)	a_B (mm)	b_B (mm)
1	525	340	3.98	315	230
2	600	340	3.98	390	230
3	675	340	3.98	465	230
4	825	340	3.98	615	230

Lim, J.B.P., Hancock, G.J., Charles Clifton, G., Pham, C.H., Das, R., 2016. DSM for ultimate strength of bolted moment-connections between cold-formed steel channel members. Journal of Constructional Steel Research 117, 196–203.

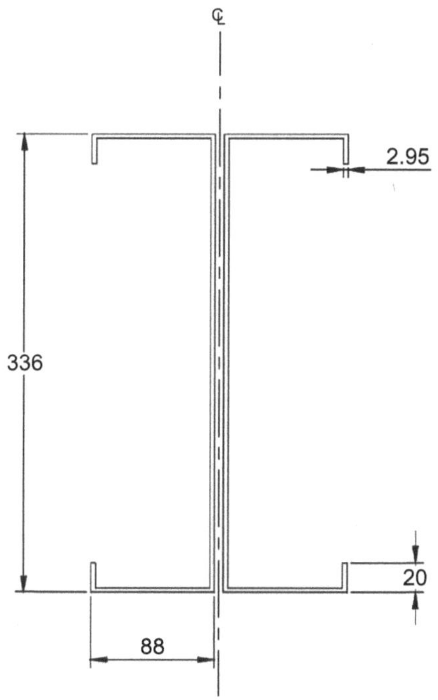

Figure 13.22 Back-to-back channel sections used for column and rafter members.
Lim, J.B.P., Nethercot, D.A., 2003. Ultimate strength of bolted moment-connections between cold-formed steel members. Thin-walled Structures 41, 1019–1039.

bending. Table 13.5 shows the ratio a_B/D and the ultimate moment M_u^{exp} for all four apex joints, where *D* is the overall depth of the channel section.

Using the effective width method, in accordance with BS 5950-5 (1998), the moment capacity of the back-to-back channel sections was calculated to be 96.8 kN m; it should be noted that distortional buckling was not considered. It can be seen from Table 13.5 that as the value of a_B/D increases, the ultimate moment

Figure 13.23 Photograph of the laboratory test setup of apex joints.
Lim, J.B.P., Nethercot, D.A., 2003. Ultimate strength of bolted moment-connections between cold-formed steel members. Thin-walled Structures 41, 1019–1039.

Table 13.5 **Experimental test results**

Test	Cold-formed steel sections		
	a_B (mm)	a_B/D	M_u^{exp} (kN m)
1	315	0.94	75.0
2	390	1.16	77.5
3	465	1.38	82.5
4	615	1.83	87.5

Lim, J.B.P., Hancock, G.J., Charles Clifton, G., Pham, C.H., Das, R., 2016. DSM for ultimate strength of bolted moment-connections between cold-formed steel channel members. Journal of Constructional Steel Research 117, 196–203.

also increases. For example, Test 1, having a ratio of a_B/D of 0.93, failed at a bending moment 23% less than the moment capacity of the back-to-back channel sections. On the other hand, Test 4, having a ratio of a_B/D of 1.83, failed at a bending moment only 10% less than the moment capacity of the back-to-back channel sections.

Lim and Nethercot (2003b) presented design charts from which the reduction in strengths can be interpolated for different bolt group lengths and different ratios of D/t. These are shown in Fig. 13.24. These design charts were shown by Lim and Nethercot (2003b) to be able to predict the moment capacity of the joints described by Chung and Lau (1999) and Wong and Chung (2002), as noted in Section 13.1.1.

13.3 Frame idealization

13.3.1 Full-scale tests

To quantify the effect of different joint details on frame behavior, Lim and Nethercot (2004a) conducted two full-scale frame tests: frames A and B. Both

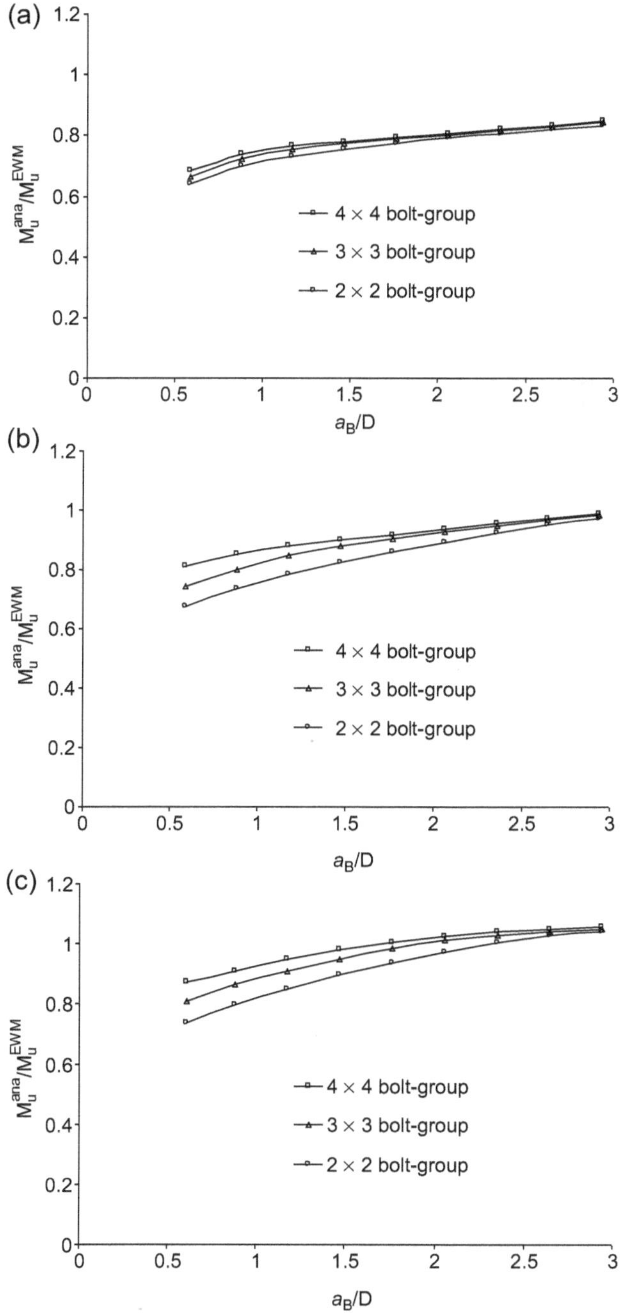

Figure 13.24 Variation of ultimate moment capacity of joint against bolt group length after Lim and Nethercot (2003b). (a) $D/t = 168$ ($D = 336$ mm and $t = 2$ mm); (b) $D/t = 112$ ($D = 336$ mm and $t = 3$ mm); (c) $D/t = 84$ ($D = 336$ mm and $t = 4$ mm).
Lim, J.B.P., Nethercot, D.A., 2003. Ultimate strength of bolted moment-connections between cold-formed steel members. Thin-walled Structures 41, 1019–1039.

frames were of 12 m span and 3 m column height with a roof pitch of 10 degree, but used different sizes of brackets and bolt arrangements (see Fig. 13.13). Figs. 13.25 and 13.26 show the general test arrangement and aerial view, respectively, of frame B.

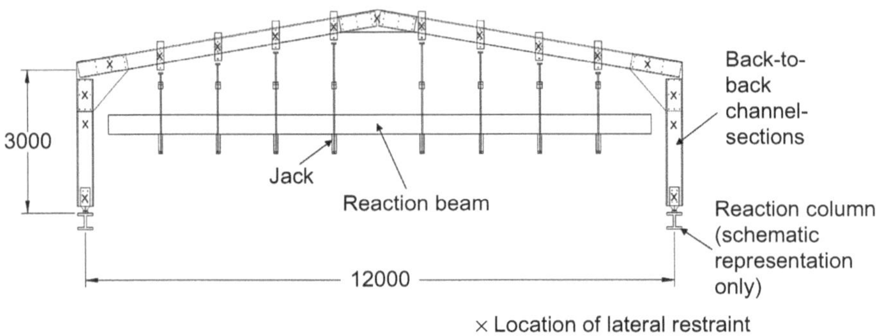

Figure 13.25 General test arrangement of frame B.
Lim, J.B.P., Nethercot, D.A., 2004. Finite element idealization of a cold-formed steel portal frame. Journal of Structural Engineering 130, 78–94.

Figure 13.26 "Aerial" view of frame B.
Lim, J.B.P., Nethercot, D.A., 2004. Finite element idealization of a cold-formed steel portal frame. Journal of Structural Engineering 130, 78–94.

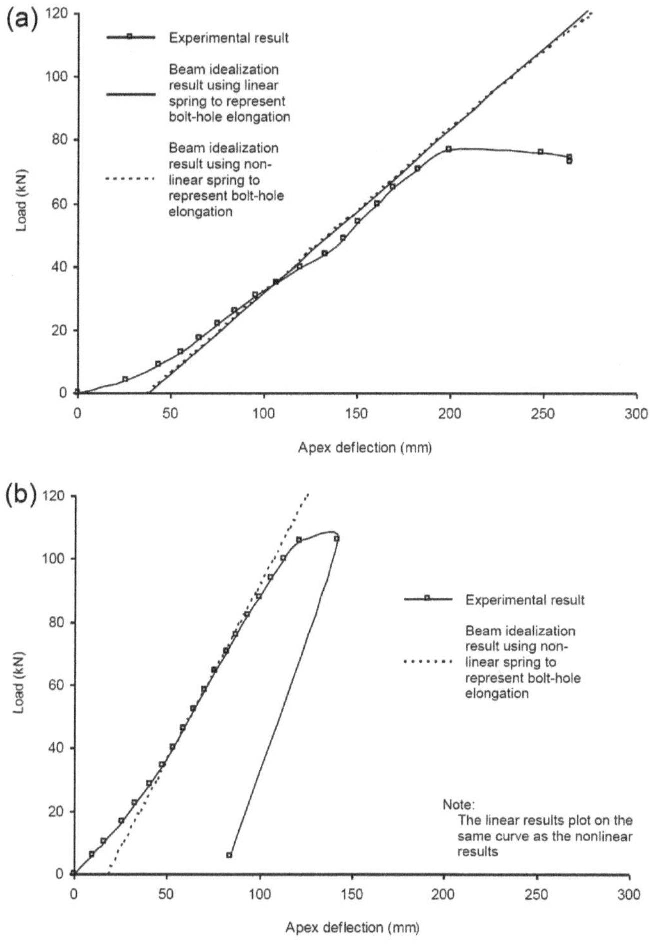

Figure 13.27 Variations of applied loads against apex deflections for frames A (a) and B (b). Lim, J.B.P., Nethercot, D.A., 2004. Finite element idealization of a cold-formed steel portal frame. Journal of Structural Engineering 130, 78–94.

Fig. 13.27 shows the experimental variation of load against apex deflection for both frames. As can be seen, the effect of increasing the bolt-group size from 315 × 230 mm (frame A) to 615 × 230 mm (frame B) increases the gradient of load against deflection by a factor of approximately three.

Fig. 13.27 also shows the gradients obtained using a simple beam idealization, described in Section 13.2.2. It should be noted that the gradients of the beam idealization, against which the experimental gradients can be compared, are plotted with an initial offset along the deflection axis to account for the deflection due to misalignment of the bolt holes; this enables the experimental and analytical results to be compared at loads when all the bolt shanks are in full bearing contact against the connected plates. The values of the initial offsets were determined from a line of best fit drawn for the load deflection curve of each frame.

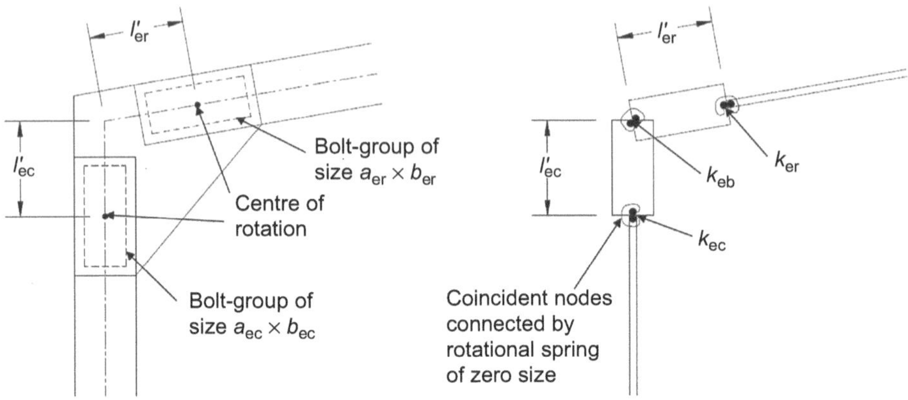

Figure 13.28 Details of beam idealization of eaves joint.
Lim, J.B.P., Nethercot, D.A., 2004. Finite element idealization of a cold-formed steel portal frame. Journal of Structural Engineering 130, 78–94.

From the results of these two tests, it can be concluded that the joints of CFS portal frames were semirigid and of finite connection length. Further details of the tests can be found in Lim and Nethercot (2004a).

13.3.2 Beam idealization for the frame

Fig. 13.28 shows details of a beam idealization for a CFS portal frame. As can be seen, the column and rafter members are each connected at the eaves joint through rotational spring elements of stiffness k_{ec} and k_{er}, respectively. Each rotational spring is of zero size and connects two coincident nodes, with one node belonging to the member and the other to the eaves bracket. The values of k_{ec} and k_{er} for the spring elements depend on the bolt group size, the number of bolts used, and the bolt-hole elongation stiffness, k_b. Essentially, each bolt group is assumed to rotate around its center of rotation, as described by Crawford and Kulak (1971) and Bryan (1993) (see Fig. 13.15).

The finite connection length of the eaves joint is idealized using two rigid beam elements. As can be seen, the column and rafter connections at the eaves have finite connection lengths of l'_{ec} and l'_{er}, respectively (see Table 13.6).

The apex joint is idealized in a similar manner to the eaves joint, as shown in Fig. 13.29. Fig. 13.30 shows details of the beam idealization of the whole portal frame. For the convenience of obtaining results and plotting graphical output, six elements are used for the column and 12 elements for the rafter; an additional element is used for each leg of the eaves and the apex brackets.

13.3.3 Joint effects on bending moments

Fig. 13.31 shows bending moment diagrams of the frame under vertical loads: three bending moment diagrams are provided for the cases of rigid-jointed frames, frame A, and frame B, respectively. All three frames are subjected to the same vertical

Table 13.6 **Parameters pertaining to frames A and B in analysis using beam idealization**

Frame	l'_{ec} (mm)	l'_{er} (mm)	l'_{ar} (mm)	$k_{ec} \times 10^3$ (kN m/rad)	$k_{er} \times 10^3$ (kN m/rad)	$k_{ar} \times 10^3$ (kN m/rad)	$k_{eb} \times 10^3$ (kN m/rad)	$k_{ab} \times 10^3$ (kN m/rad)
A	400.1	400.1	287.5	2.4	2.4	2.4	70.0	39.4
B	550.1	550.1	446.5	6.8	6.8	6.8	93.6	32.6

Lim, J.B.P., Nethercot, D.A., 2004. Finite element idealization of a cold-formed steel portal frame. Journal of Structural Engineering 130, 78–94.

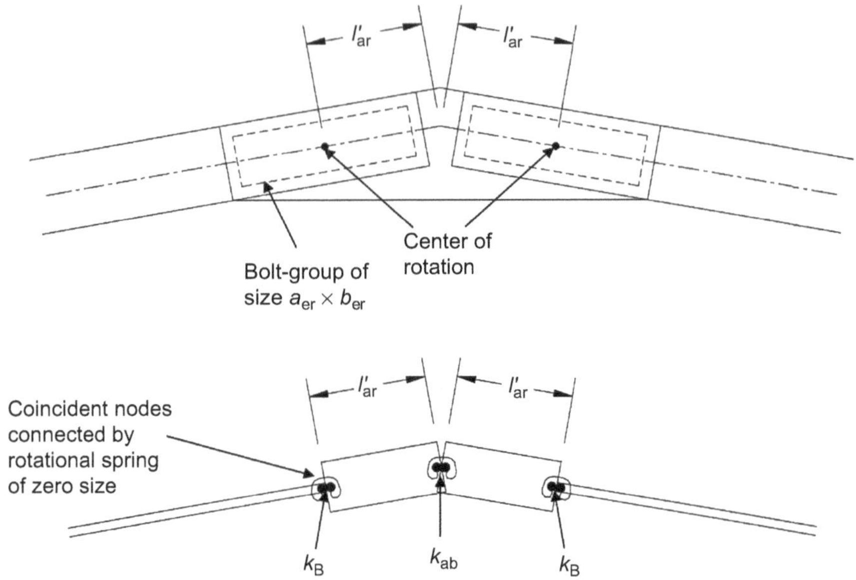

Figure 13.29 Details of beam idealization of an apex joint.
Lim, J.B.P., Nethercot, D.A., 2004. Finite element idealization of a cold-formed steel portal frame. Journal of Structural Engineering 130, 78–94.

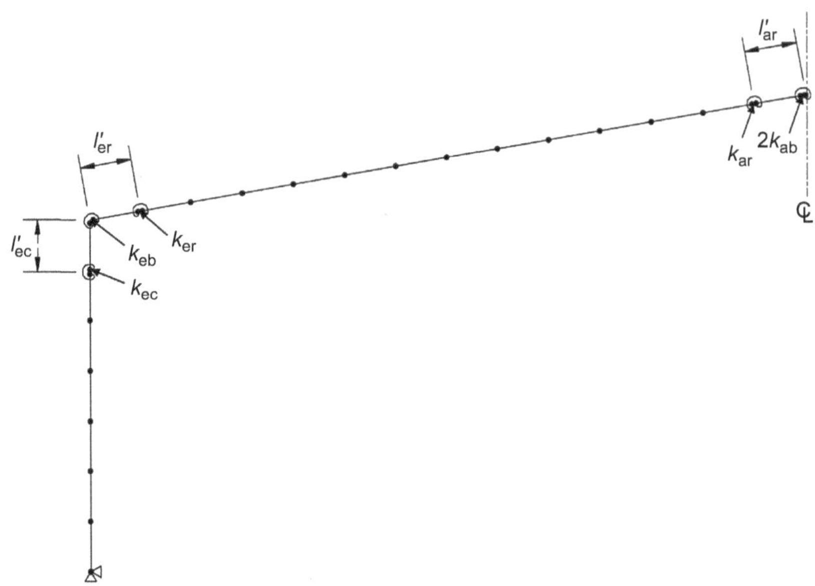

Figure 13.30 Details of beam idealization of a portal frame.
Lim, J.B.P., Nethercot, D.A., 2004. Finite element idealization of a cold-formed steel portal frame. Journal of Structural Engineering 130, 78–94.

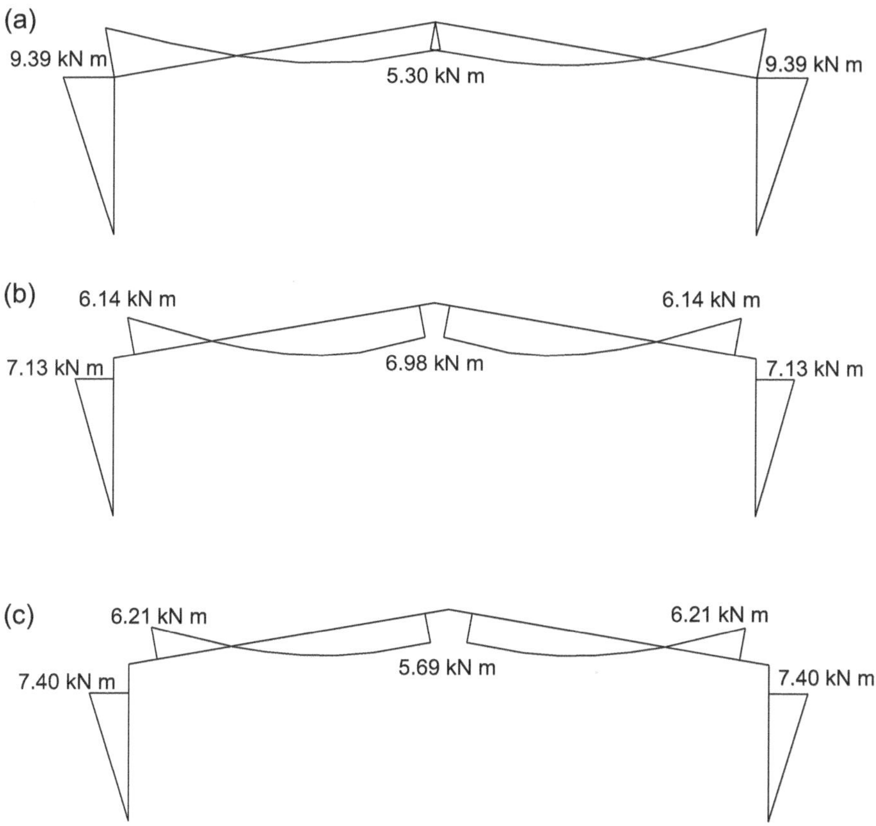

Figure 13.31 Comparison of bending moment diagram of frames under vertical load: (a) rigid frame; (b) frame A; (c) frame B.
Wrzesien, A.M., Lim, J.B.P., Nethercot, D.A., 2012. Optimum joint detail for a general cold-formed steel portal frame. Advances in Structural Engineering 15, 1623–1639.

load. While the magnitude of this vertical load is not important, as the analysis is elastic and it is the relative differences in bending moments that are important, the magnitude of the vertical load corresponds to the serviceability load of a rigid-jointed frame. As can be seen, the bending moment at the eaves of frames A and B is less than that of the rigid-jointed frame, while the bending moment at the apex is larger than that of the rigid-jointed frame for vertical load cases.

13.3.4 Concluding remarks

The results of full-scale tests show that semirigid design should be used for CFS portal frames, as the rotational flexibility of the joints has a large effect on both the bending moment distribution of the frame and frame deflections. It may be noted, however, that semirigid joint design has the advantage that the ratio of the bending moment at the eaves to the bending moment at the apex can be controlled by the designer by choosing suitable connection rotational stiffness and connection lengths for the joints. On the other hand, the joint detail also needs to be chosen to control deflections under serviceability load.

The effect of cladding, however, has been ignored in the tests and subsequent analysis. Section 13.4 describes the effect of cladding on the frame, which can have an important effect on the behavior of small-span frames.

13.4 Behavior of clad frames

13.4.1 Clad frame tests

The previous section described the effect of joint detail on the behavior of unclad CFS portal frames. This work ignores the effect of the sheeting, which can be important, as demonstrated for small-span frames. In this section, six full-scale laboratory tests conducted on CFS portal framing systems with cladding are described; these tests investigate the effects of joint flexibility and stressed-skin diaphragm action. Full details of this work can be found in Wrzesien et al. (2015).

The buildings tested were of 6 m span, 3 m height, 10 degree pitch, and 3 m frame spacing (see Fig. 13.32). Two different joints were considered, joint types 1 and 2 (see Fig. 13.13); the stiffness of joint 2 was approximately twice that of joint 1.

The results of the clad frame tests were intended to represent the behavior of a building of 9 m overall length with two braced gable frames and internal frames, and spacing of 3 m between the frames. Owing to symmetry and also due to laboratory space limitations, only one gable frame and one internal frame were used for each clad frame test. Load cells located in the gable braces allowed the load distribution between frames to be determined. Fig. 13.33(a) shows the nominal dimensions of the channel sections used for the frames in all six tests. In the internal frames the channel sections are placed back to back, and in the gable frames (including the gable posts) single channel sections are used. Fig. 13.33(b) shows the nominal dimensions of the single-skin profile used for the roof sheeting. The roof profile has a depth of 30 mm and a nominal thickness of 0.7 mm. Fig. 13.33(c) shows the nominal dimensions of the top-hat purlins whose thickness is 1.0 mm.

Firstly, tests on frames without roof sheeting were conducted. Vertical loading was applied in two tests (see Fig. 13.34(a)), and horizontal loading in four tests (see Fig. 13.34(b)). For horizontal loading only, the frame tests were carried out with and without roof sheeting to determine the effect of stressed-skin action. The component tests were described separately for both the roof panel (Wrzesien et al., 2009) and the beam-to-column connections (Wrzesien et al., 2015).

13.4.2 Clad frame test results

From the full-scale tests on the bare frames, it was observed that the failure load of buildings with joints 1 and 2 failed at approximately the same load (see Figs. 13.35 and 13.36). In terms of stiffness, the frame with joint 2 was approximately 60% and

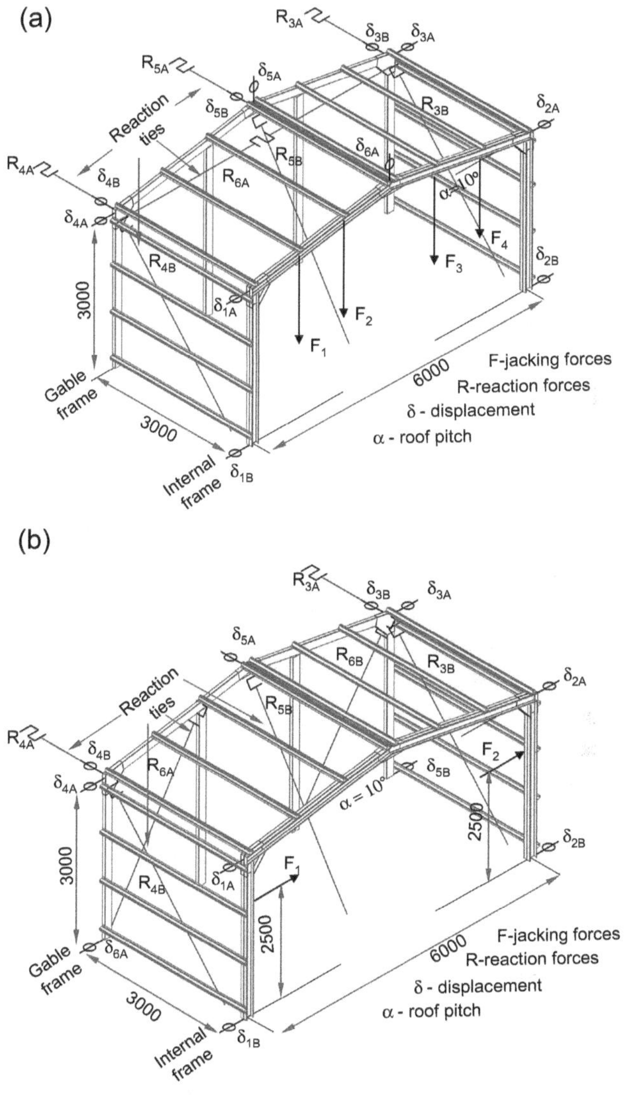

Figure 13.32 General arrangement of full-scale test frame: (a) vertical load; (b) horizontal load, $H_F = 2500$ and 2400 mm for frame A and frame B respectively.
Wrzesien, A.M., Lim, J.B.P., Xu, Y., Macleod, I.A., Lawson, R.M., 2015. Effect of stressed skin action on the behaviour of cold-formed steel portal frames. Engineering Structures 105, 123–136.

30% stiffer than the frame with joint 1 under vertical load and horizontal load, respectively. It should be noted that for the tests conducted without sheeting, no load transfer to the gable frames from the internal loaded frames was recorded. The results of the tests on the internal frame can therefore be assumed to be identical to a bare frame test.

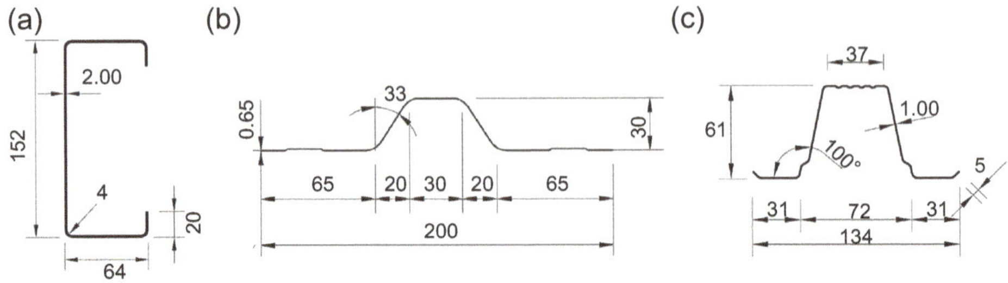

Figure 13.33 Nominal dimensions of components used for building tests: (a) column/rafter; (b) sheeting profile; (c) purlin.
Wrzesien, A.M., Lim, J.B.P., Xu, Y., Macleod, I.A., Lawson, R.M., 2015. Effect of stressed skin action on the behaviour of cold-formed steel portal frames. Engineering Structures 105, 123–136.

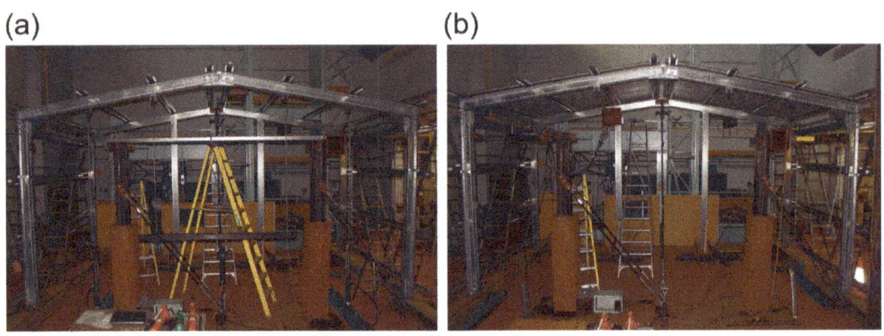

Figure 13.34 Photograph of full-scale portal frame building tests: (a) building with no sheeting under vertical load; (b) building with roof sheeting under horizontal load.
Wrzesien, A.M., Lim, J.B.P., Xu, Y., Macleod, I.A., Lawson, R.M., 2015. Effect of stressed skin action on the behaviour of cold-formed steel portal frames. Engineering Structures 105, 123–136.

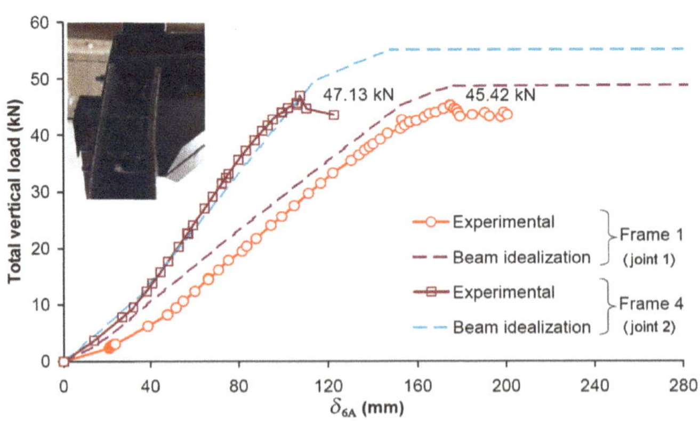

Figure 13.35 Variation of load against apex deflection for frames loaded in vertical direction.
Wrzesien, A.M., Lim, J.B.P., Xu, Y., Macleod, I.A., Lawson, R.M., 2015. Effect of stressed skin action on the behaviour of cold-formed steel portal frames. Engineering Structures 105, 123–136.

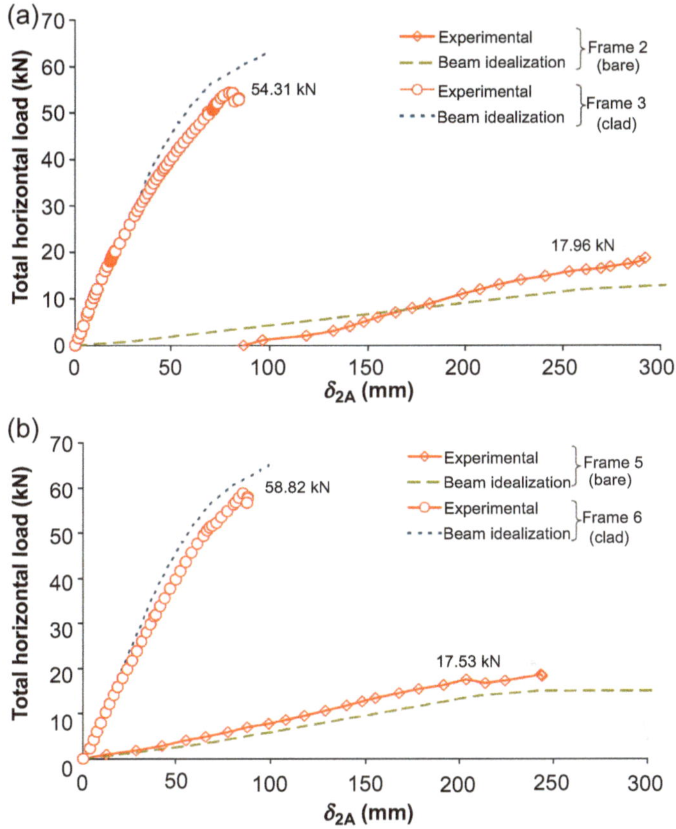

Figure 13.36 Variation of load against deflection for bare and clad building loaded in horizontal direction: (a) joint 1; (b) joint 2.
Wrzesien, A.M., Lim, J.B.P., Xu, Y., Macleod, I.A., Lawson, R.M., 2015. Effect of stressed skin action on the behaviour of cold-formed steel portal frames. Engineering Structures 105, 123–136.

However, the failure load increased by a factor of almost three compared with the failure load of the frame with no roof sheeting (see Fig. 13.36). Furthermore, the stiffness of the frame increased by a factor of almost 10 relative to the bare frame. In both tests, failure of the roof panels occurred with severe deformations of the purlins and gable rafters (as shown in Fig. 13.37).

When the experimentally derived data were used to design buildings, it was found that the cladding stiffness should not be ignored in practice as the loading attracted to the gables may be underestimated by as much as a factor of seven. This has an important effect on the gable frames, which are usually designed for a tributary wind load acting on half a bay according to the bare frame model.

The design showed that a lighter internal frame can be designed by including diaphragm action in the three-dimensional stressed-skin analysis. For a typical building of 6 m span and 12 m length consisting of three internal frames and two end gables, plane

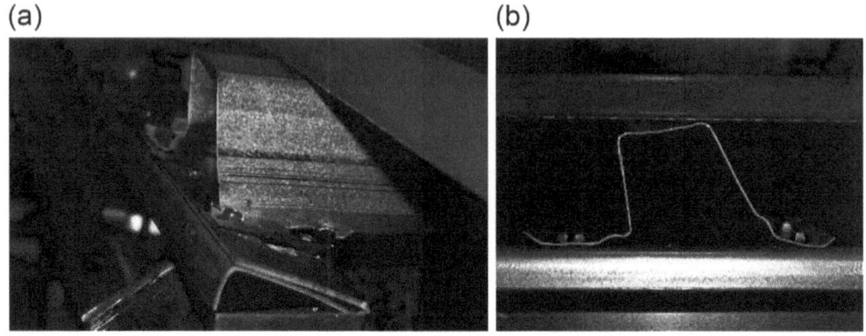

Figure 13.37 Mode of failures observed in clad frame tests: (a) failure of purlin connection and deformation of the gable frame rafter; (b) shear deformation due to diaphragm action. Wrzesien, A.M., Lim, J.B.P., Xu, Y., Macleod, I.A., Lawson, R.M., 2015. Effect of stressed skin action on the behaviour of cold-formed steel portal frames. Engineering Structures 105, 123−136.

frame analysis results in 981 kg of steel in the framework, and when stressed-skin action is considered, the overall weight of the steel framework can be reduced by 42%. In such designs, the roof cladding becomes an important structural component in providing lateral stability.

The following general conclusions were drawn.

- The bare frame model identifies a wind load combination as being critical; allowing for diaphragm action shows that the critical load combination depends on the length-to-span ratio of the building.
- The bare frame design model may not be safe for designing clad structures, as it ignores the additional force in the edge purlins, end cladding, and gables.
- To avoid failure of the end gables, they should be designed to resist wind forces higher than those acting on a single bay between frames.

As can be seen from the building tests, the stiffness of the joints has a small effect on the "true" horizontal deflections of a clad portal frame building. Increasing the rotational stiffness of the joints has a large effect on vertical deflections, eg, increasing the stiffness by the factor of two (from joints 1 and 2) offers approximately 40% reduction in apex deflections.

However, a three-dimensional stressed-skin analysis allows the designer to follow "true" load paths and produce a design which is often more economical. In such designs, however, roof sheeting may become a key structural component responsible for the building's stability, so it follows that the stability of the structure can be compromised if the roof sheeting is removed or the stiffness of the roof panel is changed (eg, by adding openings or replacing sheeting profiles).

13.5 Fire design

In terms of portal frames in fire boundary conditions, a common recommendation is to protect the columns against fire and leave the rafters unprotected. Reference is often also made to the SCI design method SCI P313 (2002) for single-story structures. However, this design method is intended for conventional hot-rolled steel portal frames. No current guidance exists specifically for the design of CFS portal frame structures in fire boundary conditions.

From the results of a full-scale fire test (Johnston et al., 2014), the base fixity and in-plane restraint from side-rails were shown to have a beneficial effect in preventing premature collapse and an undesirable outward collapse mechanism. As the fixity of CFS portal frame bases cannot be easily quantified or regulated, it is recommended by Johnston et al. (2015a,b) that side-rails are protected along their entire length, in addition to columns, to help prevent sway failure and premature collapse (see Fig. 13.38).

A full-scale site fire test conducted by Johnston et al. (2014) demonstrated an asymmetric inward failure at 714°C. Although some outward lateral movement of the columns was observed, an eventual inward failure mechanism was seen (see Fig. 13.39(a)). Failure of the site test frame was due to member buckling rather than failure of the connections (see Fig. 13.39(b)).

Figure 13.38 Image of the sway failure captured by shell idealization of the portal frame presented in Fig. 13.25 at elevated temperatures.
After Johnston, R.P.D., Sonebi, M., Lim, J.B.P., Armstrong, C.G., Wrzesien, A.M., Abdelal, G., Hu, Y., 2015b. The collapse behaviour of cold-formed steel portal frames at elevated temperatures. Journal of Structural Fire Engineering 6, 77–101.

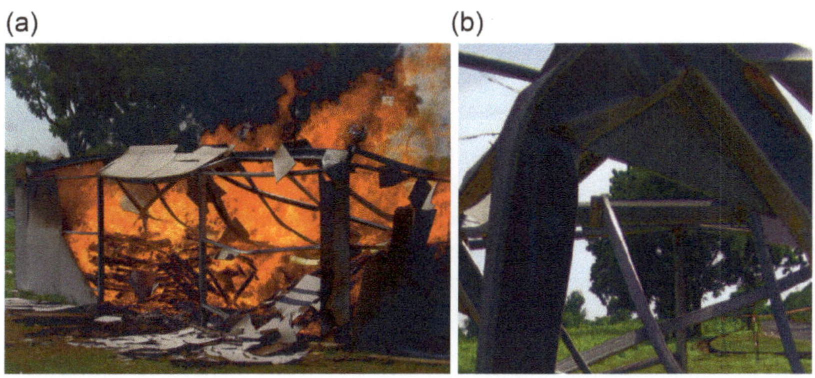

Figure 13.39 Photograph from the full-scale site test after Johnston et al. (2014): (a) frame collapse; (b) buckling of the column in the vicinity of the eaves joint.
Johnston, R.P.D., Lim, J.B.P., Lau, H.H., Xu, Y., Sonebi, M., Armstrong, C.G., Switzer, C., Mei, C. C. 2014. Cold-formed steel portal frames in fire: full-scale testing and finite element analysis. The Structural Engineer (Online), 92.

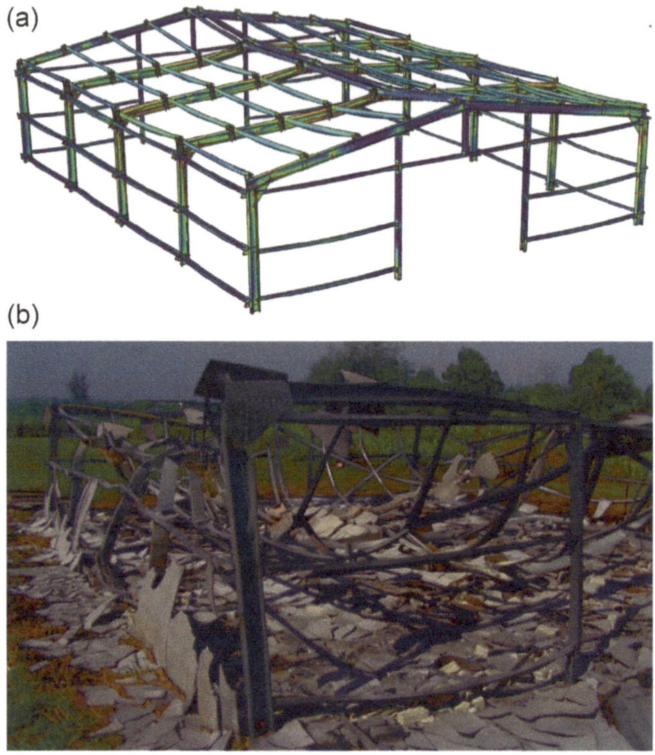

Figure 13.40 Site test results versus NLFEA shell model after Johnston et al. (2015a): (a) finite element model of the building prior to collapse (predicted temperature of 704°C); (b) collapse of the building (recorded temperature of 714°C).
Johnston, R.P.D., Lim, J.B.P., Lau, H.H., Xu, Y., Sonebi, M., Armstrong, C.G., Mei, C.C., 2015. FE investigation of cold-formed steel portal frames in fire. Proceedings of the ICE – Structures and Buildings [Online].

Johnston et al. (2015b) also recommended that different fire scenarios be considered by designers to ensure a conservative approach. Nonlinear elastoplastic shell models demonstrated a good agreement with physical testing, more so than beam models, and may therefore be used to predict the response of CFS portal frames subject to fire (see Fig. 13.40). SCI P313 guidance for hot-rolled steel portal frames should not be used for CFS portal frames in fire boundary conditions.

The results from experimental and nonlinear finite element analysis (NLFEA) testing at elevated temperatures by Johnston et al. (2014), and Johnston et al. (2015a,b) concluded the following.

- The secondary structural elements had a beneficial effect on the behavior of CFS portal frames in fire, in terms of giving a higher failure temperature and preventing undesirable collapse mechanisms in fire. Adequate provision should be made where the continuity of secondary members is broken up to facilitate openings.

- Base fixity influences the collapse mode and failure temperature. However, the fixity would need to be accurately quantified for CFS portal frames due to the pinned or partial rigidity of base connections. In lieu of necessary research, it was recommended that the side-rails, columns, and associated connections to columns and fire-rated cladding are suitably protected along their entire length.
- NLFEA shell models were validated against site test results, and therefore can be used to assist designers in predicting the behavior of CFS portal frames in fire.
- A like-for-like 22 m span CFS alternative to the hot-rolled steel portal frame example presented in the SCI P313 design guidance failed with an asymmetric collapse mechanism at a temperature of 595–601°C, depending on the base condition. This is comparable to the failure temperature of 890°C specified for the 22 m span hot-rolled frame in the SCI guidance. Results demonstrated that without the inclusion of in-plane restraint, the 22 m CFS frame with fully pinned bases failed with an undesirable outward sway collapse mechanism at 595°C. Although not representative of a real fire, the ISO (International Organization for Standardization) standard curve is commonly used in practice. Therefore, based on the difference in collapse temperature, it is recommended that practicing engineers consider the difference in the time domain for collapse in fire.
- The use of the SCI P313 guidance document for the design of single-story hot-rolled portal frames in fire boundary condition is potentially unconservative for CFS portal frames. Design recommendations and a simplified mathematical model are presented in Johnston (2015).

13.6 Conclusions

This chapter reviews different arrangements in bolted-moment connections between CFS members. As can be seen, almost all the joints comprise channel sections for the members, connected through gusset plates bolted to the webs of the channel sections. The general arrangement involves a bolt group in a regular grid with each bolt passing through the webs of the channel sections. Variations to this arrangement include connections with bolts through the flanges of the channel sections, or more bolts located away from the center of rotation of the bolt group.

Using a general joint arrangement comprising a 3 × 3 array of bolts, a method for determining the rotational stiffness of the joints is described, which can then be incorporated into a frame analysis. In some cases, semirigid joints of finite connection lengths having a practical joint detail can actually result in a frame with a higher load-carrying capacity than an equivalent rigidly jointed frame. Full-scale frame tests have demonstrated the accuracy of the beam idealization for a bare frame of 12 m span.

Full-scale building tests are also described, conducted to investigate the effect of roof sheeting for small-span buildings. It is shown that stressed-skin diaphragm action is important to consider, particularly if the objective is to prevent tearing of cladding at serviceable loads.

Finally, fire design criteria for such buildings are considered. As can be seen, the structural system maintains stability and is stronger than the sum of its individual components. A fire-protected lower side-rail is recommended to improve the stability of the system, specifically the stability of the collapsing walls in fire.

References

Baigent, A.H., Hancock, G.J., 1978. The Behaviour of Portal Frames Composed of Cold-formed Members. University of Sydney, School of Civil Engineering, Sydney.

Baigent, A.H., Hancock, G.J., 1982. Structural analysis of assemblages of thin-walled members. Engineering Structures 4, 207–216.

Bryan, E.R., 1993. The design of bolted joints in cold-formed steel sections. Thin-walled Structures 16, 239–262.

BS 5950-5, 1998. Structural Use of Steelwork in Building Part 5: Code of Practice for Design of Cold Formed Thin Gauge Sections. British Standards Institution, London.

BS EN 1993-1–8, 2005. Eurocode 3-Design of Steel Structures. Part 1-8: Design of Joints. European Committee for Standardization, Brussels.

Chung, K.F., Lau, L., 1999. Experimental investigation on bolted moment connections among cold formed steel members. Engineering Structures 21, 898–911.

Crawford, S.F., Kulak, G.L., 1971. Eccentrically loaded bolted connections. Journal of the Structural Division, ASCE 97, 765–783.

Dubina, D., Stratan, A., Ciutina, A., Fulop, L., Zsolt, N., 2004. Monotonic and cyclic performance of joints of cold formed steel portal frames. Loughborough, UK. In: Loughlan, J. (Ed.), 4th International Conference on Thin-walled Structures, pp. 381–388.

Dundu, M., Kemp, A.R., 2006. Plastic and lateral-torsional buckling behaviour of single cold-formed channels connected back-to-back. Journal of Structural Engineering 132, 1223–1233.

Hancock, G.J., 1985. Portal frames composed of cold-formed channel-sections. In: Narayanan, R. (Ed.), Chapter 8, Steel Framed Structures. Elsevier Applied Science Publishers, London.

Johnston, R.P.D., Lim, J.B.P., Lau, H.H., Xu, Y., Sonebi, M., Armstrong, C.G., Mei, C.C., 2015a. FE investigation of cold-formed steel portal frames in fire. Proceedings of the ICE – Structures and Buildings [Online]. Available: http://www.icevirtuallibrary.com/content/article/10.1680/stbu.14.00082.

Johnston, R.P.D., Lim, J.B.P., Lau, H.H., Xu, Y., Sonebi, M., Armstrong, C.G., Switzer, C., Mei, C.C., 2014. Cold-formed steel portal frames in fire: full-scale testing and finite element analysis [Online] The Structural Engineer 92.

Johnston, R.P.D., Sonebi, M., Lim, J.B.P., Armstrong, C.G., Wrzesien, A.M., Abdelal, G., Hu, Y., 2015b. The collapse behaviour of cold-formed steel portal frames at elevated temperatures. Journal of Structural Fire Engineering 6, 77–101.

Kirk, P., 1986. Design of a cold formed section portal frame building system. St. Louis, Missouri, USA. In: 8th International Specialty Conference on Cold-formed Steel Structures, pp. 295–310.

Kwon, Y.B., Chung, H.S., Kim, G.D., 2006. Experiments of cold-formed steel connections and portal frames. Journal of Structural Engineering 132, 600–607.

Lim, J.B.P., Hancock, G.J., Charles Clifton, G., Pham, C.H., Das, R., 2016. DSM for ultimate strength of bolted moment-connections between cold-formed steel channel members. Journal of Constructional Steel Research 117, 196–203.

Lim, J.B.P., Nethercot, D.A., 2002. F. E.-assisted design of the eaves bracket of a cold-formed steel portal frame. Steel & Composite Structures 2, 411–428.

Lim, J.B.P., Nethercot, D.A., 2003a. Serviceability design of a cold-formed steel portal frame having semi-rigid joints. Steel & Composite Structures 3, 451–474.

Lim, J.B.P., Nethercot, D.A., 2003b. Ultimate strength of bolted moment-connections between cold-formed steel members. Thin Walled Structures 41, 1019–1039.

Lim, J.B.P., Nethercot, D.A., 2004a. Finite element idealization of a cold-formed steel portal frame. Journal of Structural Engineering 130, 78–94.

Lim, J.B.P., Nethercot, D.A., 2004b. Stiffness prediction for bolted moment-connections between cold-formed steel members. Journal of Constructional Steel Research 60, 85–107.

Mäkeläinen, P., Kankaanpää, J., 1996. Structural design study on a light-gauge steel portal frame with cold-formed sigma sections. St. Louis, Missouri, USA. In: 13th International Specialty Conference on Cold-formed Steel Structures, pp. 349–371.

Mills, J., LaBoube, R., 2004. Self-drilling screw joints for cold-formed channel portal frames. Journal of Structural Engineering 130, 1799–1806.

Phan, D.T., Lim, J.B.P., Tanyimboh, T.T., Wrzesien, A.M., Sha, W., Lawson, R.M., 2015. Optimal design of cold-formed steel portal frames for stressed-skin action using genetic algorithm. Engineering Structures 93, 36–49.

Rhodes, J., Burns, R., 2006. Development of a portal frame system on the basis of component testing. In: 18th International Specialty Conference in Cold-formed Steel Structures (Orlando, Florida, USA).

SCI P313, 2002. In: Simms, W.I., Newman, G.M. (Eds.), Single Storey Steel Framed Buildings in Fire Boundary Conditions. The Steel Construction Institute, Ascot, UK.

Vlasov, V.Z., 1961. Thin-walled Elastic Beams. Israel Program for Scientific Translation, Jerusalem.

Wong, M.F., Chung, K.F., 2002. Structural behaviour of bolted moment connections in cold-formed steel beam-column sub-frames. Journal of Constructional Steel Research 58, 253–274.

Wrzesien, A.M., Lim, J.B.P., Lawson, R.M., 2009. The ultimate strength and stiffness of modern roof systems with hat-shaped purlins. Hong Kong, China. In: Chan, S.L. (Ed.), Sixth International Conference on Advances in Steel Structures, pp. 480–487.

Wrzesien, A.M., Lim, J.B.P., Xu, Y., Dundu, M., Macleod, I., Lawson, R.M., 2012. Stressed skin effects on cold-formed steel portal frames with semi-rigid joints – experimental study. In: The 6th International Conference on Coupled Instabilities in Metal Structures (Glasgow, UK).

Wrzesien, A.M., Lim, J.B.P., Xu, Y., Macleod, I.A., Lawson, R.M., 2015. Effect of stressed skin action on the behaviour of cold-formed steel portal frames. Engineering Structures 105, 123–136.

Yu, W.K., Chung, K.F., Wong, M.F., 2005. Analysis of bolted moment connections in cold-formed steel beam–column sub-frames. Journal of Constructional Steel Research 61, 1332–1352.

Zadanfarrokh, F., Bryan, E.R., 1992. Testing and design of bolted connections in cold-formed steel sections. St. Louis, Missouri, USA. In: 11th International Specialty Conference on Cold-formed Steel Structures, pp. 625–662.

Zaharia, R., Dubina, D., 2006. Stiffness of joints in bolted connected cold-formed steel trusses. Journal of Constructional Steel Research. 62, 240–249.

Zbirohowski-Koscia, K., 1967. Thin-walled Beams – From Theory to Practice. Crosby Lockwood, London.

Index

'*Note:* Page numbers followed by "f" indicate figures, "t" indicate tables.'

Kolibri 360 GmbH
Ferdinand-Jühlke-Straße 7
99095 Erfurt, Deutschland
produktsicherheit@kolibri360.de